T0350385

Applied Big Data Analytics and Its Role in COVID-19 Research

Peng Zhao
Intelligentrabbit, LLC, USA

Xi Chen
Beijing University of Civil Engineering and Architecture, China

A volume in the Advances in Data Mining and Database Management (ADMDM) Book Series

Published in the United States of America by
IGI Global
Engineering Science Reference (an imprint of IGI Global)
701 E. Chocolate Avenue
Hershey PA, USA 17033
Tel: 717-533-8845
Fax: 717-533-8661
E-mail: cust@igi-global.com
Web site: http://www.igi-global.com

Library of Congress Cataloging-in-Publication Data

Names: Zhao, Peng, 1985- author. | Chen, Xi, 1982- author.
Title: Applied big data analytics and its role in COVID-19 research / by Peng Zhao, and Xi Chen.
Description: Hershey, PA : Engineering Science Reference, [2022] | Includes bibliographical references and index. | Summary: "This book provides emerging research on the development and implementation of real-world cases in big data analytics for various industrial and public sections including healthcare, business, social media, and government by highlighting topics such as data processing, deep learning, statistical inference, data visualization, and decision support systems"-- Provided by publisher.
Identifiers: LCCN 2021051170 (print) | LCCN 2021051171 (ebook) | ISBN 9781799887935 (hardcover) | ISBN 9781799887942 (paperback) | ISBN 9781799887959 (ebook)
Subjects: LCSH: COVID-19 (Disease)--Research. | Big data. | Data mining.
Classification: LCC RA644.C67 Z435 2022 (print) | LCC RA644.C67 (ebook) | DDC 614.5/92414--dc23/eng/20211128
LC record available at https://lccn.loc.gov/2021051170
LC ebook record available at https://lccn.loc.gov/2021051171

This book is published in the IGI Global book series Advances in Data Mining and Database Management (ADMDM) (ISSN: 2327-1981; eISSN: 2327-199X)

British Cataloguing in Publication Data
A Cataloguing in Publication record for this book is available from the British Library.

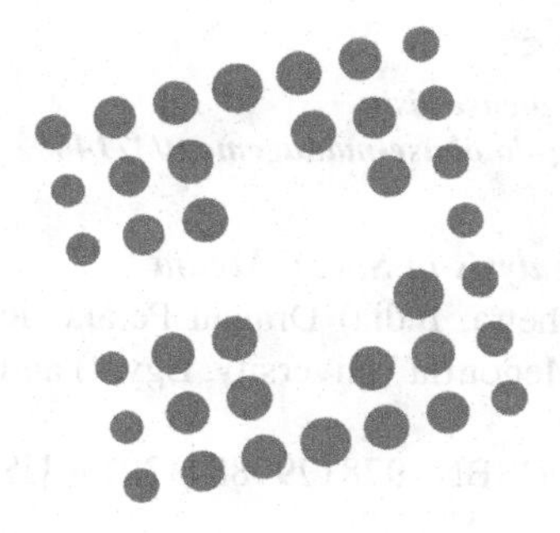

Advances in Data Mining and Database Management (ADMDM) Book Series

ISSN:2327-1981
EISSN:2327-199X

Editor-in-Chief: David Taniar, Monash University, Australia

MISSION

With the large amounts of information available to organizations in today's digital world, there is a need for continual research surrounding emerging methods and tools for collecting, analyzing, and storing data.

The **Advances in Data Mining & Database Management (ADMDM)** series aims to bring together research in information retrieval, data analysis, data warehousing, and related areas in order to become an ideal resource for those working and studying in these fields. IT professionals, software engineers, academicians and upper-level students will find titles within the ADMDM book series particularly useful for staying up-to-date on emerging research, theories, and applications in the fields of data mining and database management.

COVERAGE

- Profiling Practices
- Neural Networks
- Enterprise Systems
- Decision Support Systems
- Factor Analysis
- Web-based information systems
- Database Security
- Database Testing
- Quantitative Structure–Activity Relationship
- Association Rule Learning

IGI Global is currently accepting manuscripts for publication within this series. To submit a proposal for a volume in this series, please contact our Acquisition Editors at Acquisitions@igi-global.com or visit: http://www.igi-global.com/publish/.

The Advances in Data Mining and Database Management (ADMDM) Book Series (ISSN 2327-1981) is published by IGI Global, 701 E. Chocolate Avenue, Hershey, PA 17033-1240, USA, www.igi-global.com. This series is composed of titles available for purchase individually; each title is edited to be contextually exclusive from any other title within the series. For pricing and ordering information please visit http://www.igi-global.com/book-series/advances-data-mining-database-management/37146. Postmaster: Send all address changes to above address.

Titles in this Series

For a list of additional titles in this series, please visit:
http://www.igi-global.com/book-series/advances-data-mining-database-management/37146

Data Mining Approaches for Big Data and Sentiment Analysis in Social Media
Brij B. Gupta (National Institute of Technology, Kurukshetra, India) Dragan Peraković (University of Zagreb, Croatia) Ahmed A. Abd El-Latif (Menoufia University, Egypt) and Deepak Gupta (LoginRadius Inc., Canada)
Engineering Science Reference • © 2022 • 313pp • H/C (ISBN: 9781799884132) • US $225.00

Blockchain Technology Applications in Businesses and Organizations
Pietro De Giovanni (Luiss University, Italy)
Business Science Reference • © 2022 • 315pp • H/C (ISBN: 9781799880141) • US $225.00

Handbook of Research on Essential Information Approaches to Aiding Global Health in the One Health Context
Jorge Lima de Magalhães (Institute of Drugs Technology - Farmanguinhos, Oswaldo Cruz Foundation, - FIOCRUZ Brazil & Global Health and Tropical Medicine – GHMT, Institute of Hygiene and Tropical Medicine, NOVA University of Lisbon, Portugal) Zulmira Hartz (Global Health and Tropical Medicine – GHMT, Institute of Hygiene and Tropical Medicine, NOVA University of Lisbon, Portugal) George Leal Jamil (Informações em Rede Consultoria e Treinamento, Brazil) Henrique Silveira (Global Health and Tropical Medicine – GHMT, Institute of Hygiene and Tropical Medicine, NOVA University of Lisbon, Portugal) and Liliane C. Jamil (Independent Researcher, Brazil)
Medical Information Science Reference • © 2022 • 390pp • H/C (ISBN: 9781799880110) • US $395.00

Ranked Set Sampling Models and Methods
Carlos N. Bouza-Herrera (Universidad de La Habana, Cuba)
Engineering Science Reference • © 2022 • 276pp • H/C (ISBN: 9781799875567) • US $195.00

For an entire list of titles in this series, please visit:
http://www.igi-global.com/book-series/advances-data-mining-database-management/37146

701 East Chocolate Avenue, Hershey, PA 17033, USA
Tel: 717-533-8845 x100 • Fax: 717-533-8661
E-Mail: cust@igi-global.com • www.igi-global.com

Table of Contents

Preface

During the months since the onset of the global pandemic in the early 2020, the novel coronavirus, namely COVID-19, has swept across the world, causing millions of deaths. At the time of writing in early spring of 2021, the pandemic is still ongoing, while the economic and social shocks associated with COVID-19 are only beginning to be realized. Such a health crisis has fundamentally affected the regular life, along with a long-time impact on the human society. There have been large amounts of academic publications proposed in multiple aspects from various visions of the pandemic, ranging from biomedical studies, clinical analytics, epidemiological researches, to social, economic, and political issues. With the advent of the cutting-edge technologies, particularly the big data analytics, researchers have made significant contributions to the study of COVID-19 from a variety of real-world problem-solving perspectives. Connecting COVID-19 research with big data analytics plays a crucial role in the advancement of industry and academia. This book, *Applied Big Data Analytics and Its Role in COVID-19 Research*, provides emerging research topics that emphasize on the development and implementation of the big data analytics in dealing with a variety of COVID-19 issues and controversies, including the evidence-based policy-making for disease control measures, the implementation and assessment of non-pharmaceutical interventions, the advances of biomedical approaches in diagnosis and treatment, and the strategical layouts in responding to social, economic, and political dynamics. The objective of this book is to investigate the role of the big data in COVID-19 research by incorporating a broad range of state-of-the-art techniques, such as big data processing, machine learning, deep learning, predictive analysis, data visualization, computer vision, image recognition, voice recognition, motion recognition, social media analysis, natural language processing (NLP), and geographic information system (GIS). Such key terms are crucial for the development and implementation of big data analytics in the COVID-19 research and the pandemic-related applications.

In a resolutely practical and data-driven universe, the way of research initiatives and industrial implementations has been changed by transforming analytical tasks, decision-making strategies, and business opportunities into the intelligence-based process. In the context of COVID-19 and the age of new normal, big data analytics has become one of the most efficient tools among various research themes and real-world applications, e.g. tracking the pandemic trend, predicting epidemiological parameters, implementing non-pharmaceutical disease control measures (i.e. the stay-at-home order, the social-distancing regulation, the face-covering policy, etc.), improving the efficiency of case identification and diagnostics, screening appropriate drugs and vaccine targets, maintaining the health of the internet information spreading and communication, monitoring public sentiments and political attitudes, and supporting decision-making processes in macroeconomic regulations, investment strategies, and business management and operations. The most significant challenge comes from the complexity and variability of the pandemic, which belongs to an interdisciplinary subject with big data analytics that can be attributed to many features and phenomena. The difficulty to implement applied methodologies for handling the COVID-19 research starts at the entry gate of the multi-dimensional investigation by understanding the role of data-driven approaches in between. This book attempts to fill all of the gaps between the domain expert knowledge and the data science method, and is expected to serve as a comprehensive reference in interdisciplinary research for COVID-19. Due to the lack of a comprehensive and complete reference book that integrates big data thinking with the COVID-19 pandemic, this book is obviously unique in its nature, contribution, value, scope, and originality by incorporating cutting-edge technologies and real-world applications, along with the current urgency of the COVID-19 research. The main objective of this book is to deliver a practical description of the COVID-19 pandemic using big data analytics with the proof from real-world examples and case studies.

With the above features and initiatives, this book is designed as a vital resource for data scientists, statisticians, epidemiologists, biomedical scientists, bioinformaticians, clinical analysts, pharmacologists, informatics experts, economists, sociologists and other researchers in the field of computational science, data mining, artificial intelligence, predictive analysis, data visualization, and computer vision. This book provides insights for public health authorities, government officials, lawmakers, and entrepreneurs to make decisions by taking the advantages of big data, thereby propelling the entire decision-making process more scientific, reasonable, and predictable in the age of COVID-19. Practitioners from computing, information technology, and big data industries, can also draw insights and inspiration from this book in terms of the development and management of the pandemic-related projects. It will be useful and applicable for developers to examine and identify major initializations and methods of computational requirements, system architectures, big data pipelines, sandbox implementations, testing and evaluation processes, and application deployments. The book also integrates both theoretical and practical configurations, which incorporate the most recent studies and applications in COVID-19, the theoretical framework and knowledge behind each topic, the cases of using big data and computational tools, the brand-new database, the open-source code, and the full cycle of the project development. Therefore, it can serve as a complementary textbook for advanced courses in big data analytics, predictive analysis, machine learning, deep learning, data mining, applied mathematics, statistical inference, bioinformatics, computational informatics, political science, econometrics, quantitive finance, business administration, and operational research. Moreover, the book provides a large number of open-questions and discussions for future research directions, which will be beneficial for senior undergraduates, postgraduates, and doctoral students in the form of the enlightenment of research opportunities. As the target audience for this book, pre-required knowledge sets, such as database management, regression analysis, statistical modeling, mathematical simulation, etc., are recommended prior to use this book. Besides, this book is coding-in-Python-intensive, despite the availabilities of the open-source code layout, therefore, it is better to be familiar with the Python programming language. The database and code can be accessable at the GitHub repository (https://github.com/pzhaoir/Applied-Big-Data-Analytics-and-Its-Role-in-COVID-19-Research). The theme of this book is designed according to the timeline of the COVID-19 pandemic.

Chapter 1 describes how big data analytics helps to track, process, and analyze COVID-19 data by incorporating the major components of visual mining methods. Topics cover a broad range of cutting-edge techniques that deal with a variety of big data problems, such as data aggregation, data preprocessing, big data pipeline construction, data workflow architecture design, visual mining, and web-based dashboard development. A comprehensive understanding of COVID-19 big data solutions will be investigated in this chapter, ranging from data aggregation, big data processing, to data visualization. Main issues about COVID-19 data gathering and preprocessing are concentrated on a series of subset topics, including data acquisition and aggregation, data cleaning and preprocessing, data collection and validation. Challenges of big data processing against COVID-19 data problems will be investigated by providing the existed potential big data solutions based on Apache ecosystems. A conceptual data workflow architecture will be examined by introducing several big data tools, such as Hadoop, MapReduce, Spark, Airflow, etc. A full cycle of COVID-19 visualizer and dashboard designs will also be illustrated by examining typical COVID-19 visual outputs using a variety of big data visualization tools, such as Tableau, R Shiny, Python visualization modules, and GIS visualization tools.

Chapter 2 presents a set of epidemiological models for disease control purposes of COVID-19, along with the empirical analysis based on the U.S. data. Data analysis and mathematical simulation results are illustrated to preview the trend of the COVID-19 outbreak under different scenarios. The effect of interventions will be compared with that of the no-actions based on a series of SIR family models that have been applied in the COVID-19 predictive analysis and simulation. A case study will be represented by using the U.S. real-time daily report data for the trend analysis and parameter estimations based on SIR, SIR-D, and SIR-F. The optimal model will be selected for simulating future situations given the no-actions scenarios, the effects under lockdown, and new medicines and vaccinations. Experimental results perfectly depict the COVID-19 pandemic outbreak in the U.S. from beginning to the second wave as of late 2020. The simulation results also provide the prevision of how the COVID-19 pandemic ends by investigating different scenarios and effects. Together with the discussion of future research directions, this chapter offers a comprehensive understanding of how epidemiological models work for predicting and simulating the COVID-19 pandemic. The conclusion indicates that the public authorities can reduce the epidemic scale based on a strict strategy projected from the simulation results.

Chapter 3 attempts to explain a broad range of forecasting techniques applied in the COVID-19 trend predictions, along with dealing with real-world problems using the U.S. real-time daily report data. Three types of forecasting methods, i.e. machine learning models, time series forecasting techniques, and deep learning algorithms, are categorized and introduced, mathematically and empirically. A series of models, such as SVM, bayesian ridge regression, decision tree, XGBoost, ARIMA, and LSTM, have been trained. A wavelet transformation will be applied to reduce the noise term and to make the input data stable. A comprehensive model evaluation and comparison will be presented with discussions of the model performance and efficiency for each model. To justify the outcomes from each model, this chapter has presented case studies of three pandemic scenarios, including the early stage, the second wave, and the real-time prediction. This chapter also illustrates the most recent studies and research trend in this field, along with the challenges existed in the predictive models for COVID-19 with the possible research in the future. The main focus of the chapter is to present how artificial intelligence helps for forecasting the cases of COVID-19, and to further assist for constructing the smart healthcare system.

Chapter 4 focuses on how big data analytics works to explore the human mobility pattern that can be used for evaluating and implementing mobility restriction effects during the COVID-19 pandemic. A data aggregation approach will be introduced by illustrating how the novel dataset is created in terms of its components, the structure of designs, and functionalities. A case study will be performed via visualizing mobility patterns for mapping nationwide human behaviors, exploring seasonal mobility features at the county-level, and segmenting driven patterns in major cities of the United States. Visual mining results indicate the overall mobility pattern overtime and across the nation, along with seeking the reasons why the stay-at-home order failed in the U.S. and the lessons learned from the past. The pandemic-mobility management system is proposed in terms of its architecture designs and functionalities by calculating the weights of involved factors, such as mobility measurements, COVID-19 cases information, and medical risk indicators, via variable importance based on machine learning algorithms. Experimentally, the pandemic-mobility management system can generate a new index that reflects the relationships between human mobility, pandemic spreading, and medical risk. Such a management system will make significant contribution in the form of reducing the spread of COVID-19 via evidence-based policy-making processes of the mobility restriction.

Chapter 5 deals with investigating how artificial intelligence assists to monitor social distancing violations using computer vision and deep learning. The algorithm is a DNN-based human detector that has been proposed to detect and track static and dynamic individuals in public places for monitoring social distancing metrics under multiple visual conditions. The system is able to operate in various challenging conditions with the outperforming model efficiency and complexity. The detector can be formalized upon a three-stage module that contains human detection, people tracking and identification, and inter-distance estimation. Such a framework has been integrated and used on CCTV surveillance cameras with the real-time performance. In this chapter, such modules are illustrated with a sample test procedure in terms of implementing a social distancing monitor under the framework of the deep learning-based model. Instead of the common-used video, a new dataset will be applied in the testing process, along with examples throughout the whole stages of the development of the monitor. An innovated social control management system, called DeepCovidPark, will be introduced and discussed in the form of future research direction, which is a real-world representation of successfully applied social distancing monitor in the context of the COVID-19 pandemic.

Chapter 6 illustrates how artificial intelligence serves as the solution toward monitoring the face covering conditions. Several typical face mask detection algorithms will be examined with an experimental testing procedure using the selected model and a new image set of training and testing algorithms for identifying multiple face covering conditions on both static images and real-time video streams. Three types of deep learning-based face mask detection models will be discussed, ranging from the CNN-based architectures, the hybrid deep learning models as the feature extractor with the classical machine learning-based classifiers, to the deep transfer learning-based methods. A deep transfer learning-based approach with fine-tuning the MobileNetV2 architecture is selected as the proposed model, which is applied to solve the real-world problems, including a basic prototype of the face mask detection, a face covering types detector, and a detector of the masked face conditions. Each face mask detector has been implemented in the form of the human-computer interaction with a real-time representation. The detector can identify whether individuals wear a face mask or not, recognize which types of the face covering a person wears, and monitor whether people wear face masks correctly or not, which will benefit the current situation by implementing the social management and control regulations during the COVID-19 pandemic.

Chapter 7 aims on illustrating how to identify COVID-19 suspected and confirmed cases rapidly and remotely using big data analytics. Three modules together offer a comprehensive understanding of the automatic system for the COVID-19 self-assessment application using a variety of data types, including clinical data, textual information, and human voice signals. The mobile app user interface (UI) has been designed for each module, providing a broad range of functionalities, such as the capability of the symptoms-based self-assessment, the functionality for communication and information sharing, and the mechanism of the voice-based self-testing. The symptoms-based self-assessment module allows users to evaluate whether they have infected the disease or not by answering several questions. Questions are generated by analyzing clinical data from hospitalized patients and suspected cases.

A smart tool will learn users' responses through a pre-trained model using machine learning. The communication and information sharing module allows users to communicate in real-time and offline. The functionality of the module allows users to post text and voice messages, share images and documents, and make real-time voice/video calls. This chapter will then analyze the beta test data that contains signal messages in regards to COVID-19. Users with typical symptoms and/or being tested positive will be identified automatically through an unsupervised learning algorithm. The voice-based self-testing module records the user's voice through coughing, breathing, and reading a few pre-defined sentences. Such recordings are processed by a deep learning model that has been trained through an existing COVID-19 voice data collection. The module can distinguish the sounds of COVID-19 suspected cases from healthy individuals. Each module will be investigated and discussed throughout the data acquisition, experimental designs, methodologies, experimental results, and potential improvements as the solution and recommendation for implementing a multi-angle assessment system for the COVID-19 self-testing capacity.

Chapter 8 investigates how to implement the current diagnosis of COVID-19 by incorporating data-driven approaches with the construction of a smart health framework. Topics cover a broad range of smart diagnosis innovations for supporting current assays and diagnostics, such as data analysis for nucleic acid test, machine learning-based serological signatures identification, medical image classification using deep learning, and decision support system for automatic diagnosis with clinical information. Each topic has been illustrated and discussed throughout methodologies, data collections, experimental designs and results, limitations and potential improvements. All applicational potentials have been examined with real-world datasets based on the existed studies. A set of machine learning algorithms, such as Naive Bayes, random forest, k-nearest neighbors, XGBoost, etc., has been incorporated for implementing intelligence-based biomedical tests, including RT-PCR, RT-LAMP, CRISPR-based diagnosis, and serological signatures identification. A pre-trained deep learning model will be investigated in terms of the network architecture, which can be used to implement the chest image classification, along with its usage with a real-world dataset. A decision support system for the COVID-19 automatic diagnosis will be proposed by using a XGBoost model and presenting the decision tree structure.

Chapter 9 provides a systematic review for investigating current research in drug discovering and vaccine development for COVID-19 throughout protein structural basis analysis and visualization, machine learning- and deep learning-based models, as well as a big data-driven approach. In the age of big data, computational approaches play a crucial role in drug discovery and vaccine development through the entire process of biomedical design. Despite the widely applied methods, many challenges still remain with the traditional data-driven approaches in terms of cost efficiency, model complexity, and reliability. With the development of AI, such bottlenecks have been broken through by using advanced machine learning and deep learning algorithms in the form of the cutting-edge technologies in drug development. This chapter will conduct a systematic survey study to investigate how big data and AI can help in drug discovering and vaccine development for COVID-19 throughout three major groups of methods, such as protein structural basis, AI-based models, and big data-driven approaches, along with the data and code accessibility for each study.

Chapter 10 proposes a cloud-based architecture to detect misleading information on COVID-19-related news and articles. The system has been illustrated through misinformation extraction, fake news detection, and ground-truth testing. A web-based application has been presented with a dashboard-like user interface design using cloud computing. A bench of word embeddings and deep learning algorithms has been investigated for determining the optimal model. The anti-misinformation system can identify the fake news in a second with a reliability study operated in a cloud computing environment. Several word embedding approaches have been chosen as the candidate feature extractors, including TF-IDF, Word2Vec, and Global Vector (GloVe), among which the most efficient methods will be selected for extracting text features based on the theoretical and experimental conclusions summarized from related studies. A variety of deep learning algorithms will be investigated in terms of the network structure for each model, such as DBN, RNN, LSTM, GRU, and CNN, whereas the candidate models will be selected as the fake news classifiers based upon the experimental results derived from recent studies. Finally, the optimal model will be deployed in a web-based UI with Amazon Web Services (AWS) using Python Flask API containerized in Docker and powered by Elastic Container Service (ECS).

Chapter 11 aims on investigating how big data analytics and AI models help in monitoring, tracking, and characterizing pandemic-related anti-Asian racism and xenophobia throughout data acquisition and preprocessing, text classification, model training and selection, real-time xenophobia monitering, and anti-Asian hater tracking and profiling. Data used in this chapter is collected from multiple social media channels, such as YouTube and Twitter, for different tasks, along with the information gathering and data cleaning process. To monitor racism and xenophobia, an optimal anti-Asian classifier has been trained before constructing the xenophobia index, which can be applied in detecting anti-Asian hate speech in real-time. Common-used word embedding methods, such as Embeddings from Language Models (ELMo) and Bidirectional Encoder Representations from Transformers (BERT), have been investigated in terms of the basic functionality, whereas BERT and GoogleNews-vectors have been applied in feature extraction, along with a set of machine learning and deep learning classifiers. To examine how big data helps in tracking and profiling anti-Asian haters, a novel dataset, namely COVID-HATE-TRACK, has been created on tracking anti-Asian haters from an automatically labeled dataset based on tweets by using machine learning algorithms, followed by characterizing haters' profiles using sentiment analysis and Apriori algorithms.

Chapter 12 provides a systematic and comprehensive survey on current applications of big data analytics that deal with a broad range of social and economic changes incorporating economic development, investment management, and business operation issues associated with COVID-19. Existing studies in this field are divided into three layers, ranging from the macroeconomic monitoring and measures, asset and investment management, to business operations and social changes, along with the discussion of the role of big data in between. Various ongoing and urgent topics, such as pandemic recession, economic uncertainty, sustainable development, labor demand and supply, energy consumption, financial market volatility, housing and rental, consumer behaviors, risk control, business operation and management, and social changes, have been incorporated with an examination on how big data works to provide insightful evidence in the decision-making of economic policies, investment strategies, and business operations.

As such, this book is proposed to help to understand the multiple aspects of the COVID-19 research topics using big data analytics. Such works will derive and represent the comprehensive investigations of the pandemic-big-data theme construction, architect, and intersection with computational, epidemiological, biomedical, pharmacological, political, economical, and social domains. The book essentially deals with the role of big data analytics in the COVID-19 research by incorporating a broad range of real-world applications using cutting-edge technologies, which have been represented for almost all research orientations of the COVID-19 pandemic. Therefore, this book will accelerate and influence the interdisciplinary research of COVID-19 towards the next generation of data-driven methodologies and ideologies.

Acknowledgment

With all my hearty wishes, I would like to thank all of those who supported and helped in making the progress of this book. Special thanks go to Drs. Shuang-Hun Chen, Mathias Hoffmann, and Shikha Takahashi for providing comments and suggestions. Many thanks with appreciation to faculties from Beijing University of Civil Engineering and Architecture, Prof. Hou Pingying, Ms. Shen Bingjie, Ms. Lin Qing, and Ms. Yuan Yi, at School of Humanities, for the assistance in wrapping up and proofreading this book. Thanks also to the Assistant Development Editor, Angelina Olivas, Director of Intellectual Property & Contracts, Jan Travers, and Book development Coordinator, Maria Rohde, at the IGI Global. A very special thanks to all those contributing to the COVID-19 Research Project, Bangcheng Li, Tristan Huang, Alex Brown, and Jingyao Guo, for providing assistance in data collections, database management, and coding debugs to make this book a reality.

Peng Zhao
Intelligentrabbit, LLC, USA

Chapter 1
Data Gathering, Processing, and Visualization for COVID-19

ABSTRACT

The novel coronavirus has been impacting human society since 2019. The global death toll has exceeded 400 million as of October 2021. To better understand the dynamic of COVID-19, big data technologies can in principle be applied to provide the overview and the preview for how it spread spatially and temporally. This chapter introduces how big data helps in terms of tracking, processing, visualizing, and analyzing COVID-19 information by illustrating the major components of data visualization designs. Topics cover a broad range of cutting-edge techniques that deal with a variety of big data problems, such as data aggregation, data preprocessing, big data pipeline construction, data workflow architecture design, visual mining, and web-based dashboard development.

INTRODUCTION

Since December 2019, a novel coronavirus, namely Severe Acute Respiratory Syndrome Coronavirus 2 (SARS-CoV-2), which causes Coronavirus Disease-19 (COVID-19), has been spreading rapidly in the whole world. Due to its high human-to-human transmission efficiency and the severe consequence of the infection, COVID-19 has become one of the most significant health

DOI: 10.4018/978-1-7998-8793-5.ch001

crises in modern history. World Health Organization (WHO) declared that COVID-19 is a global pandemic on March 11, 2020. According to WHO, a total of 236,599,025 individuals have been confirmed as infected cases worldwide, including 4,831,486 deaths, as of 6:49 CEST, October 8, 2021 (WHO, 2020). The highest numbers of COVID-19 cases have been reported in the United States of America (44.2M), followed by India (33.9M), Brazil (21.5M), the United Kingdom (8.05M), and Russia (7.58). Till date, the COVID-19 pandemic is ongoing despite the availability of medications and high vaccination rates in some regions, due to a series of COVID-19 variants, such as B.1.1.7 (Alpha), B.1.351 (Beta), B.1.617.2 (Delta), etc. (Rubin, 2021; Bernal et al., 2021). Fighting against the COVID-19 pandemic has become the main theme in the past two years. Research communities, public health organizations, government authorities, and industrial sectors have been involved to overcome the challenges of the pandemic, while many innovative disease control measures have been proposed by using cutting-edge technologies, such as big data analytics, artificial intelligence (AI), Internet of Things (IoT), and 5G (Rodríguez-Rodríguez et al., 2021; Siriwardhana et al., 2020).

Various state-of-the-art technologies have been applied to deal with the current urgency of the pandemic, while big data is the foundation that supports many other applications, such as epidemiological models, pandemic trend predictions, and machine learning tasks. One such noteworthy component is tracking major measurements of COVID-19 through visual mining tools, while big data processing plays a crucial role among the full cycle of the project development. Big data analytics is an automated process which uses a set of techniques or tools to access large-scale datasets for extracting useful information and insights. Such a process involves a series of customized and proprietary steps. It requires specific knowledge to handle and operate the workflow properly. Due to the nature of 4Vs (Volumes, Variety, Velocity and Veracity), a robust, reliable and fault-tolerant data processing pipeline is essential for any big data project. The sustainable automation can consolidate all tasks throughout ETL, data warehousing, and data visualization. The upstream vendor data will be ingested into a data lake, where source data is maintained and go through the data processing of cleaning, scrubbing, normalization and insight extraction. Prior to the visual mining and data analytical tasks, data should be preprocessed towards a user-friendly manner.

A massive amount of data visualizers and dashboards has been generated to track and analyze the feature of COVID-19 spreading patterns. Data can be classified into temporal measurements (i.e. infection, recovery, death counts, hospitalizations, test results over time), geographical information (i.e. disease measurements at different geo-spatial levels), and demographical data (such as disease measurements per race). However, gathering appropriate data for tracking the pandemic facets of COVID-19 is still challenging due to differences of regional reported records and the complexity of data formats. Data visualization, on the other hand, has become one of the most powerful tools for tracking data related to COVID-19, along with many web-based visualizers and dashboards. Prominent example dashboards include Coronavirus COVID-19 Global Cases by the Center for Systems Science and Engineering (CSSE) at Johns Hopkins University (JHU), WHO Coronavirus Dashboard, Coronavirus Data in the United States by New York Times (NYT), the U.S. Centers for Disease Control and Prevention (CDC) COVID Data Tracker, etc., in which big data techniques have been widely applied. It is important to compare different regions in terms of identifying potential emerging hotspots and assessing possible effects of disease control strategies by understanding the dynamics of the pandemic.

Although such dashboards present insightful data visualization in regards to the novel coronavirus, they do not particularly emphasize on the operational mechanism that illustrates the full cycle of the visualizer design and dashboard development. As a result, it is difficult for research communities to develop new visual representations as per updated information streams and analytical requirements for tracking COVID-19 dynamics. Motivated by the currently urgent situation of COVID-19 and the lessons learned from the pandemic, this chapter deals with how big data analytics works in tracking, processing, visualizing COVID-19 data. The objectives of the chapter are:

- illustrating the role of big data in tracking and visualizing COVID-19 epidemiological data using state-of-the-art big data analytical tools;
- providing a comprehensive understanding for COVID-19 big data processing by introducing cutting-edge big data solutions;
- investigating how big data works throughout the full cycle of the COVID-19 data visualization development by examining major components of the big data pipeline.

BACKGROUND

Big data plays a crucial role in gathering, processing, and visualizing COVID-19 data. Thanks to the fast revolutionary of information technologies and systems, the avalanche-like growth of data has prompted the emergence of new models and technical methods for distributed data processing, including MapReduce, Dryad, and Spark (Khan et al., 2014). For processing large-scale datasets, ad-hoc systems have been proposed for distributed computing in terms of cutting-edge data workflow architectures (Gonzalez et al., 2014). Over the last few years, open-source platforms have become increasingly popular as many big data tools are easily available. Most recent big data applications are concentrated on hybrid approaches using open-source tools, e.g. Apache ecosystems. Suleykin & Panfilov (2019) proposed a big data pipeline using Apache Hadoop, Postgre SQL, and Apache Airflow. Such a system architecture can be performed with stages through multiple storage spaces for industrial KPIs of millions records. Such ideologies can be applied in COVID-19 research in terms of processing big data with multiple open-source tools. COVID-19 datasets are represented in different formats for different regions around the world, with which only big data management tools are suitable, due to many options that can be used for processing in the form of database models, such as SQL, NoSQL, MongoDB, Cassandra, etc.

On the other hand, COVID-19 visualizers and dashboards have been widely applied in tracking the pandemic. Except for popular dashboards that can be accessible from the websites, several COVID-19 data visualization outputs have also been proposed by research communities and data science groups. Yang et al. (2020) developed an interactive tracking, visualizing, and analyzing platform for COVID-19 in terms of presenting the global map of COVID-19 infected cases, illustrating the pandemic trends at multi-grained levels, and integrating a multiple vision of the disease spreading dynamics. Similarly, Pan et al. (2020) implemented a novel data visualization platform for tracking COVID-19 globally, including a comprehensive database of world-wide COVID-19 cases, heatmaps and detailed trends of confirmed cases per region, the public concern flow based on daily tweets collections, and the breaking news integrations. Some COVID-19 visualizers have been deployed into a web-based server throughout data collection, big data processing, and data visual presentation as many developers are seeking an automatic solution in the form of big data analytics as services (Idogawa et al., 2020). Moreover, with the development of the vaccine design and the vaccination progressive, more dashboards have been implemented for visualization purposes as per the current urgency in disease control measures (Shrotri et al., 2021; Shaheen et al., 2021). As such, big data-based visualization approaches make huge contributions in tracking and analyzing the COVID-19 pandemic, thereby providing many insights for supporting the evidence-based policy-making in disease control measures.

MAIN FOCUS OF THE CHAPTER

Issues, controversies, problems

In the recent age of big data, a large amount of data has been created from a vast amount of data sources. Useful information and valuable knowledge can be derived from such raw data embedded in multiple big data providers. Analyzing healthcare and epidemiological data is one of the most demanded tasks for knowledge extraction and information retrieval in the context of COVID-19. Insights discovered from such epidemiological measurements help research communities and public authorities to get a better understanding of the novel coronavirus in terms of the inspiration in making rapid reactions to track, monitor, and fight the disease. As per a famous quote "a picture is worth a thousand words", data visualization makes it efficient to analyze the data, in which big data analytics plays a fundamental role among data gathering and preprocessing, big data pipeline and data workflow construction, and visualizer design and dashboard development. Despite various COVID-19 visualizers and dashboards, little is discussed on how big data works throughout the whole developing cycle of the COVID-19 visualization process.

Features of the Chapter

In this chapter, a comprehensive understanding of COVID-19 big data solutions will be investigated in terms of how big data helps in gathering, processing, and visualizing pandemic-related data. This chapter composes three components, i.e. data aggregation, big data processing, and data visualization for COVID-19. Main issues about COVID-19 data gathering and preprocessing are concentrated on a series of subset topics, including data acquisition and aggregation, data cleaning and preprocessing, data collection and validation. Challenges of big data processing against COVID-19 data problems will be investigated by providing existing and potential big data solutions based on Apache ecosystems. A conceptual data workflow architecture will be examined by introducing several big data tools, such as Hadoop, MapReduce, Spark, Airflow, etc. A full cycle of COVID-19 visualizer and dashboard designs will also be illustrated through examining typical COVID-19 visual products using a variety of big data visualization tools, such as Tableau, R Shiny, Python visualization modules, and GIS visualization tools.

SOLUTIONS AND RECOMMENDATIONS

Data Gathering and Preprocessing

Data Acquisition

Data gathering is the foundation in constructing a COVID-19 visualizer, where identifying data sources plays one of the most critical roles in the data visualization pipeline. A variety of publicly available datasets can be applied, of which most of them are updated frequently, and county- or state-level records of COVID-19 confirmed cases and deaths are incorporated in a specific country or region. Thanks to the contribution of the data providers and the global data science community, various data sources regarding COVID-19 cases are offering multiple precision with concentrated on the first glance of the pandemic. Common-used datasets for visualization purposes have been released by universities' research teams, public health authorities, news paper institutions, multi-platform publishers, and non-profit organizations. Most of them can be acquired from official websites and Github repositories.

The first research group that aggregated COVID-19 data was the JHU CSSE, which published the total cases, recovered counts, and deaths of COVID-19 at the country, state, and county levels (Dong et al., 2020). Such a dataset has been releasing to the public in a sizable and ongoing manner. Later on, The NYT released the COVID-19 tracking data that was collecting and tracking novel coronavirus cases and deaths on the county level since January 2020 (The New York Times, 2021). Confirmed cases in NYT dataset were defined as individuals who tested positive and were attributed to the county in which the man or woman was treated and recorded on the date that the case was announced to the public. The Atlantic, a magazine and multi-platform publisher, incubated the COVID Tracking Project that provided a grassroots effort in measuring and tracking the pandemic trend in the United States (Miller & Curry, 2020). Daily updates for test results, hospitalizations, patients in ICUs, and deaths were reported per state, whereas some of such data records are not available for every state due to a high amount of variability.

Besides, similar measurements for tracking COVID-19 cases have been also released by non-profit organizations, e.g. the USAFacts updates county-level daily data in the morning of the following day (USAFacts, 2020).

From the data preprocessing perspective, the first step towards constructing the COVID-19 tracker and visualizer relies on fundamentally understanding attributes and contents of the data sources in terms of types of measurements associated with geographical levels. The differences among the above-mentioned datasets are summarized in Table 1, based upon how the data are collected and compiled. Such four data sources apply different approaches to aggregate COVID-19 data, while most of them incorporate primary measurements regarding the novel coronavirus as per different geographical levels. On the other hand, validating the data quality plays a crucial role in any data-driven projects, thereby requiring to compare the consistency of those sources with the standard database. The U.S. CDC releases COVID-19 data for confirmed cases throughout the entire country (Centers for Disease Control and Prevention, 2020). Such an authorized database can be used to validate the data aggregation among other data sources.

Table 1. The summarized comparison among major data sources for tracking COVID-19

Source	Attributes	Geographical Levels	Data Availability
JHU	infected, recovered, and death counts; tested counts; probable infected and death counts	national, state, and county (only national-level data available for recovered counts; only state-level data available for tested counts)	available at GitHub https://github.com/CSSEGISandData/COVID-19
NYT	infected and death counts; place of fatality; probable infected and death counts	national, state, and county	available at GitHub https://github.com/nytimes/covid-19-data
The Atlantic	infected, recovered, and death counts; tested and hospitalized counts; place of fatality; probable infected and death counts	national and state	available at GitHub https://github.com/COVID19Tracking
USAFacts	infected and death counts; probable death counts	national, state, and county	can be downloaded from the website https://usafacts.org/visualizations/coronavirus-covid-19-spread-map/

Abnormal Data Detection and Repairing

Abnormal records usually appear among datasets that are ready to use in the next stage of data analysis and visualization. As one of the most important procedures in data preprocessing, abnormal data detection and repairing will offer the robust maintenance for obtaining better visual and analytic results. Common issues among COVID-19 datasets have been commonly discussed by observing several types of abnormalities, such as the order dependency violation, delayed updates on specific dates, abnormal data points and periods (Wang et al., 2020a). Order dependency is widely applied in relational databases, particularity, it is important to validate the characteristics for the cumulative time series data. The order dependency violation in COVID-19 data happens when the cumulative count of a variable on a certain day is bigger or equal to that on the next day. Moreover, delayed updates are usually happened on weekends and/or holidays as the situation that there are significantly fewer daily new test results reported on weekends and holidays. Such an effect is very common in COVID-19 datasets at both state and county levels due to the limited workforce in local healthcare systems. Such a problem can be detected by testing the intra-week seasonality. Furthermore, Single anomaly points and abnormal periods are everywhere in the raw data of COVID-19. The unexpected spikes at one time spot appear in the cumulative time series plot as the large batch of reported data was released and the reporting standards was changed. Many data-driven projects are experiencing a continuous abnormal period due to the increasing speed in COVID-19 measurements. Such changes can affect the pattern in visualization, whereas they are essential in time series and can be detected by segmenting the structural change.

To deal with the data abnormality issue, several basic validation procedures should be implemented prior to the data repairing. Firstly, researchers need to understand the nature of each measurement in COVID-19 datasets, e.g. the number of infected cases and death counts should be non-negative integers, along with utilizing the dependence structure among observations. Secondly, the population should be adjusted to normalize data to major measurements per specific amount of residents. Finally, it is necessary to compare major metropolitan areas with that of the aggregated counties into the combined statistical areas as the spread of COVID-19 also has geo-spatial patterns. Except for the basic validation procedures, a variety of data repairing approaches can be applied to handle the problems mentioned regarding data abnormality issues. The main idea is to estimate the correct values at the abnormal data points.

To deal with time series models for count data, the generalized linear model (GLM) can be applied in terms of modeling the samples conditionally on the historical data. The integer autoregressive moving average (INARMA) model is another option for analyzing count time series, whereas the GLM method yields a more convenient way in predictions, and can be easily implemented in computational tools, such as R and Python, which perform the model estimation through the quasi-conditional maximum likelihood function. Some studies combined GLM and exponential predictors in estimating time series data counts by incorporating individual county linear, exponential, and exponential epidemic predictors (Altieri et al., 2020). Moreover, the discrete-time spatial epidemic model has been widely used with combinations of infected, susceptible, and removed cases (Wang et al., 2020b).

Web Scraping for Data Collection

As many research studies have been proposed to the effort and prevention of COVID-19 outbreak, scholarly production datasets focused on the pandemic have been conducted, providing the insightful overview of scientific research studies, to identify information for data science communities, public authorities, and epidemiologies, who are the most active in the task force to fight the novel coronavirus. Such research records from existing articles can be collected in terms of metadata gathering and processing from different publication databases, such as arXiv, bioRxiv, PubMed, Scopus, and Google Scholar. Such data can be extracted by using the web scraping techniques and be summarized and preprocessed through text analysis and data wrangling. Such an ideology can be incorporated with bibliometrics and scientometrics, in which data science makes much more contributions in generating the scholarly production dataset report for COVID-19 research analysis. Santos et al. (2020) proposed a dataset that can be used by other researchers to implement automatic mechanisms for extracting insights from abstracts and keywords of publish studies in regards to COVID-19. Various entities and relationships can be also extracted through applying the complex network analysis technique involved in this dataset, along with identifying most influential scholars and research groups in initiatives of COVID-19 studies. Despite the limited scale of collected data, the pipeline and source codes are publicly accessible from GitHub, thereby providing many opportunities in creativities of further implementations. Inspired by the same idea, this chapter applies web scraping methods to deploy a quick screening tool that can detect and extract general information of COVID-19 datasets from relevant studies on Google Scholar. Sample searching results are shown in table 2. Such datasets can be used as the supplementary datasets for multiple research objectives.

Table 2. The summarized information among supplementary data sources for tracking COVID-19. Note: searching results represent the numbers of total article related to the dataset as of October 6, 2021.

Source	General Information and Usages	Searching Results	Data Availability
Amazon Web Services (AWS) data lake for analysis of COVID-19 data	a centralized repository with continuously updated datasets regarding the spread of COVID-19 and the characteristics of SARS-CoV-2. Dashboards can be created based on data using visualization tools. Research communities can use this data repository to perform epidemiology studies of COVID-19	258	https://dj2taa9i652rf.cloudfront.net/
CDC COVID-19 Cases, Data, and Surveillance	The data hub can help researchers to explore the COVID-19-related trends for generating the guidance on public health actions. This dataset is collected through reliable steps, including counting for cases and deaths, reviewing through jurisdictional websites, sharing by COVID Data Tracker. Visualization tools can be used for data analysis, predictions, and dashboard designs in terms of representations of COVID-19 spread patterns	416	https://www.cdc.gov/coronavirus/2019-ncov/cases-updates/
Google Cloud Platform (GCP)	provides a repository of public datasets for COVID-19. The datasets are collected from multiple data sources, including the New York Times, Google, OpenStreetMap, European Center for Disease Prevention and Control, and Global Health Data from the World Bank. The data can be queried through BigQuery's Sandbox.	393	https://console.cloud.google.com/marketplace/browse?filter=category:covid19
DXY	a GitHub repository that is maintained by many volunteers, which aggregates multiple data sources, such as local media, government reports, and health organizations, to provide cumulative COVID-19 cases in near real-time.	282	https://github.com/BlankerL/DXY-COVID-19-Data
1point3acres COVID-19 tracker	a real-time update and digestible information hub designed for COVID-19 data sharing and visualization, proposed and maintained by international students from China.	82	https://coronavirus.1point3acres.com/

Despite availabilities of various data sources, one of the most primary issues is that users are difficult to access the original raw data and customize such data to satisfy specific research proposals in categories, data structures, and resource selections. Some data sources are publicly read-only, thereby it is challenging to extract useful data from the original background database. To deal such challenges, web scraping can be applied in developing a toolset to extract, refine, unify, and store COVID-19 data at different scales of current available data sources. One of the most adoptable tools is COVID-Scraper, which is an open-source tool for automatically scraping and processing multi-scale spatiotemporal COVID-19 data (Lan et al., 2021). Such a toolset can collect, filter, organize, preprocess, and store COVID-19 records from multiple types of spatiotemporal data sources, along with up-to-date data records available in GitHub repository and a cloud-based database for the public. Besides, web scraping techniques have been applied in constructing an expert system, which helps research communities to conduct independent studies of COVID-19, e.g. the risk diagnosis and mitigation system of COVID-19 (Mufid et al., 2020). Such a system can perform an early detection feature of COVID-19 using the rule-based approach, along with information obtained by web scraping from official websites of the task force for the acceleration of dealing with the novel coronavirus. As such, web scraping techniques can be applied as a powerful tool to collect extra datasets for establishing supplementary databases in COVID-19 research.

Big Data Tools and Workflows

Processing Big Data for COVID-19 Challenges

Information technology plays a crucial role in COVID-19 research, of which it can map and integrate decision-making processes efficiently. Insights derived from data sources related to COVID-19 can be visualized and analyzed by using big data analytics, which handles a large scale of data with the combination of highly complex and optimized algorithms. In the era of the Industry 4.0, big data analytics becomes the core technology in processing, recording, storing, and analyzing the information from the whole society. Particularly, when the COVID-19 pandemic disrupts the regular life, public authorities are seeking efficient ways in dealing with the health crisis by using big data, whereas it is challenging to implement such a cutting-edge technology in terms of big data processing, data storage, and data pipeline deployment. The challenges of continuous changes in data structure have to be conquered by much effort, especially during the current COVID-19 pandemic, with its incoming unprecedented data processing needs. Several big data platforms can be applied to process vast amounts of data collected from COVID-19 data sources in a cloud-based parallel computing environment.

The survey of recent studies in discussing big data challenges of COVID-19 suggests that a variety of big data and database management tools, such as Hadoop, MapReduce, and Spark, have been incorporated in collecting, storing, processing, visualizing, and analyzing COVID-19 datasets, along with database management systems (DBMSs), such as SQL and NoSQL.

Apache Hadoop is a set of open-source toolsets that is designed for problem-solving initiatives of massive amounts of data and computation (Vavilapalli et al., 2013). The base Hadoop framework is composed of major modules, such as Hadoop Common, Hadoop Distributed File System (HDFS), YARN, MapReduce, and Ozone. Current studies using Apache Hadoop for COVID-19 are mainly concentrated on processing big data with HDFS and MapReduce (Azeroual & Fabre, 2021; Vijay & Nanda, 2020). HDFS has been widely used to solve various big data challenges, ranging from data collection, storage, sharing, to data analytics. HDFS can handle large amounts of structured and unstructured data efficiently as multiple computing nodes are working simultaneously. MapReduce, on the other hand, can be used to process and generating big data with a parallel and distributed data modeling on a cluster (Dean & Ghemawat, 2008). To implement a MapReduce task, JobTracker works for assigning the task to each node, while TaskTracker will accept the task. Both of them form a complete MapReduce framework to handle large amounts of data.

Apache Spark is a unified computing engine that is composed of a set of libraries for parallel data processing on computer clusters (Zaharia et al., 2016). Spark is the most efficient open-source engine for implementing big data task in the form of a standard tool for many developers and data scientists. Spark can process 100 times faster than Hadoop, along with supporting multiple programming languages, such as Java, Python, Scala, and R. Spark also supports the integration with HDFS, HBase, Canssandra, Amazon S3, Hive, etc. Spark Core is the key foundation in dealing with a variety of big data tasks, including dispatching and scheduling distributed tasks, managing data storage, commanding basic I/O functionalities, and conducting fault tolerance strategy. Resilient Distributed Dataset (RDD) is proposed for parallel operations, including transformations and actions. Transformation operations work for mapping, filtering, joining, and unionizing on RDD. As such, Spark has been applied as one of the most powerful tools in solving COVID-19 data problems (Elmeiligy et al., 2020; Khashan et al., 2020).

From the DBMS perspective, managing COVID-19 data requires a powerful tool that handles complex data models. NoSQL provides such a mechanism for storage and retrieval of data that is modeled in both tabular and non-tabular relations applied in relational databases. NoSQL can be used in many typical use cases, such as caching layer for data, storing and mining non-transactional data, serving as a store for pre-aggregated data, etc. (Mazumder, 2016). Over one hundred NoSQL databases are available across open-sourced big data platforms, while several popular databases, such as MongoDB, Cassandra, and HBase, are widely applied in dealing with large scale and medium size databases. NoSQL has been applied to deal with DBMS for COVID-19, e.g. discussions of the general methodology in data selection and security using NoSQL for COVID-19 DBMS (ElDahshan et al., 2020).

Alternative Options and Potential Solutions

Expect for the exiting big data tools for COVID-19 DBMS, alternative options can be also adopted in the form of potential solutions. Many other similar tools have not been incorporated in current studies. Although many options, such as Hive, Impala, Cloudera, etc., can also be applied as potential solutions, this chapter chooses three big data tools which are listed as follows:

- Apache Sqoop is a widely popular component of the Hadoop ecosystem, which is a tool to import the structured data from relational database systems into Hadoop for MapReduce tasks (Vohra, 2016). Importing data with Sqoop goes through two main steps. The first step is to collect the necessary metadata, and the second step is to submit a map job to the cluster. The imported data is usually stored in text or in Sequence and Avro files as binary data. For the process of exporting data, Sqoop shifts data from HDFS to RDBMS, which is processed through checking for the database and transferring the data. In addition, Sqoop can be utilized by a variety of structured data sources, such as Oracle, Postgres, and can directly load data from Hive and HBase, which make transferring data efficiently and cost-effectively.
- Hadoop Hue is another open-source user interface (UI), through which users can manage HDFS and MapReduce applications without utilizing the command line (Sharma & Navdeti, 2014). Users choose Hue as an optimal choice to operating Hadoop cluster, due to its advantages, including Hadoop API access, presence of HDFS file browser, editor for Hive and Pig query, Hadoop shell access, etc. One of the most important feature of Hue is the capability to access the HDFS browser. Users can control the works on HDFS through its interface. Users can

apply SQL Hive queries through the editor and then the browser will present the data analytical results.

- Apache Airflow is an open-source workflow platform that can manage and monitor the data workflows (Mitchell et al., 2019). Directed Acyclic Graph (DAG) enables the Airflow to operate workflow schedule. Airflow uses SerialExecutor to assign a task to workers, to determine number of tasks to work on simultaneously, and to update the progress status of the tasks. Airflow deploys a database backend to MySQL or PostgresSQL using Hooks to connect and store the configuration and status information for all DAGs and tasks. Moreover, Airflow can track original data with lineage support through inlets and outlets of the tasks to check the workflow over time. Operators are building blocks that deal with the real work and conduct tasks in a specific order as per the functional structure of Airflow. The major operational functions of Airflow can be fulfilled by three categories, such as action, transfer, and sensor. Airflow also allows the user to generate new operators through defining the parameters for the new operator and adding config values in executing codes.

Each of these tools has its own merits, thereby it is important to select the appropriate one to deal with the COVID-19 big data problems according to different objectives of the research needs.

Big Data Pipeline and Workflow

Implementing a big data pipeline allows data flows to work automatically and intelligently, thereby reducing the cost in processing big data for COVID-19 research. A variety of data problems, such as data processing, application integration, and data storing and warehousing, can be incorporated in workflow technologies, which have been widely applied in describing and automating big data pipelines. Many big data tools have been developed based upon cutting-edge data workflows that allow users to implement them on multiple types of computational systems, particularity on distributed ones. As the complexity of big data processing is increasing, different data formats are transmitting and integrating throughout almost all big data projects, along with dealing with other data complexities, such as volume growth, velocity, and variety. Nowadays, as the open-source platforms are becoming popular and powerful, ready-to-use primitives have been incorporated as an important feature of existing big data systems for distributed data processing, which is the abstraction of the data engineering objectives that deliver details of the implementation of computational demands (Suleykin & Panfilov, 2019).

Prior to the big data era, traditional ways of big data pipeline and data workflow are challenging in extracting raw data into a data analytic. Such data models rely on processing the data files which satisfy the size limit. The job scheduling tool, such as Autosys, cannot implement UIs to monitor each step of data processing. Each different type of raw data has its own challenge, which makes it difficult to standardize the data pipeline. Therefore, traditional data pipelines and workflows may not be suitable to meet current big data challenges of COVID-19 research. Constructing a reliable, robust, and fault-tolerant data pipeline plays a critical role in the data processing and storage framework, which enables of addressing big data loads and transactions with data from multiple sources for COVID-19 research. To satisfy such the urgent demand, this chapter presents a novel data workflow architecture in terms of a conceptual representation in establishing a big data pipeline for processing COVID-19 data. The proposed architecture conceptually illustrates the full cycle of the data flow transaction from external raw data extraction to the visual mining and data analytics using the hybrid architecture designs of big data tools.

The proposed workflow consolidates the ETL process, data analytics and data warehousing onto a state-of-the-art big data platform, i.e. Apache Airflow. Once the data is extracted from external sources, the data engine will load datasets into the data lake as-is from original databases, e.g. ftp/sftp, webservice, AWS S3, etc., followed by the automated process of cleansing, scrubbing, normalizing and data quality checking steps, which can be triggered accordingly. Such a novel data workflow architecture can be applied in handling a variety of COVID-19 big data problems with the fully automated features of the system. The whole ETL process applies Apache Airflow to ingest various feed and data validation, along with data quality checking, and finally produces data asset using Apache Spark. However, such a system requires the pre-actions to setup Airflow and Cloudera clusters, thereby may cause some unexpected difficulties in initializing the system architecture.

Data Visualization

COVID-19 Data Visualizers and Dashboards

As the novel coronavirus has been spreading globally, research communities and health authorities are making efforts on tracking and visualizing the pandemic patterns over time. A large number of COVID-19 data visualizers and dashboards have been proposed, whereas most of them share the same properties in terms of visual methods, functionalities, and dashboard designs. Almost all of the current existing dashboards designed for tracking and visualizing COVID-19 are fully-interactive, real-time, and informative. Various types of dashboards have been developed, while most of them are focusing on the actual COVID-19 measurements, such as confirmed cases and deaths at the beginning of the pandemic (Leung et al., 2020). With the vaccine development, more and more dashboards and data visualizers have been proposed in dealing with the vaccination progressive globally (Mathieu et al., 2021). Despite the availability and variety of COVID-19 data visualization, some issues are still remaining during the visual techniques and dashboard designs, from perceptual, scientific, and technical perspectives.

The majority of the existing COVID-19 visualizers incorporated GIS data by applying either a bubble map or a choropleth map to represent the total numbers of infected cases and deaths at different geographical levels. The total number of COVID-19 measurements for each region is represented by the radius of the bubble, in an absolute value or a relative adjustment with respect to population. However, many bubbles overlap as the severity of the pandemic in many regions can be indicated by the sizes of the bubbles, thereby making it difficult to visualize the accurate representations of the disease in dense regions. Alternatively, choropleth maps represent diversities in shading, coloring, saturating, or symboling within predefined regions to represent the COVID-19 measurements. Although a choropleth map can solve the overlapping problem, smaller regions may not be able to visible due to the sizes of geographical natures. Except for visualizing spatial data, temporal records of the COVID-19 outbreak data has also been incorporated in data visualization as many COVID-19 visualizers explore the pandemic trends for different regions. The temporal information of COVID-19 is usually represented by time-series plots, line charts, column diagrams, stacked column charts, areas under curve, and stacked areas under curve. Moreover, some COVID-19 visualizers applied a hybrid method for handling the spatiotemporal information from the COVID-19 pandemic data, which can be represented by line graphs, e.g. daily new infected cases over time for all states and counties can be indicated by a line graph with multiple lines, where each one represents the trend of the territory.

Besides, one of the most significant research trends in COVID-19 data visualization relies on using hybrid big data analytical tools and platforms. For example, Villanustre et al. (2021) developed a visualization framework to model the COVID-19 spread by using an innovative big data analytics tool, namely LexisNexis HPSS, which is an open-source platform that incorporates a system architecture implemented on commodity computing clusters to provide high-performance, data-parallel processing for tracking and analyzing COVID-19 data. Similarly, a conceptual system design for Geo-Online Exploratory Data Visualization (Geo-OEDV) of COVID-19 has been proposed by incorporating several cartographic and information tools, such as Information and Communication Technologies (ICT), Web Geographical Information Systems (WebGIS), web-based dashboards, infographics, and statistical modelings (Bernasconi & Grandi, 2021). Such attempts have implemented the full cycle of big data pipelines, from raw data to visual mining, thereby providing valuable experience in constructing COVID-19 dashboards in terms of the automatically operational systems.

Data Visualization Tools

Applying data visualization tools plays a fundamental role in producing COVID-19 visualizers and dashboards, thereby identifying the appropriate software is essential on the top of the project initialization. Tableau is one of the most popular business intelligence (BI) tools that can be applied in many analytical tasks and visual mining projects, along with several major components, such as Tableau Public, Tableau Desktop, Tableau Server, and Tableau Online. Tableau plays an important role in visualizing COVID-19 data due to its advantages, such as the completely interactive UI, the easy-to-use and easy-to-learn property, high-performance in real-time visualizations, and easy-to-share dashboards. Tableau has been applied in current studies of data analytics and visualization for COVID-19 in terms of data aggregation, reporting, and dashboard designs (Akhtar et al., 2020). However, Tableau is not attractive for many other dashboard designers due to its disadvantages, such as output limitations, the weakness in ETL, highly formatted reports, and difficulties in data preparation. With the improvement of Tableau, such problems can be solved by incorporating other programming softwares, such as R and Python.

R Shiny is a RStudio package that can be applied to build interactive web applications with R programming. It is the most popular tool that has been widely used in data reporting and data data visualization, where users can implement HTML elements to stylize the content of the application. R Shiny allows users to build a web-based visualizer in the real-time manner without JavaScript. An interactive UI theme working with any R environment provides pre-defined and customized output widgets for delivering almost all plots and diagrams that are demanded for tracking and visualizing COVID-19 data as per printed outputs in R objects. Several visualization tools for COVID-19 have been proposed by using R Shiny, e.g. an open-source analytical tools for exploring the COVID-19 outbreak with representations of interactive web apps under the R Shiny framework (Wu et al., 2020); a real-time application, namely the COVID-19 Watcher, for checking and visualizing the NYT and COVID Tracking Project datasets by using ggplot2 and R Shiny (Wissel et al., 2020).

Over the last decade, Python becomes the hotspot in data science communities and has rapidly adopted in implementing analytical tasks due to a variety of visualization liberties, such as Matplotlib, Seaborn, Plotly, Gleam, geoplotlib, etc. Despite being over a decade age, Matplotlib is still the dominated library for data visualization, of which many other Python libraries are initialized based on wrappers over matplotlib. Visual mining task can be implemented by running a few lines of code with Seaborn, which harnesses the advantages of matplotlib, while plotly offers many advanced visual outputs, comparing to basic visualization modules in Python. Moreover, Geoplotlib provides a toolbox for generating maps and handling GIS data through a variety of map types, such as choropleths, heatmaps, and dot density maps. Besides, Gleam allows users to produce visual outputs into interactive web applications that work with any Python data visualization library. As such, Python provides many options for producing COVID-19 visualizers and designing dashboards. For example, COVID-TRACK is a web application designed by using Python, which delivers a platform for tracking, testing, incidence, hospitalizations, and deaths in regards to COVID-19 (Zohner & Morris, 2021).

As observing pandemic information on maps has become one of the most intuitive ways, many COVID-19 visualizers and dashboards apply GIS softwares that are designed for visualizing and analyzing geographical statistics other than using R or Python. ArcGIS is such a GIS analytical and visualization platform that can be applied in handling professional geo-visual analytics. It has been widely used by research communities and public health authorities to develop and illustrate groundbreaking research for COVID-19. Prominent examples include the JHU dashboard designed by Esri ArcGIS Online services (Kamel Boulos & Geraghty, 2020) and the WHO COVID-19 Dashboard by ArcGIS (Costa et al., 2020). Some research groups also apply ArcGIS services to illustrate dynamic spatiotemporal changes of COVID-19, e.g. the visual mining project of medical resource deficiencies in the U.S. under COVID-19 pandemic by using ArcGIS Pro Dashboards (Sha et al., 2020). Similar to ArcGIS, QGIS is another option for producing COVID-19 maps with GIS data visualization. QGIS has also been incorporated in the current COVID-19 visualizer and dashboard designs, e.g. the thematic mapping application to analysis pandemic situation in large cities using QGIS-based dashboard designs (Kuznetsov et al., 2021).

FUTURE RESEARCH DIRECTIONS

Although this chapter provides a comprehensive understanding of the COVID-19 tracking and visualizing platform throughout data aggregation, big data processing, and data visualization, some issues that have not yet involved can be discussed in the form of future research directions. Existing visualizers and dashboards are mainly relied on COVID-19 datasets associated with disease measurements, whereas mining extra factors related to COVID-19 will provide depth in exploring the feature of the pandemic. Such factors can be extracted from a variety of representations, such as public sentiments, socioeconomic measurements, government reactions, etc., which need to build novel databases by performing more visual analytics in the future. Moreover, most existing dashboards represent descriptive analysis for COVID-19, while adding additional functionalities, such as pandemic simulations, estimations, and predictions, may provide previews for decision makers. Such a concept requires a series of data-driven approaches, such as epidemiological models, forecasting techniques, and machine learning-based predictions, which will be illustrated in the next two chapters, and further, may be implemented in the futures studies.

CONCLUSION

This chapter provides a comprehensive understanding of COVID-19 data problems by emphasizing on how big data helps in gathering, processing, and visualizing pandemic-related data. Major COVID-19 data sources have been reviewed, along with several common methods been illustrated for data cleaning, preprocessing, and data collection and validation. Big data pipelines have been investigated by examining cutting-edge big data tools and data workflows, which can be applied to propose an automatic and efficient solution for COVID-19 big data processing. A conceptual framework of data workflow architecture has been proposed and discussed in terms of a hybrid method that incorporates a set of state-of-the-art big data solutions. COVID-19 visualizers have also been examined by illustrating data visualization approaches and tools for COVID-19 dashboard designs. As a result, this chapter verifies the importance of using big data in tracking and analyzing COVID-19 for decision-making processes during the pandemic.

REFERENCES

Akhtar, N., Tabassum, N., Perwej, A., & Perwej, Y. (2020). Data analytics and visualization using Tableau utilitarian for COVID-19 (Coronavirus). *Global Journal of Engineering and Technology Advances*.

Altieri, N., Barter, R. L., Duncan, J., Dwivedi, R., Kumbier, K., Li, X., . . . Yu, B. (2020). *Curating a COVID-19 data repository and forecasting county-level death counts in the United States*. arXiv preprint arXiv:2005.07882.

Azeroual, O., & Fabre, R. (2021). Processing Big Data with Apache Hadoop in the Current Challenging Era of COVID-19. *Big Data and Cognitive Computing*, *5*(1), 12. doi:10.3390/bdcc5010012

Bernal, J. L., Andrews, N., Gower, C., Gallagher, E., Simmons, R., Thelwall, S., ... Ramsay, M. (2021). Effectiveness of Covid-19 vaccines against the B. 1.617. 2 (delta) variant. *New England Journal of Medicine*.

Bernasconi, A., & Grandi, S. (2021). A Conceptual Model for Geo-Online Exploratory Data Visualization: The Case of the COVID-19 Pandemic. *Information (Basel)*, *12*(2), 69. doi:10.3390/info12020069

Centers for Disease Control and Prevention. (2020). *Centers for Disease Control and Prevention Coronavirus disease 2019 (COVID-19) 2020.* Available at https://www.cdc.gov/coronavirus/2019-ncov/cases-updates/cases-in-us.html#cumulative

Costa, J. P., Grobelnik, M., Fuart, F., Stopar, L., Epelde, G., Fischaber, S., ... Davis, P. (2020). Meaningful big data integration for a global COVID-19 strategy. *IEEE Computational Intelligence Magazine*, *15*(4), 51–61. doi:10.1109/MCI.2020.3019898

Dean, J., & Ghemawat, S. (2008). MapReduce: Simplified data processing on large clusters. *Communications of the ACM*, *51*(1), 107–113. doi:10.1145/1327452.1327492

Dong, E., Du, H., & Gardner, L. (2020). An interactive web-based dashboard to track COVID-19 in real time. *The Lancet. Infectious Diseases*, *20*(5), 533–534. doi:10.1016/S1473-3099(20)30120-1 PMID:32087114

ElDahshan, K. A., AlHabshy, A. A., & Abutaleb, G. E. (2020). Data in the time of COVID-19: A general methodology to select and secure a NoSQL DBMS for medical data. *PeerJ. Computer Science*, *6*, e297. doi:10.7717/peerj-cs.297 PMID:33816948

Elmeiligy, M. A., Desouky, A. I. E., & Elghamrawy, S. M. (2020). *A multi-dimensional big data storing system for generated Covid-19 large-scale data using Apache Spark.* arXiv preprint arXiv:2005.05036.

Gonzalez, J. E., Xin, R. S., Dave, A., Crankshaw, D., Franklin, M. J., & Stoica, I. (2014). Graphx: Graph processing in a distributed dataflow framework. In *11th USENIX Symposium on Operating Systems Design and Implementation (OSDI 14)* (pp. 599-613). USENIX.

Idogawa, M., Tange, S., Nakase, H., & Tokino, T. (2020). Interactive web-based graphs of coronavirus disease 2019 cases and deaths per population by country. *Clinical Infectious Diseases*, *71*(15), 902–903. doi:10.1093/cid/ciaa500 PMID:32339228

Kamel Boulos, M. N., & Geraghty, E. M. (2020). Geographical tracking and mapping of coronavirus disease COVID-19/severe acute respiratory syndrome coronavirus 2 (SARS-CoV-2) epidemic and associated events around the world: How 21st century GIS technologies are supporting the global fight against outbreaks and epidemics. *International Journal of Health Geographics*, *19*(1), 8–8. doi:10.118612942-020-00202-8 PMID:32160889

Khan, N., Yaqoob, I., Hashem, I. A. T., Inayat, Z., Mahmoud Ali, W. K., Alam, M., Shiraz, M., & Gani, A. (2014). Big data: Survey, technologies, opportunities, and challenges. *TheScientificWorldJournal*, *2014*, 2014. doi:10.1155/2014/712826 PMID:25136682

Khashan, E. A., Eldesouky, A. I., Fadel, M., & Elghamrawy, S. M. (2020). *A big data based framework for executing complex query over Covid-19 datasets (Covid-QF)*. arXiv preprint arXiv:2005.12271.

Kuznetsov, I., Panidi, E., Kikin, P., Kolesnikov, A., Korovka, V., & Galkin, V. (2021). Issues of geographic information systems and thematic mapping application to analysis of epidemiological situation in large cities. *The International Archives of the Photogrammetry, Remote Sensing and Spatial Information Sciences*, *43*, B4–B2021. doi:10.5194/isprs-archives-XLIII-B4-2021-287-2021

Lan, H., Sha, D., Malarvizhi, A. S., Liu, Y., Li, Y., Meister, N., Liu, Q., Wang, Z., Yang, J., & Yang, C. (2021). COVID-Scraper: An open-source toolset for automatically scraping and processing global multi-scale spatiotemporal COVID-19 records. *IEEE Access: Practical Innovations, Open Solutions*, *9*, 84783–84798. doi:10.1109/ACCESS.2021.3085682 PMID:34812396

Leung, C. K., Chen, Y., Hoi, C. S., Shang, S., Wen, Y., & Cuzzocrea, A. (2020, September). Big data visualization and visual analytics of COVID-19 data. In *2020 24th International Conference Information Visualisation (IV)* (pp. 415-420). IEEE. 10.1109/IV51561.2020.00073

Mathieu, E., Ritchie, H., Ortiz-Ospina, E., Roser, M., Hasell, J., Appel, C., ... Rodés-Guirao, L. (2021). A global database of COVID-19 vaccinations. *Nature Human Behaviour*, 1–7.

Mazumder, S. (2016). Big data tools and platforms. In *Big data concepts, theories, and applications* (pp. 29–128). Springer. doi:10.1007/978-3-319-27763-9_2

Miller, K., & Curry, K. (2020). *The COVID tracking project*. Available at https://github.com/COVID19Tracking

Mitchell, R., Pottier, L., Jacobs, S., da Silva, R. F., Rynge, M., Vahi, K., & Deelman, E. (2019, December). Exploration of workflow management systems emerging features from users perspectives. In *2019 IEEE International Conference on Big Data (Big Data)* (pp. 4537-4544). IEEE. 10.1109/BigData47090.2019.9005494

Mufid, M. R., Basofi, A., Mawaddah, S., Khotimah, K., & Fuad, N. (2020, September). Risk Diagnosis and Mitigation System of COVID-19 Using Expert System and Web Scraping. In *2020 International Electronics Symposium (IES)* (pp. 577-583). IEEE. 10.1109/IES50839.2020.9231619

Pan, Z., Mehta, D., Tiwari, A., Ireddy, S., Yang, Z., & Jin, F. (2020, December). An Interactive Platform to Track Global COVID-19 Epidemic. In *2020 IEEE/ACM International Conference on Advances in Social Networks Analysis and Mining (ASONAM)* (pp. 948-951). IEEE. 10.1109/ASONAM49781.2020.9381436

Raine, S., Liu, A., Mintz, J., Wahood, W., Huntley, K., & Haffizulla, F. (2020). Racial and ethnic disparities in COVID-19 outcomes: Social determination of health. *International Journal of Environmental Research and Public Health*, *17*(21), 8115. doi:10.3390/ijerph17218115 PMID:33153162

Rodríguez-Rodríguez, I., Rodríguez, J. V., Shirvanizadeh, N., Ortiz, A., & Pardo-Quiles, D. J. (2021). Applications of artificial intelligence, machine learning, big data and the internet of things to the COVID-19 pandemic: A scientometric review using text mining. *International Journal of Environmental Research and Public Health*, *18*(16), 8578. doi:10.3390/ijerph18168578 PMID:34444327

Rubin, R. (2021). COVID-19 Vaccines vs Variants—Determining How Much Immunity Is Enough. *Journal of the American Medical Association*, *325*(13), 1241–1243. doi:10.1001/jama.2021.3370 PMID:33729423

Santos, B. S., Silva, I., da Câmara Ribeiro-Dantas, M., Alves, G., Endo, P. T., & Lima, L. (2020). COVID-19: A scholarly production dataset report for research analysis. *Data in Brief*, *32*, 106178. doi:10.1016/j.dib.2020.106178 PMID:32837978

Sha, D., Miao, X., Lan, H., Stewart, K., Ruan, S., Tian, Y., Tian, Y., & Yang, C. (2020). Spatiotemporal analysis of medical resource deficiencies in the US under COVID-19 pandemic. *PLoS One*, *15*(10), e0240348. doi:10.1371/journal.pone.0240348 PMID:33052956

Shaheen, A. W., Ciesco, E., Johnson, K., Kuhnen, G., Paolini, C., & Gartner, G. (2021). Interactive, on-line visualization tools to measure and drive equity in COVID-19 vaccine administrations. *Journal of the American Medical Informatics Association: JAMIA*, *28*(11), 2451–2455. doi:10.1093/jamia/ocab180 PMID:34480569

Sharma, P. P., & Navdeti, C. P. (2014). Securing big data hadoop: A review of security issues, threats and solution. *International Journal of Computer Science and Information Technologies*, *5*(2), 2126–2131.

Shrotri, M., Swinnen, T., Kampmann, B., & Parker, E. P. (2021). An interactive website tracking COVID-19 vaccine development. *The Lancet. Global Health*, *9*(5), e590–e592. doi:10.1016/S2214-109X(21)00043-7 PMID:33667404

Siriwardhana, Y., Gür, G., Ylianttila, M., & Liyanage, M. (2020). *The role of 5G for digital healthcare against COVID-19 pandemic: Opportunities and challenges*. ICT Express.

Suleykin, A., & Panfilov, P. (2019, January). Implementing Big Data Processing Workflows Using Open Source Technologies. In *Proceedings of the 30th DAAAM International Symposium* (pp. 394-404). 10.2507/30th.daaam.proceedings.054

The New York Times. (2021). *Coronavirus (Covid-19) Data in the United States*. Retrieved from https://github.com/nytimes/covid-19-data

USAFacts. (2020). *Coronavirus Locations: COVID-19 Map by County and State*. Available at https://usafacts.org/visualizations/ coronavirus-covid-19-spread-map/

Vavilapalli, V. K., Murthy, A. C., Douglas, C., Agarwal, S., Konar, M., Evans, R., ... Baldeschwieler, E. (2013, October). Apache hadoop yarn: Yet another resource negotiator. In *Proceedings of the 4th annual Symposium on Cloud Computing* (pp. 1-16). 10.1145/2523616.2523633

Vijay, V., & Nanda, R. (2020). MRC-COVID (Map Reduce with Cache) System for Big data Analytics. *Inter J All Res Edu Sci Meth, 12*(8).

Villanustre, F., Chala, A., Dev, R., Xu, L., Furht, B., & Khoshgoftaar, T. (2021). Modeling and tracking Covid-19 cases using Big Data analytics on HPCC system platform. *Journal of Big Data*, *8*(1), 1–24. doi:10.118640537-021-00423-z PMID:33425651

Vohra, D. (2016). Apache Sqoop. In *Practical Hadoop Ecosystem* (pp. 261–286). Apress. doi:10.1007/978-1-4842-2199-0_5

Wang, G., Gu, Z., Li, X., Yu, S., Kim, M., Wang, Y., ... Wang, L. (2020a). *Comparing and integrating US COVID-19 daily data from multiple sources: a county-level dataset with local characteristics*. Academic Press.

Wang, L., Wang, G., Gao, L., Li, X., Yu, S., Kim, M., . . . Gu, Z. (2020b). *Spatiotemporal dynamics, nowcasting and forecasting of COVID-19 in the United States.* arXiv preprint arXiv:2004.14103.

WHO. (2020). *WHO Coronavirus (COVID-19) Dashboard.* Available at https://covid19.who.int/

Wissel, B. D., Van Camp, P. J., Kouril, M., Weis, C., Glauser, T. A., White, P. S., Kohane, I. S., & Dexheimer, J. W. (2020). An interactive online dashboard for tracking COVID-19 in US counties, cities, and states in real time. *Journal of the American Medical Informatics Association: JAMIA*, *27*(7), 1121–1125. doi:10.1093/jamia/ocaa071 PMID:32333753

Wu, T., Hu, E., Ge, X., & Yu, G. (2020). *Open-source analytics tools for studying the COVID-19 coronavirus outbreak.* MedRxiv; doi:10.1101/2020.02.25.20027433

Yang, Z., Xu, J., Pan, Z., & Jin, F. (2020, December). Covid19 tracking: An interactive tracking, visualizing and analyzing platform. In *2020 IEEE/ACM International Conference on Advances in Social Networks Analysis and Mining (ASONAM)* (pp. 941-943). IEEE. 10.1109/ASONAM49781.2020.9381414

Zaharia, M., Xin, R. S., Wendell, P., Das, T., Armbrust, M., Dave, A., Meng, X., Rosen, J., Venkataraman, S., Franklin, M. J., Ghodsi, A., Gonzalez, J., Shenker, S., & Stoica, I. (2016). Apache spark: A unified engine for big data processing. *Communications of the ACM*, *59*(11), 56–65. doi:10.1145/2934664

Zohner, Y. E., & Morris, J. S. (2021). COVID-TRACK: World and USA SARS-COV-2 testing and COVID-19 tracking. *BioData Mining*, *14*(1), 1–15. doi:10.118613040-021-00233-2 PMID:33472672

ADDITIONAL READING

Arfat, Y., Usman, S., Mehmood, R., & Katib, I. (2020). Big data tools, technologies, and applications: A survey. In *Smart Infrastructure and Applications* (pp. 453–490). Springer. doi:10.1007/978-3-030-13705-2_19

Baviskar, M. R., Nagargoje, P. N., Deshmukh, P. A., & Baviskar, R. R. (2021). A Survey of Data Science Techniques and Available Tools. *International Research Journal of Engineering and Technology.*

Bragazzi, N. L., Dai, H., Damiani, G., Behzadifar, M., Martini, M., & Wu, J. (2020). How big data and artificial intelligence can help better manage the COVID-19 pandemic. *International Journal of Environmental Research and Public Health, 17*(9), 3176. doi:10.3390/ijerph17093176 PMID:32370204

Dasgupta, N. (2018). *Practical big data analytics: Hands-on techniques to implement enterprise analytics and machine learning using Hadoop, Spark, NoSQL and R*. Packt Publishing Ltd.

Du, D. (2018). *Apache Hive Essentials: Essential techniques to help you process, and get unique insights from, big data*. Packt Publishing Ltd.

Miller, J. D. (2017). *Big data visualization*. Packt Publishing Ltd.

Mittal, M., Balas, V. E., Goyal, L. M., & Kumar, R. (Eds.). (2019). *Big data processing using spark in cloud*. Springer. doi:10.1007/978-981-13-0550-4

Rao, T. R., Mitra, P., Bhatt, R., & Goswami, A. (2019). The big data system, components, tools, and technologies: A survey. *Knowledge and Information Systems*, *60*(3), 1165–1245. doi:10.100710115-018-1248-0

Wu, J., Wang, J., Nicholas, S., Maitland, E., & Fan, Q. (2020). Application of big data technology for COVID-19 prevention and control in China: Lessons and recommendations. *Journal of Medical Internet Research*, *22*(10), e21980. doi:10.2196/21980 PMID:33001836

Zhou, C., Su, F., Pei, T., Zhang, A., Du, Y., Luo, B., ... Xiao, H. (2020). COVID-19: challenges to GIS with big data. *Geography and Sustainability, 1*(1), 77-87.

KEY TERMS AND DEFINITIONS

Apache Airflow: An open-source big data management platform, proposed by Airbnb as one of the most efficient data solutions to manage the industrial-level data workflow challenges.

Apache Spark: An open-source analytical engine for big data processing with an interface for programming entire clusters of implicit data parallelism and fault tolerance.

Big Data Pipeline: A sequence of data processing components connected in series, where the output of one part is the input of the next one, in which the pipeline can be operated in parallel or in time-sliced manner.

Data Acquisition: A process to deal with physical conditions and transform the sample data into numerical values that can be read by computer.

Data Visualization: An efficient method to communicate with data and information through graphical representations.

Data Workflow: A set of operations that processes information and data from raw to processed.

ETL: Stands for extract, transform, and load, which is the general procedure of delivering data from one data source to another.

Hadoop Ecosystem: A big data platform which offers a variety of services to solve the big data problems with four main components, such as HDFS, MapReduce, YARN, and Hadoop Common.

Chapter 2
Simulating and Preventing COVID-19 Using Epidemiological Models

ABSTRACT

With the global spreading of COVID-19, disease control has become a critical problem and an overwhelming challenge for our healthcare system. The decision-making of the control is mostly difficult because the disease is highly contagious, the policy-making procedures inappropriate, as well as the medical treatments and vaccines insufficient. Computational approaches such as mathematical modeling and simulation can assist to measure and prevent the pandemic. This chapter presents a set of SIR-based models for disease control in the context of COVID-19 with the empirical analysis based on the U.S. data. Data analysis and mathematical simulation results are illustrated to preview the progress of the outbreak and its future given different types of scenarios. The effect of interventions has been compared with that of the no-actions. The conclusion indicates that the public authorities can reduce the epidemic scale based on a strict strategy projected from the simulation results.

INTRODUCTION

The whole world has been facing the biggest risk in the shape of COVID-19 global pandemic. The increasing number of infected and deceased patients

DOI: 10.4018/978-1-7998-8793-5.ch002

makes a huge impact on the society. It is urgent to uncover the natural progression of the novel coronavirus. Generally, a disease follows the host formation and progression, such as exposure to the infection, host formation, and the spread of the infection. As an infectious disease, the cause of COVID-19 can be defined as an epidemiological triad, which is a traditional model involved three components: the agent, the host, and the environment. The agent carries the infection that is transmitted to the host under a certain environment. Cutting off the connection between those factors can disrupt the proliferation of COVID-19 effectively. Interrupting factors, such as agent-host, environment-host, and agent-environment, can be conceptualized into three scenarios: community, hospital, and aerosol-generating medical procedure (Tsui et al., 2020). In this scenario, prediction tools can serve to determine different objectives such as the pandemic trend, the medical equipment supplies and distributions, and the degrees of policy execution (i.e. quarantine regulations, lockdown orders, travel restriction policies). Strong policies are needed to be administered and executed in order to control the pandemic. On the other hand, the disease control policy requires rationality and reliability. Data-driven approaches are helpful for the policy-making process in terms of accurate parameter estimations and mathematical simulations. Further, policy-makers need to rebuild the infrastructure in terms of a quick and accurate responses, which is an important component of the smart healthcare system. Data mining methods along with the epidemiological models always make the contribution to the smart healthcare system (Jain & Bhatnagar, 2017).

Responses from a smart healthcare system require formal procedures for measuring and tracking the outbreak of a new disease. Epidemiological models with mathematical simulations can be served to manage the impacts and reduce the risk. A successful simulation analysis for an infectious disease relies on measuring influential factors including the infectious duration, the opportunity of contact, the transmission probability, and the susceptibility. (Kucharski, 2020). The new virus will die off after major actions have been taken for susceptible carries, otherwise, the disease may eventually become an epidemic. Relationships between susceptible, infected, and recovered population are important to uncover the transmission pattern of an infectious disease (Hethcote, 2000). The susceptible-infected-removed (SIR) model has been considered as the most popular stochastic simulations in the field of epidemiology because of its efficiency and simplicity. Firstly, it can predict the number of susceptible, infected, and recovered individuals at any given time. Secondly, its simplicity allows the computational process easier by estimating a small number of parameters. Various SIR models have been used to reveal the outbreaks' pattern of other diseases such as SARS, H1N1, and MERS (Zhang, 2007; Coburn et al., 2009; Kwon & Jung, 2016).

Motivated by the current urgent demand of epidemiological simulation and prediction, this chapter presents how SIR-based models work with stochastic simulations for the disease control in the context of the COVID-19 pandemic. A brief introduction of SIR family models will be given, followed by an exploratory analysis of the U.S. data from the Johns Hopkins University. Case studies will be conducted to illustrate model usages in terms of parameter estimations and stochastic simulations for the disease control measures. The objectives of this chapter are listed as follows:

- reviewing the most recent studies and applications for COVID-19 predictions and simulations based on the SIR family of models.
- illustrating several SIR-based models, such as the basic SIR, SIR-D, SIR-F, SEIR, and eSIR, in terms of model structures and ordinary differential equations.
- providing a real-world case study in regards to estimate parameters with trend analysis using SIR, SIR-D, and SIR-F.
- simulating future situations in the U.S. under different scenarios and effects, including the no-actions scenario, the effect under new medicines, the effect under vaccinations, and the effect under lockdowns.
- discussing potential solutions that can be adopted for SIR-based predictions and simulations, i.e. the mixed model approach.

BACKGROUND

Similar to other infectious virus, COVID-19 follows a natural progressive feature in terms of contact infections, which is possible to preview the transmission process overtime. A family of SIR models are potentially applicable in the disease control of the COVID-19 pandemic. Chen et al. (2020) applied a SIR model to analyze COVID-19 pandemic features with asymptomatic persons. Similar studies using a basic SIR model have been presented in terms of tracking and simulating the progression of COVID-19 in Cameroon, Colombia, India, and Sweden (Nguemdjo et al., 2020; Manrique-Abril et al., 2020; Simha et al., 2020; Qi et al., 2020). Two modified SIR models, SIR-F and SIR-D, have been used in predicting the endpoint of the pandemic and in simulating the future under possible effects. A SIR-F model architecture has been proposed to predict the spread of COVID-19 and its simulate prevention effects (Ndiaye et al., 2020; Alanazi et al., 2020). Similarly, a SIR-D model has been used in COVID-19 predictions across nations (Rajesh et al., 2020; Pinter et al., 2020).

For the COVID-19 pandemic, related studies concentrated on estimations and simulations using SIR-based models are limited in the United States. Existing studies have been presented in terms of parameter estimations by fitting either a national-level or a states-level data. Shapiro et al. (2020) estimated the infection rate and the reproduction number in the U.S. using a simple SIR model. Similarly, the time-varying parameters have been estimated for the 12 countries including the United States using the SIR epidemic models (Marinov & Marinova, 2020). The SIR models have also been applied to fit the regional data such as New York and New Jersey with a concentration on the transmission rate and the pandemic trend (Ambrosio & Aziz-Alaoui, 2020). A classical susceptible-exposed-infected-removed (SEIR) model, a modified SIR model, has been applied in forms of estimating the infectivity, forecasting the spread, modeling the dynamics, simulating control scenarios, and making management strategies for COVID-19 (Tang et al., 2020; Wu et al., 2020; He et al., 2020; López & Rodo, 2021; Radulescu & Cavanagh, 2021). An extended susceptible-infected-removed (eSIR) model can be adopted, covering a wide range of time-varying models with adjustable effects overtime (Wang et al., 2020). It has also been applied to predict the pandemic trend in Italy comparing with that of China (Wangping, et al., 2020).

Despite the limited research in this field, many industrial applications have been presented in the United States. The COVID-19 Hospital Impact Model for Epidemics (CHIME) has been applied to form an industrialized decision-making dashboard by University of Pennsylvania, which is designed to assist hospitals and public health officials to manage hospital capacity during the COVID-19 pandemic (CHIME, 2020). The CHIME App provides estimations of total new and census inpatient hospitalizations, ICU admissions, and patients requiring ventilation based on a SIR model. Similarly, the Severe COVID-19 Model & Mapping Tool, introduced by Columbia University and Mount Sinai, has been designed to predict the number of severe cases, hospitalizations, emergence care and ICU usages, and deaths for 3 up to 6 weeks using the SIR-based models (Li et al., 2020).

MAIN FOCUS OF THE CHAPTER

Issues, Controversies, Problems

SIR family models are powerful to estimate infected, recovery, and fatal cases of COVID-19, which can provide insights for pandemic predictions and simulations. Most existing studies are concentrated on the early stage of the pandemic across countries and regions such as China, Italy, and EU. Experimental results vary over researches, and some of them are far away from the reality due to the difficulties of the predictive modeling in the early stage, such as unknown transmission mechanism, insufficient reported cases, and wrong assumptions for the spread of the disease in different communities. On the other hand, some researchers believe that complicated models can yield better predictive results because additional factors have been involved, i.e. time-varying features and adjustable effects. However, such a conclusion may not be holding in other countries or regions. Existing studies for the United State are rare in the early stage of the pandemic. Overall, most of them focus on time-dependent parameter estimations and trend predictions, however, effects under different scenarios and effects have not been investigated. Although several real-world applications have been proposed in the form of decision-making dashboards and visual mining interactions, the underlying model mechanisms have not been illustrated with unknown parameter estimations. To settle the controversies and solve the above-mentioned problems, this chapter is designed to offer a comprehensive understanding of the SIR-based models in terms of their model structures and ordinary differential equations, along with a real case study for predicting the pandemic trend and simulating the situations under different scenarios and effects in the United States.

Features of the Chapter

This chapter explains a broad range of SIR family models that have been applied in the COVID-19 predictive analysis and simulation, followed by a case study using the U.S. real-time daily report data for the trend analysis and parameter estimations using a basic SIR, SIR-D, and SIR-F. The SIR-F model has been selected for simulating future situations given the no-actions scenarios, the effects under lockdown, and new medicines and vaccinations. Experimental results well depict the COVID-19 pandemic outbreak in the United State from beginning to the second wave as of late 2020. The simulation results also provide the prevision of how the COVID-19 pandemic ends by investigating different scenarios and effects. Together with the discussion of future research directions, this chapter offers a comprehensive understanding of how epidemiological models work for predicting and simulating the COVID-19 pandemic.

SOLUTIONS AND RECOMMENDATIONS

SIR Family Models

The Basic SIR Model

SIR model is one of the simplest mathematical models, which can describe the spread of a disease and help people to understand the outbreak of infectious diseases. Kermack and McKendrick first proposed the original formulation of epidemic model (Capasso & Serio, 1978). Its fundamental theory is to forecast the number and the distribution of cases of a certain infectious disease as it is transmitted through a population over time. The population is assigned to compartments with labels as S, I or R (Susceptible, Infectious, or Recovered). In the SIR model, the susceptible are remain people, after excluding those persons, who are confirmed to be virus carriers through testing, from the total population. The susceptible can be transited to the infected cases with the contact rate β. The model structure can be described as follow:

$$S \xrightarrow{\beta I} I \xrightarrow{\gamma} R$$

Where S represents susceptible and is equal to the difference between total population and the confirmed cases. I represents the infected and is equal to the number of the confirmed cases minus the recovered and the death. R represents the recovered or the fatal and is equal to the summation of the recovered and the deaths. β and γ are effective contact rate [1/min] and recovery rate [1/min], respectively. The ordinary differential equations (ODEs) of the SIR model are shown as below:

$$\frac{dS}{dT} = -N^{-1},$$

$$\frac{dI}{dT} = N^{-1}\beta SI - \gamma I > 0\,,$$

$$\frac{dR}{dT} = \gamma I\,,$$

Where *N* indicates the total population and is the summation of *S*, *I*, and *R*. *T* indicates the elapsed time from the beginning of the virus spreading. The functions can be simplified as follows:

$$\frac{dx}{dt} = -\rho xy\,,$$

$$\frac{dy}{dt} = \rho xy - \sigma y\,,$$

$$\frac{dz}{dt} = \sigma y\,,$$

$$R_0 = \rho\sigma^{-1} = \beta\gamma^{-1}.$$

Where R_0 is the reproductive rate, a possibility for an infected person to infect other people. If R_0 is greater than 1, a high probability for an outbreak is indicated to occur.

The SIR-D Model

SIR-D model involves the number of the susceptible, infected, dead, and recovered. Compared to SIR model, SIR-D model considers human factors, including age, sex, social behavior, and location. Among these factors, transmission rate during the pandemic is largely affected by social behavior. Accordingly, in SIR-D model, a susceptible person, who is in contact with an infected person, may get infected. The transmission of the infected is either to recover from the disease or to die after getting infected. The ODEs of the SIR-D model is indicated as follow:

$$\frac{dS}{dT} = -\frac{\beta}{N} IS \,,$$

$$\frac{dI}{dT} = \frac{\beta}{N} IS - \gamma_R I - \gamma_D I \,,$$

$$\frac{dR}{dT} = \gamma_R I \,,$$

$$\frac{dD}{dT} = \gamma_D I \,.$$

Where S represents the susceptible population, I represents the infected, R represents the recovered, and D represents the dead. SIR-D model assumes total population $N=S+I+R+D$ is constant and not changed during pandemic. β is the transmission rate and indicates the growth rate of the number of the infected. γR is the recovery rate and indicates the growth rate of the number of the recovered. The death rate is γD, which shows the change in the number of deaths. In the SIR-D model, reproduction rate can be defined as R0= β/(γR+γD).

The SIR-F Model

The SIR-F model is potentially useful in predicting and simulating the pandemic, in which "F" stands for "fatal with the confirmation". In the beginning of COVID-19 outbreak, the virus was unprecedented and many doctors and medical professionals never knew the pathological characteristics of COVID-19, therefore, many cases were confirmed after the death. In the SIR-F model, the total population is divided into four parts: the susceptible, the infected, the recovered, and the fatal. The susceptible individuals are the remaining part of subtracting the confirmed cases from the total population. However, in fact the coronavirus carriers without knowing even by themselves exist in the group of the susceptible. Among these virus carriers, some can be self-healed by their own immune system, while others may be confirmed as positive after they die. SIR-F model involves S* to describe this actual circumstance, as shown as below in the model structure:

$$S \xrightarrow{I} S^* \xrightarrow{1} F,\ S \xrightarrow{I} S^* \xrightarrow{1-1} I,\ I \xrightarrow{\gamma} R \text{ and } I \xrightarrow{2} F$$

Where S represents susceptible, S* represents the confirmed and the uncategorized, I represents the confirmed and categorized as infected, and F represents the confirmed as fatal. The parameter β is the effective contact rate [1/min], α*1* is the direct fatality probability of S*, α*2* is the mortality rate of the infected cases [1/min], and γ is the recovery rate [1/min]. The total population N is equal to the sum of S, *I, R,* and F. The ordinary differential equations of SIR-F model are presented in appendix. Importantly, S* plays a role of a supplementary compartment in this model to formalize two death circumstances: the transitions from the susceptible directly to the fatal, and those from the susceptible indirectly to the fatal through infected. The ODEs are presented as follow:

$$\frac{dS}{dT} = -N^{-1}\beta S\,,$$

$$\frac{dI}{dT} = -N^{-1}\left(1-\alpha_1\right)\beta SI - \left(\gamma - \alpha_2\right)I\,,$$

$$\frac{dR}{dT} = \gamma I\,,$$

$$\frac{dF}{dT} = N^{-1}\alpha_1\beta SI + \alpha_2 I\,.$$

Where *N=S+I+R+F* is the total population and *T* is the elapsed time from the starting date. The functions can be simplified as follows:

$$\frac{dx}{dt} = -\rho xy\,,$$

$$\frac{dy}{dt} = \rho\left(1-\theta\right)xy - \left(\sigma +\right)y\,,$$

$$\frac{dz}{dt} = \sigma y\,,$$

$$\frac{dw}{dt} = \rho\theta xy + y\,,$$

$$R_0 = \rho\left(1-\theta\right)\left(\sigma +\right)^{-1} = \beta\left(1-\alpha_1\right)\left(\gamma+\alpha_2\right)^{-1},$$

where R_0 is the reproductive rate.

The SEIR Model

Complex disease processes are oversimplified due to the simplicity of the SIR model. It does not consider the incubation period when people are exposed and became infected and contagious in the context of COVID-19. The factor regards those who have been exposed but not yet contagious, should be taken into account of the modeling and simulating the pandemic pattern. The goal of using a SEIR model is to calculate the number of infected, recovered, and dead cases on the basis of the number of contacts, probability of disease transmission, incubation period, recovery rate, and the fatality rate. In the SEIR model, the total population is divided into susceptible, exposed, infected and recovered. In most cases, the individuals may not be infectious after being infected. Based on this scenario, the SEIR model considers the incubation duration, in order to simulate the virus transmission. The model structure of the SEIR model is illustrated as follow:

$$S \xrightarrow{\beta} E \xrightarrow{\sigma} I \xrightarrow{\gamma} R$$

Where S is the susceptible, E is the exposed, I is the infected, and R is the recovered. β, σ, and γ symbolize the infectious rate, the incubation rate, and the recovery rate, respectively. The ODEs of SEIR model are demonstrated as follows:

$$\frac{dS}{dt} = \varphi - \mu S - \frac{\beta SI}{N},$$

$$\frac{dE}{dt} = \frac{\beta SI}{N} - \left(\mu+\sigma\right)E\,,$$

$$\frac{dI}{dt} = \sigma E - \left(\gamma + \mu + \alpha\right) I ,$$

$$\frac{dR}{dt} = \gamma I - \mu R .$$

Where $N=S+E+I+R$ is the total population. φ, μ, and α represent the per-capita birth rate, the per-capita natural death rate, and the average fatality caused by virus, correspondingly.

The eSIR Model

The time-varying parameters and strategy adjustments are also important to be considered in of improving the accuracy of estimations and simulations. The transmission rate is unchangeable in SIR, however, it may not be the case in actual situations as the speed of transmission changes over time through different interventions. The main idea of the eSIR model is to add a transmission modifier to enables a time-varying probability of the transmission rate. Different from the SIR model with constant transmission and removal rates, the eSIR model considers the changes of transmission rate over time. Carrying out various forms of human interventions is a strong influential factor on the transmission rate over time, such as wearing masks, taking in-home isolation, city blockade, etc. Moreover, the virus could go through mutation to improve its spreading capacity by hiding in more virus carriers. Therefore, according to these influences, the eSIR model allows a time-varying probability. The model structure can be defined as:

$$S \xrightarrow{(t)} IR$$

Where S represents the susceptible, I represents the infected, and R represents the removed, which includes the recovered and the death. Assuming at a time t, the transmission modifier (t) can be defined as below:

$$(t) = \left\{1 - p^{S}\left(t\right)\right\}\left\{1 - p^{I}\left(t\right)\right\} \in \left[0,1\right]$$

Where $p^S(t) \in [0,1]$ is the probability of a susceptible person being isolated at home, and $p^I(t)$ is the probability of an infected person being quarantined in hospital. The disease transmission can be given as follow:

$$\beta\left\{1-p^S\left(t\right)\right\}\theta_t^S\left\{1-p^I\left(t\right)\right\}\theta_t^I = \beta\pi\left(t\right)\theta_t^S\theta_t^I$$

Where θ_t^S is the probability of a person having the susceptible risk and θ_t^I is the probability of a person being confirmed as the infected. The ODEs of eSIR model are presented as follows:

$$\frac{d\theta_t^S}{dt} = -\beta\pi\left(t\right)\theta_t^S\theta_t^I,$$

$$\frac{d\theta_t^I}{dt} = \beta\pi\left(t\right)\theta_t^S\theta_t^I - \gamma\theta_t^I,$$

$$\frac{d\theta_t^R}{dt} = \gamma\theta_t^I.$$

Where β is the effective contact rate. $\beta\pi(t)$ *is* the product of contact rate and transmission modifier, which can describe the real time transmission rate corresponding to each level of the quarantine measures.

Additionally, the eSIR model also can take the quarantine compartment into account. The eSIR model structure with time-varying quarantine rate is illustrated as below:

$$SIR, S \xrightarrow{\varphi(t)} Q$$

Where Q represents the quarantine and $\varphi(t)$ is the quarantine rate at a time t. The ODEs with the consideration of quarantine measures are as below:

$$\frac{d\theta_t^Q}{dt} = \varphi\left(t\right)\theta_t^S,$$

$$\frac{d\theta_t^S}{dt} = -\beta\theta_t^S\theta_t^I - \varphi(t)\theta_t^S,$$

$$\frac{d\theta_t^I}{dt} = \beta\theta_t^S\theta_t^I - \gamma\theta_t^I,$$

$$\frac{d\theta_t^R}{dt} = \gamma\theta_t^I.$$

Where θ_t^Q is the probability of a person being in quarantine. Since θ_t^Q, θ_t^S, θ_t^I, and θ_t^R are the independent random events, the summation of them is equal to 1.

Trend Analysis and Parameters Estimation

Data Sources and Descriptive Analysis

To implement the projection and simulation models in gaining a better understanding of how the virus may spread in a community, the population data, population pyramid and COVID-19 data are needed. The most popular daily updated COVID-19 open access dataset is provided by the Johns Hopkins University (JHU). This COVID-19 dataset contains the number of the confirmed, the death, the recovered, the incident rate, and the case fatality ratio from about 194 countries since January 2020. Besides, the population data is needed to calculate the number of susceptible cases. The dataset can be retrieved from COVID-19 Data Hub (Guidotti & Ardia, 2020). Since various age groups may have a huge mobility difference of social activities, which is also an essential factor affecting the spread of the virus among the communities in the population, the population pyramid data is used to estimate the number of days for going out in average. In this case study, only US-related data is considered. A statistical COVID-19 data summary for US from January, 2020 to September, 2020 is in Table 1.

Table 1. Statistical COVID-19 data summary of the United States

	Mean	Std	Min	25%	50%	75%	max
Confirmed	2450130	2237711	12	374894.8	1826900	4376956	6938731
Infected	1602381	1355917	9	343483	1267300	2884712	4109962
Fatal	94372.73	68561.46	0	14102.25	109088.5	150582.8	202009
Recovered	753376.1	827824.9	3	17309.5	453539.5	1341662	2626760

Using JHU COVID-19 data, the data visualization for the U.S. is presented in Figure 1. Figure 1A has shown that the cases of the infected, the fatal and the recovered have been continuously increasing since January, 2020 until now. In Figure 1B, the ratio of the recovered cases to the confirmed is almost always greater than either the ratio of the fatal to the confirmed cases or the fatal to the removed cases during the whole time period. The curve smoothness in kernel density estimation of Figure 1C has proved the reliability of making inferences about the population, based on the current data sample.

Figure 1. COVID-19 data visualization in the U.S. using JHU datasets. (A) the total number of cases; (B) the fatal per confirmed rate, the recovered per confirmed rate, and the fatal per total rate; (C) kernel density estimation.

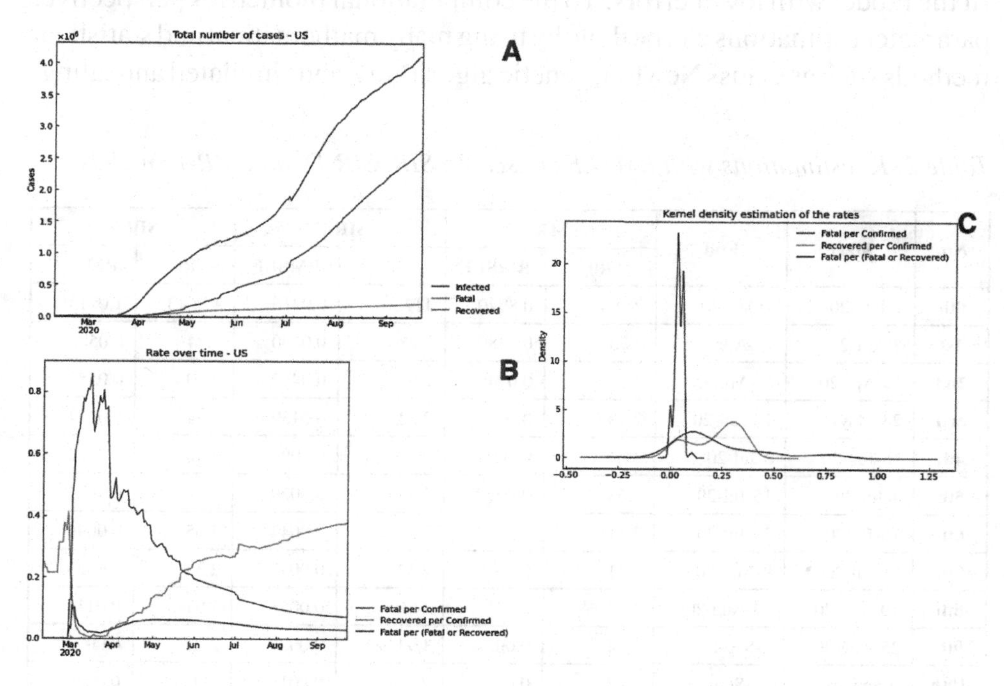

Predicting the Pandemic Trend

R_0 is the one of the most important parameters in a SIR-based model, which can be defined as the number of secondary infections produced by each infected person in the susceptible population. An outbreak will occur in the future if R_0 is bigger than 1, therefore it is an effective indicator for monitoring the pandemic trend and choosing possible interventions. Several outbreak-related indicators, such as the turning points, the initial infectious rate of the pandemic, the final proportion of the susceptible population that will be infected population, the equilibrium ratio of susceptible individuals in the population, and the critical vaccination threshold, can be acquired by estimating R_0. Table 2 provides estimated R_0 over three different models including SIR, SIR-D, and SIR-F given 11 periods of time. The root mean squared logarithmic error (RMSLE) measurement has been calculated in order to compare the precise observed and estimated values of the parameter across the models. According to the results shown in Table 2, estimated R_0s are significant at the first period when the pandemic is reaching. The values have decreased over time and finally convergent around 2. The time-varying R_0 estimations matches the pattern of the COVID-19 pandemic in the United States. In the SIR-based model, it is challenging to determine the optimal parameters that fit the model with lower errors. To the computational biometrics perspective, parameter estimations can be done by using many mathematical and statistical methods such as Gauss Newton, genetic algorithms, and simulated annealing.

Table 2. R_0 estimations with RMSLEs based on SIR, SIR-D, and SIR-F models

No.	Start	End	SIR		SIR-D		SIR-F	
			R0	RMSLE	R0	RMSLE	R0	RMSLE
0th	9-Feb-20	7-Apr-20	25.91	0.8920	117.27	0.9216	122.13	1.0713
1st	8-Apr-20	28-Apr-20	4.73	0.0380	7.28	0.0490	7.74	0.0583
2nd	29-Apr-20	22-May-20	2.30	0.0340	2.68	0.0412	2.20	0.0580
3rd	23-May-20	17-Jun-20	2.10	0.0119	2.32	0.0139	2.34	0.0172
4th	18-Jun-20	3-Jul-20	3.39	0.0094	3.57	0.0094	5.16	0.0272
5th	4-Jul-20	15-Jul-20	3.68	0.0029	3.85	0.0029	4.24	0.0075
6th	16-Jul-20	27-Jul-20	3.11	0.0047	3.25	0.0044	3.65	0.0096
7th	28-Jul-20	9-Aug-20	2.21	0.0055	2.32	0.0053	2.36	0.0123
8th	10-Aug-20	24-Aug-20	1.73	0.0055	1.80	0.0053	1.91	0.0111
9th	25-Aug-20	8-Sep-20	1.84	0.0024	1.91	0.0026	2.21	0.0093
10th	9-Sep-20	23-Sep-20	2.08	0.0019	2.17	0.0019	2.07	0.0040

The susceptible (S)-recovered (R) trend analysis can provide a better understanding of the endpoint of the COVID-19 epidemic based on the correlation analysis between susceptible cases and recovered counts. The pandemic will be ended when a decrease in the number of susceptible cases is exponentially related to recovery individuals, which means a herd of the population will never get infected again. Figure 2 illustrates a S-R trend analysis for the U.S. from February 9, 2020 to September 23, 2020. The whole time span has been divided into 11 phases. It can be observed that a negative correlation exists between the susceptible persons and the recoveries. The shape of the fitted line is not straight, indicating the changes of model parameters at some time-points.

Figure 2. S-R trend analysis

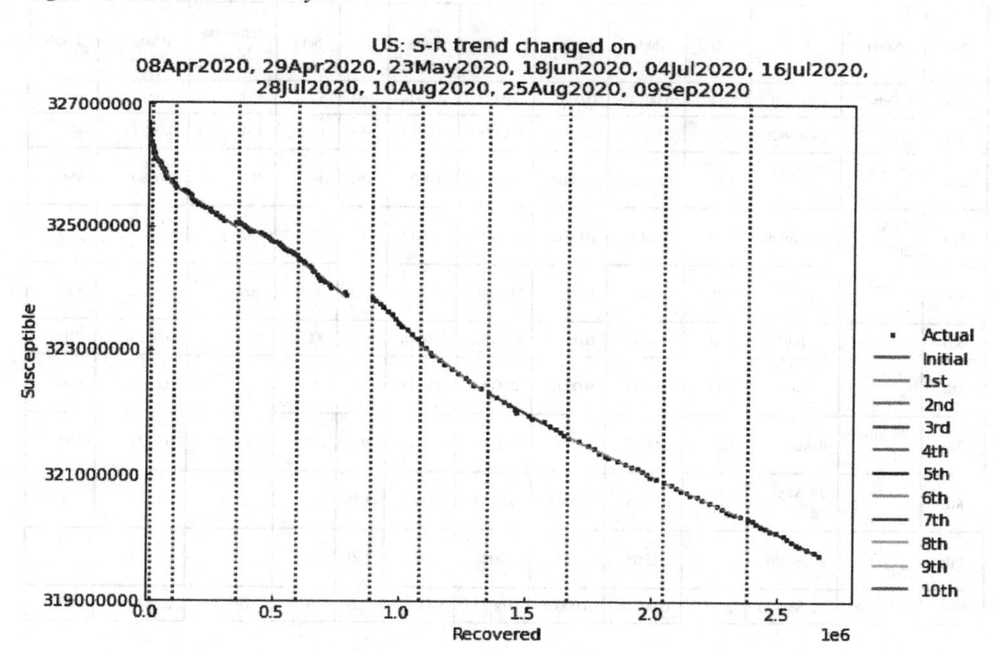

Parameter Optimization

Estimating parameters of a SIR-based model is one of the most important steps of predicting the unknown scenarios in the future. According to the above-mentioned results, the parameter estimations of SIR-based models are similar, however, the hyper-parameter optimization has not been given for each model. The hyper-parameters, such as contact rate β, fatality probability α_1, mortality rate of infected cases α_2, and recovery rate γ, can be estimated through different phases by using computational tools. In this chapter, a SIR-F model has been chosen as the example of the hyper-parameter optimization with the estimating process using a Python package called CovsirPhy (CovsirPhy Development Team, 2020). This Python library is designed for COVID-19 data analysis with phase-dependent SIR-derived ODE models. After performing the SIR-F model with the COVID-19 data of the United States, the parameter estimations, including Rt, theta, kappa, rho, sigma, tau, have been presented with RMSLE as shown in Table 3.

Table 3. Parameter estimation of SIR-F model

No.	Start	End	Rt	theta	kappa	rho	sigma	tau	1/ beta [day]	1/gamma [day]	RMSLE	Trials
0th	9-Feb-20	7-Apr-20	52.49	0.0014	0.0005	0.0487	0.0004	360	5	627	1.0713	867
1st	8-Apr-20	28-Apr-20	5.59	0.0356	0.0006	0.0128	0.0017	360	19	151	0.0583	448
2nd	29-Apr-20	22-May-20	1.86	0.0020	0.0004	0.0053	0.0024	360	46	103	0.0580	955
3rd	23-May-20	17-Jun-20	2.13	0.0001	0.0002	0.0040	0.0017	360	62	145	0.0172	340
4th	18-Jun-20	3-Jul-20	4.86	0.0013	0.0001	0.0062	0.0012	360	40	209	0.0272	547
5th	4-Jul-20	15-Jul-20	4.04	0.0004	0.0001	0.0076	0.0018	360	32	138	0.0075	710
6th	16-Jul-20	27-Jul-20	3.48	0.0004	0.0001	0.0064	0.0017	360	39	143	0.0096	889
7th	28-Jul-20	9-Aug-20	2.25	0.0007	0.0001	0.0044	0.0019	360	57	134	0.0123	457
8th	10-Aug-20	24-Aug-20	1.86	0.0015	0.0001	0.0034	0.0018	360	72	139	0.0111	236
9th	25-Aug-20	8-Sep-20	2.1	0.0008	0.0001	0.0028	0.0013	360	89	197	0.0093	272
10th	9-Sep-20	23-Sep-20	2.01	0.0078	0.0000	0.0025	0.0012	360	101	210	0.0040	607

Simulation With Possible Effects

The parameters of SIR-F model can be used to fulfill future prediction of COVID-19 outbreak situation in the United States given three different effects, including the no-actions scenario, the development of new medicines, and the lockdown policy.

The No-actions Scenario

The no-actions scenario means that no effect, such as lockdown policies, medicines, or vaccinations, will affect the population. Therefore, in the early stage of virus spreading, the number of the infected will be the highest, which has been presented in Figure 3. The increasing numbers of both infected and recovered individuals will speed up in the near future. The number of the infected will reach the highest point until January 2023, indicating that the turning point will appear after three years since the pandemic begins. The number of fatal individuals will be continuously increasing with a slow growth rate. On February 10th, 2028, the last day that the SIR-F model predicted, the number of infected cases will be approaching to zero, meaning that the pandemic will be ended and it has been keeping spreading for more than eight years. According to the SIR-F model prediction, the no-actions scenario will generate a huge impact to the whole healthcare system in the United States, which may be the worst case in the context of COVID-19. Therefore, it is urgent for the governments and health authorities to issue prevention measures as soon as possible.

Figure 3. Case predictions of the no-actions scenario until 2028

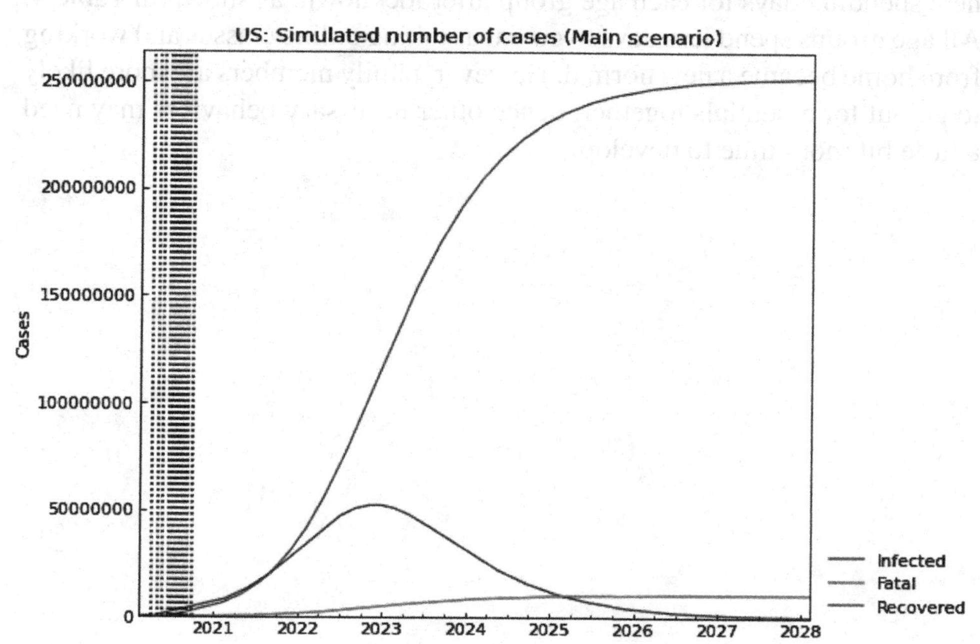

The Lockdown Scenario

The contact rate β can be reduced by proposing a lockdown policy that will minimize the probability of contacts between the susceptible and confirmed patients. In the United States, most states have been forced to lockdown when the former White House administration announced the national emergency in the middle of March, 2020. The contact rate has been dropped significantly between the first phase and the second phase, as shown in Table 3, with a decreased reproduction number. The susceptible cases that go out (g_s) and the number of contacts in a minute (c) are two factors in the mathematical formulation of the contact rate. The population pyramid data has been provided in Table 4 with the estimated spending days that go out for each age group before and after the lockdown in the United States. Initially before the lockdown, younger students spend more days in schooling, while college students do not need to go out because of the relatively flexible learning time and space. School teachers, office professionals, industry works, and government officers are more likely spending more days for work, while retired people may choose to spend the rest of their lives in a nursing home or to be taken care by their family in their own home. Situation will be changed after the lockdown. An estimated value of g_s can be calculated and also be applied to estimate the new spending days for each age group after lockdown, as shown in Table 4. All age groups spend less days to go out as taking online classes and working from home became a new normal. However, family members are more likely to go out for essentials together, hence other necessary behaviors may need a little bit more time to develop.

Table 4. Spending days before and after lockdown. Note: (.) donate the number of days spending for going out after the lockdown.

Age Groups	Period of Life	School	Office	Others	Population	Portion
0 - 2	nursery	3(0)	0(0)	0(0)	11732949	0.03829
3 - 5	nursery school	4(0)	0(0)	1(1.7)	11869337	0.038735
6 -10	elementary school	5(0)	0(0)	1(1.7)	20408368	0.066601
11 -13	middle school	5(0)	0(0)	1(1.7)	12656652	0.041304
14 -18	high school	6(0)	0(0)	1(1.7)	21176720	0.069109
19 -25	university/work	3(0)	3(1)	1(1.7)	31355852	0.102328
26 -35	work	0(0)	6(1)	1(1.7)	45837957	0.149589
36 -45	work	0(0)	5(1)	1(1.7)	41096911	0.134117
46 -55	work	0(0)	5(1)	1(1.7)	41210566	0.134488
56 -65	work	0(0)	5(1)	1(1.7)	41162200	0.13433
66 -75	retired	0(0)	0(0)	1(1.7)	15742842	0.051376
76 -85	retired	0(0)	0(0)	1(1.7)	9179360	0.029956
86 -95	retired	0(0)	0(0)	1(1.7)	2996040	0.009777

Figure 4 demonstrates the pandemic prediction using SIR-F model with new contact rate caused by lockdown effect. The turning point appears around the middle of December, 2020 as the infected cases reach the peak. The number of infected individuals increases towards the peak with a high growth rate and decreases with a gradient descent after late December. Recoveries rapidly jump after November and the fatal rate remains stable. The simulation result perfectly predicts the real situation in the United States. However, the predicted turning point arrives earlier than the reality, which means the real situation is worse than the simulation. It can be explained by several reasons, for example, citizens do not practice the stay-at-home order; some states governments emphasize the economy reopening rather than the disease control measures.

Figure 4. Case predictions of lockdown scenario until 2024

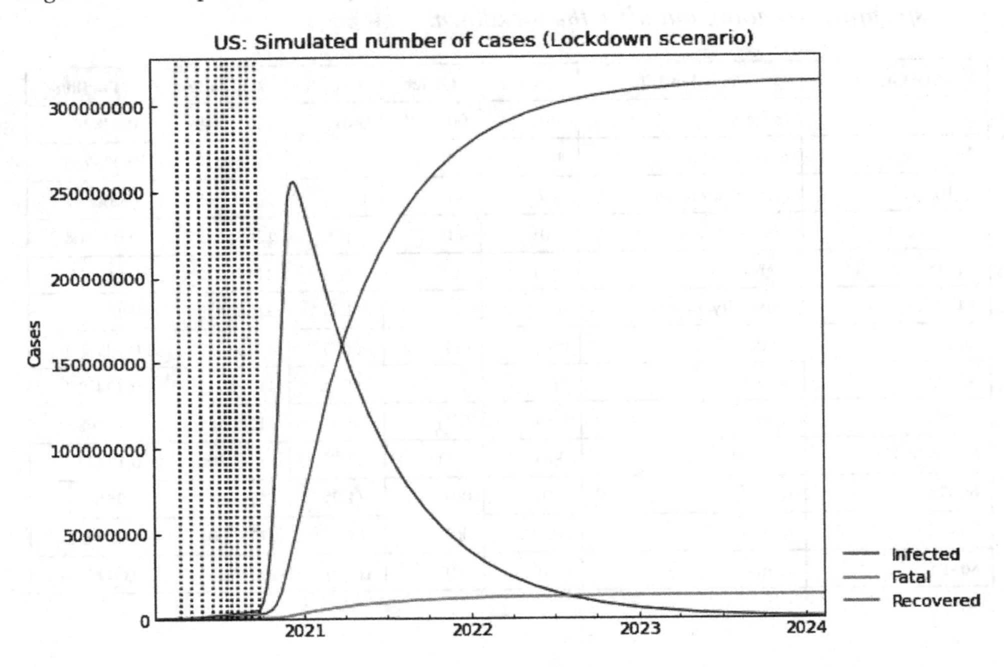

The Vaccination Scenario

It is no question that getting vaccinated will help for controlling the spreading of the virus. The progress of getting fully vaccinated will force a decreasing infection rate. Therefore, the effect of the vaccination can reduce the spreading speed, consequently, end the pandemic earlier. Moderna, Pfizer-BioNtech, and Johnson & Johnson's Janssen are three major vaccines authorized and recommended by the U.S. Centers for Disease Control and Prevention, however, such vaccines have not been released yet in September, 2020. As of February, 2021, the United States reached the first milestone as more citizens had received at least one dose of a COVID-19 vaccine than those had tested positive. To achieve better and reasonable simulation results, the data range applied for producing effects under the vaccination scenario has been updated by adding new data from September, 2020 to February, 2021. Figure 5 presents the simulation results for the effect of the vaccination scenario until February 27th, 2024. The number of infected will descend with a relatively rapid rate and then it will be down to zero after February, 2024. The recovery rate keeps growing during the predicted period.

Figure 5. Case predictions of vaccination scenario until 2024

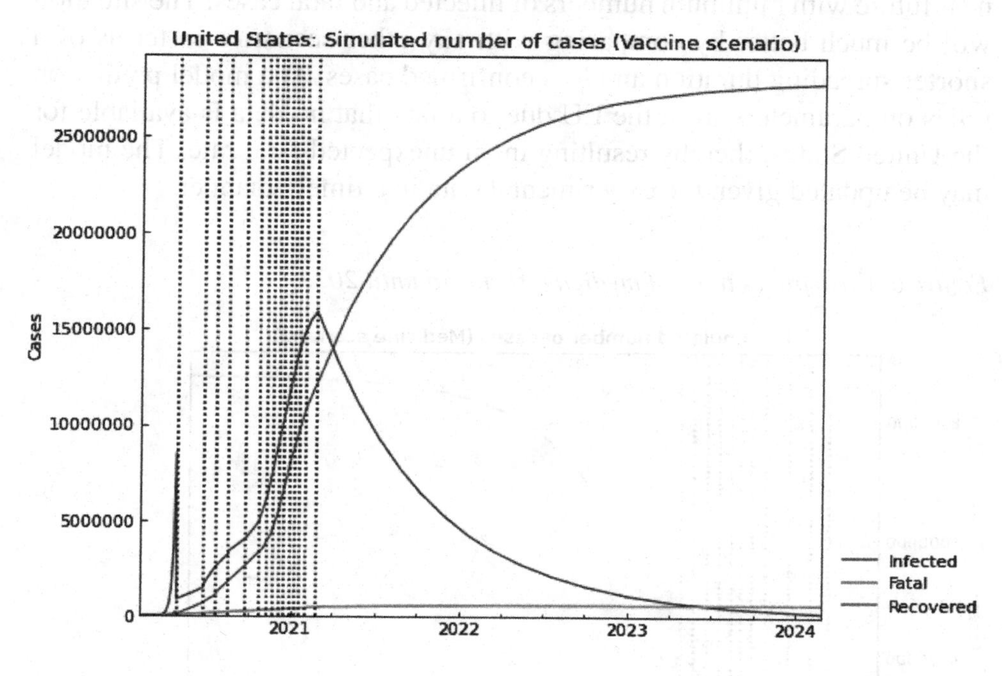

The New Medicine Scenario

The development of new medicines can push the recovery rate γ to increase, therefore, the COVID-19 outbreak will be under control immediately. Remdesivir (USAN) and Favipiravir (AVIGAN) are two candidate medicines of COVID-19, of which 47% patients have been discharged and 13% of them deceased in a 10-day course of Remdesivir administration with median follow-up of 18 days, according to a clinical study among 53 patients (Grein et al., 2020). The new model parameters can be estimated according to the discharged percentage and total observation minutes from this clinical study result. Figure 6 shows the predicted number of the infected, the fatal, and the recovered cases until April 10, 2022 under the effect of new medicine scenario using SIR-F model. Based on the new medicine-influenced model, the number of the infected appears a significant plummet on September 23, 2020 and decreases over time thereafter. Despite a significant jump of the number of the fatal cases, the growth rate of the recoveries outperforms. Obviously, the COVID-19 pandemic will be under control by the development of new medicines. On the last day of the prediction, the number of infected cases will be approaching to zero, indicating that the pandemic will be ended in the near future with minimum numbers of infected and fatal cases. The situation will be much better by comparing with any other scenarios in terms of a shorter spreading duration and less confirmed cases. The model prediction relies on parameters from the EU due to a fact that no data is available for the United States, thereby resulting in an unexpected fatal rate. The model may be updated given the experimental data in a timely manner.

Figure 6. Case predictions of medicine scenario until 2022

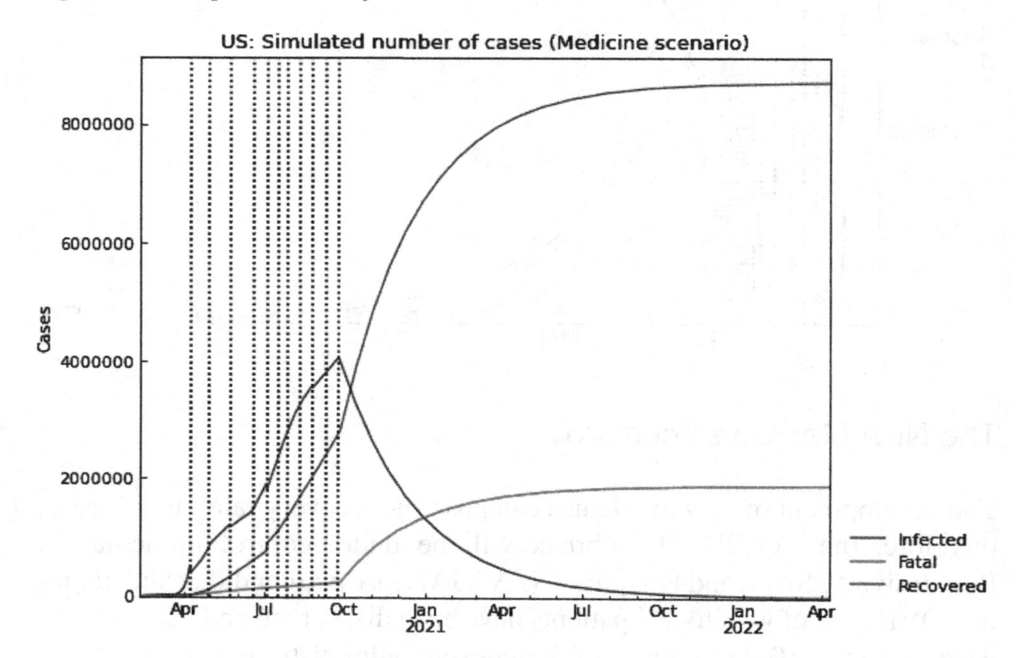

FUTURE RESEARCH DIRECTIONS

Existing studies using SIR and SEIR models rely on an assumption that the exposed population can not be infectious, however, it does not fit for the COVID-19 data in some regions. The problem can be solved by introducing time-varying features into the model. Law et al. (2020) proposed a time-varying SIR model to track transmission dynamics in the early stage of the COVID-19 pandemic. With the development of big data and machine learning techniques, it can be solved by using AI-based approaches in establishing overarching information and in predicting the advancement of infections. For instance, a pandemic trend has been estimated with an exact range of its duration by combining a SEIR model and a machine learning technology in China (Yang et al., 2020). Similar mixed models have been applied for analyzing the disease, forecasting the trend, characterizing the epidemic, estimating the transmission risk, and simulating the effects of intervention measures in the context of COVID-19 (Zhang et al., 2020; Hengjian & Tao, 2020; Hou et al., 2020; Tang et al., 2020; Stochiţoiu et al., 2020).

CONCLUSION

The SIR model is one of the most classical epidemiological simulation models, which plays an important role in biological science. In this chapter, SIR-based models have been proposed and implemented with parameter estimations and mathematical simulations under different scenarios. The primary goal of this chapter is to examine epidemiological models in terms of providing the decision-making evidence for the public healthcare system by navigating actions required to the next step of the onset pandemic. The estimated reproduction results are presented with the accuracy metric RMLSE by fitting the real data in the U.S. using the basic SIR, SIR-D, and SIR-F models. The model has been proved to be successful in producing the natural dynamic of COVID-19. The prevention measures under different scenarios have been examined comparing with the no-actions situation. The simulation results indicate that the interventions of lockdown and isolation are not enough to stop the pandemic, while COVID-19 will be end by the development of new medicines and vaccines.

REFERENCES

Alanazi, S. A., Kamruzzaman, M. M., Alruwaili, M., Alshammari, N., Alqahtani, S. A., & Karime, A. (2020). Measuring and preventing COVID-19 using the SIR model and machine learning in smart health care. *Journal of Healthcare Engineering*, *2020*, 2020. doi:10.1155/2020/8857346 PMID:33204404

Ambrosio, B., & Aziz-Alaoui, M. A. (2020). On a coupled time-dependent SIR models fitting with New York and New-Jersey states COVID-19 data. *Biology (Basel)*, *9*(6), 135. doi:10.3390/biology9060135 PMID:32599867

Capasso, V., & Serio, G. (1978). A generalization of the Kermack-McKendrick deterministic epidemic model. *Mathematical Biosciences*, *42*(1-2), 43–61. doi:10.1016/0025-5564(78)90006-8

Chen, Y.-C., Lu, P.-E., Chang, C.-S., & Liu, T.-H. (2020). *A time-dependent SIR model for COVID-19 with undetectable infected persons*. https://arxiv.org/abs/2003.00122

CHIME v1.1.5 (2020). *COVID-19 Hospital Impact Model for Epidemics (CHIME)*. University of Pennsylvania.

Coburn, B. J., Wagner, B. G., & Blower, S. (2009). Modeling influenza epidemics and pandemics: Insights into the future of swine flu (H1N1). *BMC Medicine*, *7*, 30. https://doi.org/10.1186/1741-7015-7-30

CovsirPhy Development Team. (2020). *CovsirPhy, Python package for COVID-19 analysis with SIR-derived ODE models*. https://github.com/lisphilar/covid19-sir

Grein, J., Ohmagari, N., Shin, D., Diaz, G., Asperges, E., Castagna, A., ... Flanigan, T. (2020). Compassionate use of remdesivir for patients with severe Covid-19. *The New England Journal of Medicine*, *382*(24), 2327–2336.

Guidotti, E., & Ardia, D. (2020). COVID-19 data hub. *Journal of Open Source Software*, *5*(51), 2376.

He, S., Peng, Y., & Sun, K. (2020). SEIR modeling of the COVID-19 and its dynamics. *Nonlinear Dynamics*, *101*(3), 1667–1680.

Hengjian, C., & Tao, H. (2020). *Nonlinear regression in COVID-19 forecasting*. Scientia Sinica Mathematica.

Hethcote, H. W. (2000). The mathematics of infectious diseases. *SIAM Review*, *42*(4), 599–653.

Hou, C., Chen, J., Zhou, Y., Hua, L., Yuan, J., He, S., ... Jia, E. (2020). The effectiveness of quarantine of Wuhan city against the Corona Virus Disease 2019 (COVID-19): A well-mixed SEIR model analysis. *Journal of Medical Virology*, *92*(7), 841–848.

Jain, A., & Bhatnagar, V. (2017). Concoction of ambient intelligence and big data for better patient ministration services. *International Journal of Ambient Computing and Intelligence*, *8*(4), 19–30.

Kermack, W. O., & McKendrick, A. G. (1991). Contributions to the mathematical theory of epidemics—I. 1927. *Bulletin of Mathematical Biology*, *53*(1-2), 33–55.

Kwon, C. M., & Jung, J. U. (2016). Applying discrete SEIR model to characterizing MERS spread in Korea. *International Journal of Modeling, Simulation, and Scientific Computing*, *7*(04), 1643003.

Law, K. B., Peariasamy, K. M., & Gill, B. S. (2020). Tracking the early depleting transmission dynamics of COVID-19 with a time-varying SIR model. *Scientific Reports*, *10*, 21721. https://doi.org/10.1038/s41598-020-78739-8

Li, R., Pei, S., Chen, B., Song, Y., Zhang, T., Yang, W., & Shaman, J. (2020). Substantial undocumented infection facilitates the rapid dissemination of novel coronavirus (SARS-CoV-2). *Science*, *368*(6490), 489–493.

López, L., & Rodo, X. (2021). A modified SEIR model to predict the COVID-19 outbreak in Spain and Italy: Simulating control scenarios and multi-scale epidemics. *Results in Physics*, *21*, 103746.

Manrique-Abril, F. G., Agudelo-Calderon, C. A., González-Chordá, V. M., Gutiérrez-Lesmes, O., Téllez-Piñerez, C. F., & Herrera-Amaya, G. (2020). SIR model of the COVID-19 pandemic in Colombia. *Revista de Salud Publica (Bogota, Colombia)*, *22*(2).

Marinov, T. T., & Marinova, R. S. (2020). Dynamics of COVID-19 using inverse problem for coefficient identification in SIR epidemic models. *Chaos. Solitons & Fractals: X*, *5*, 100041.

Morley, V. J. (2021). The Rules of Contagion: Why Things Spread—and Why They Stop. *Emerging Infectious Diseases*, *27*(2), 675.

Ndiaye, B. M., Tendeng, L., & Seck, D. (2020). *Comparative prediction of confirmed cases with COVID-19 pandemic by machine learning, deterministic and stochastic SIR models.* arXiv preprint arXiv:2004.13489.

Nguemdjo, U., Meno, F., Dongfack, A., & Ventelou, B. (2020). Simulating the progression of the COVID-19 disease in Cameroon using SIR models. *PLoS One*, *15*(8), e0237832. https://doi.org/10.1371/journal.pone.0237832

Pinter, G., Felde, I., Mosavi, A., Ghamisi, P., & Gloaguen, R. (2020). COVID-19 pandemic prediction for Hungary; a hybrid machine learning approach. *Mathematics*, *8*(6), 890.

Qi, C., Karlsson, D., Sallmen, K., & Wyss, R. (2020). *Model studies on the COVID-19 pandemic in Sweden.* arXiv preprint arXiv:2004.01575.

Radulescu, A., & Cavanagh, K. (2020). *Management strategies in a SEIR model of COVID 19 community spread.* arXiv preprint arXiv:2003.11150.

Rajesh, A., Pai, H., Roy, V., Samanta, S., & Ghosh, S. (2020). *CoVID-19 prediction for India from the existing data and SIR (D) model study.* medRxiv.

Shapiro, M. B., Karim, F., Muscioni, G., & Augustine, A. S. (2020). *Are we there yet? An adaptive SIR model for continuous estimation of COVID-19 infection rate and reproduction number in the United States.* medRxiv.

Simha, A., Prasad, R. V., & Narayana, S. (2020). *A simple stochastic sir model for covid 19 infection dynamics for Karnataka: Learning from Europe.* arXiv preprint arXiv:2003.11920.

Stochiţoiu, R. D., Rebedea, T., Popescu, I., & Leordeanu, M. (2020). *A self-supervised neural-analytic method to predict the evolution of covid-19 in Romania.* arXiv preprint arXiv:2006.12926.

Tang, B., Wang, X., Li, Q., Bragazzi, N. L., Tang, S., Xiao, Y., & Wu, J. (2020). Estimation of the transmission risk of the 2019-nCoV and its implication for public health interventions. *Journal of Clinical Medicine*, *9*(2), 462.

Tsui, B. C., Deng, A., & Pan, S. (2020). Coronavirus Disease 2019: Epidemiological Factors During Aerosol-Generating Medical Procedures. *Anesthesia and Analgesia.*

Wang, L., Zhou, Y., He, J., Zhu, B., Wang, F., Tang, L., ... Song, P. X. (2020). An epidemiological forecast model and software assessing interventions on the COVID-19 epidemic in China. *Journal of Data Science: JDS*, *18*(3), 409–432.

Wangping, J., Ke, H., Yang, S., Wenzhe, C., Shengshu, W., Shanshan, Y., ... Yao, H. (2020). Extended SIR prediction of the epidemics trend of COVID-19 in Italy and compared with Hunan, China. *Frontiers in medicine*, *7*, 169.

Wu, J. T., Leung, K., & Leung, G. M. (2020). Nowcasting and forecasting the potential domestic and international spread of the 2019-nCoV outbreak originating in Wuhan, China: A modelling study. *Lancet*, *395*(10225), 689–697.

Yang, Z., Zeng, Z., Wang, K., Wong, S. S., Liang, W., Zanin, M., ... He, J. (2020). Modified SEIR and AI prediction of the epidemics trend of COVID-19 in China under public health interventions. *Journal of Thoracic Disease, 12*(3), 165.

Zhang, W., Liu, J., Zhang, C., Sun, Y., & Huang, H. (2020). Characteristics of COVID-2019 in areas epidemic from imported cases. *International Journal of Public Health, 65*(6), 741–746.

Zhang, Z. (2007). The outbreak pattern of SARS cases in China as revealed by a mathematical model. *Ecological Modelling, 204*(3), 420–426. https://doi.org/10.1016/j.ecolmodel.2007.01.020

ADDITIONAL READING

Anand, A., Singh, A. K., Lv, Z., & Bhatnagar, G. (2020). Compression-then-encryption-based secure watermarking technique for smart healthcare system. *IEEE MultiMedia, 27*(4), 133–143. doi:10.1109/MMUL.2020.2993269

Banerjee, A., Srivastava, Y., & Ganguli, S. (2021). Big Data Handling for Smart Healthcare System: A Brief Review and Future Directions. *Big Data Analytics and Intelligent Techniques for Smart Cities*, 93-115.

Bhardwaj, R., & Datta, D. (2021). Development of Epidemiological Modeling RD-Covid-19 of Coronavirus Infectious Disease and Its Numerical Simulation. In *Analysis of Infectious Disease Problems (Covid-19) and Their Global Impact* (pp. 245–277). Springer. doi:10.1007/978-981-16-2450-6_12

Cui, Y., Ni, S., & Shen, S. (2021). A network-based model to explore the role of testing in the epidemiological control of the COVID-19 pandemic. *BMC Infectious Diseases, 21*(1), 1–12. doi:10.118612879-020-05750-9 PMID:33435892

Holmdahl, I., & Buckee, C. (2020). Wrong but useful—What covid-19 epidemiologic models can and cannot tell us. *The New England Journal of Medicine, 383*(4), 303–305. doi:10.1056/NEJMp2016822 PMID:32412711

Liu, X. X., Fong, S. J., Dey, N., Crespo, R. G., & Herrera-Viedma, E. (2021). A new SEAIRD pandemic prediction model with clinical and epidemiological data analysis on COVID-19 outbreak. *Applied Intelligence, 51*(7), 1–37. doi:10.100710489-020-01938-3 PMID:34764574

Machi, D., Bhattacharya, P., Hoops, S., Chen, J., Mortveit, H., Venkatramanan, S., . . . Marathe, M. V. (2021). Scalable epidemiological workflows to support covid-19 planning and response. medRxiv. doi:10.1109/IPDPS49936.2021.00072

Martcheva, M. (2015). *An introduction to mathematical epidemiology* (Vol. 61). Springer. doi:10.1007/978-1-4899-7612-3

Zhang, X. Y., Huang, H. J., Zhuang, D. L., Nasser, M. I., Yang, M. H., Zhu, P., & Zhao, M. Y. (2020). Biological, clinical and epidemiological features of COVID-19, SARS and MERS and AutoDock simulation of ACE2. *Infectious Diseases of Poverty*, *9*(1), 1–11. doi:10.118640249-020-00691-6 PMID:32690096

KEY TERMS AND DEFINITIONS

Epidemiological Modeling: A modeling analysis applied for infectious diseases through mathematical and statistical models.

eSIR: An infectious disease dynamic extended model that considers the numbers of the susceptible, the infected, and the removed with a certain infectious disease in a constant population over time.

Mathematical Simulation: A process that analyze and recognize performance of systems and conduct optimization according to practical applications.

Predictive Analysis: An analytical technique that makes simulations and forecasting in regards to uncertainties and unknown events using a variety of mathematical processes, such as statistical modeling, data mining, machine learning, etc.

SEIR: An epidemiological model that considers the numbers of the susceptible, the exposed individuals, the infected, and the recovered with a certain infectious disease in a constant population over time.

SIR: An epidemiological model that considers the numbers of the susceptible, infected cases, and recovered people with a certain infectious disease in a constant population over time.

SIR-D: An epidemiological model that considers the numbers of the susceptible, the infected, the recovered and the deaths with a certain infectious disease in a constant population over time.

SIR-F: An epidemiological model that considers the numbers of the susceptible, the infected, the recovered and the fatal with the confirmation with a certain infectious disease in a constant population over time.

Smart Healthcare System: A health service system that involves wearable devices, Internet of Theory, and mobile internet to make connections among people, facilities, and healthcare stations.

Chapter 3
Forecasting Techniques for the Pandemic Trend of COVID-19

ABSTRACT

Forecasting the trend of the COVID-19 pandemic has been crucial for controlling the spread and making related disease control policies. Various forecasting techniques can be served thereby assisting in strengthening the healthcare system to fight the pandemic. With the development of big data and machine learning techniques, prediction models become more accurate in yielding preparations against risks and threats. In this chapter, three types of forecasting methods, machine learning models, time series forecasting techniques, and deep learning algorithms, are categorized and introduced, mathematically and empirically. To justify the outcomes from each model, this chapter has presented case studies of three pandemic scenarios, including the early stage, the second wave, and the real-time prediction, with real data for the United States. Model comparisons and evaluations have been also illustrated to forecast the number of possible causes. Various existing studies about pandemic predictions are included in the current research by big data analytics.

INTRODUCTION

Over the last decade, data-driven approaches have proved to be problem-solving kernels for many complicated and sophisticated real-world applications. Such methods are relied on statistical inferences, data analysis processes,

DOI: 10.4018/978-1-7998-8793-5.ch003

machine learning algorithms, and deep learning architectures, which follow the programming reinforcements based on decision-making requiems, such as predictive analysis, data mining, and what-if analysis. The forecasting technique, as one of the most powerful and critical research domains in data-driven decision science, has been widely employed to guide the future course of disease projection and prognosis. Various statistical modeling methods and classical machine learning algorithms have been applied in medical and epidemiological areas to predict the future conditions of a specific disease, along with big data analytics for the healthcare system (Chen et al., 2017). With the rapid rise of deep learning during the past few years, a variety of studies have been performed for disease predictions, disease risk factors, and disease-genetic associations (Zhou et al., 2018; Luo et al., 2019; Ali et al., 2020). Particularly, most recent studies are focused on the real-time forecasting of the spreading features for novel coronavirus and the prediction of Covid-19 outbreak and its early response (Grasselli et al., 2020). Such a prediction system can be used in constructing a smart healthcare framework that handles the current situation by instructing the public authorities to make early prevention measures and interventions to control the new disease effectively.

With the number of Covid-19 cases has been increasing rapidly, the public authorities and individuals keep eyes on the peak and the duration of Covid-19. Policy-makers are mostly concerned with the number of infected cases and deaths in the next few weeks. Although the pandemic will eventually end, key questions of the public healthcare system remain. Can the public health system issue a pandemic alarm based on daily updated information? The question can be answered by forecasting the trend of Covid-19 using time series models based on the daily reported data. In the early stage of the outbreak, no historical data is available to guide model building, however, as the pandemic spreads and volumes of data are amassed, new insights from the prediction models have been captured. Forecasting techniques are critical in terms of better understanding the current health situation and projections of the pandemic trend. Applying appropriate models and increasing their accuracy consistently can enhance the decision-making process of the public health in order to prepare for answering the shock from this ongoing global pandemic. From the computational perspectives, fitting a forecasting model only takes a few seconds, which can be easily applied to the research studies and industrial applications. Data used for the model building relies on daily reports that can be reached from many data sources. The accuracy of the forecasting depends on updating the models in a daily manner for model development, parameter setting, and estimating prediction intervals (Petropoulos & Makridakis, 2020).

Visual mining and forecasting of the number of Covid-19 cases have become the major component for promoting the application of the computational power and resource in smart healthcare management. Several curve-fitting techniques have been used to build the predictive model that serves as the forecasting engine in the dashboards, such as the confirmed and forecasted case data model introduced by the Los Alamos National Laboratory (Los Alamos, 2020), an interactive real-time Covid-19 cases tracker called CovidCounties (Arneson et al., 2020), the worldwide coronavirus outbreak data analysis and prediction system named as CoronaTracker (Hamzah et al., 2020). However, their forecasting results vary over time and regions as they deployed different predictive models with various kinds of input variables and parameters. As the predictive results from such studies/applications are far away from the actual values, it is important to check which models are the most effective for forecasting the Covid-19 cases. Moreover, the model fitting and predicting time are also required to be taken into account in terms of the model efficiency and complexity.

To overcome the current challenges in forecasting Covid-19 cases, this chapter aims to investigate those predictive models that have been widely used. Several forecasting techniques are reviewed and examined theoretically and empirically. State-of-the-art supervised learning models are selected to perform the curve-fitting and modeling, along with the model performance evaluation and comparison among their model efficiency and complexity. The candidate models are trained using the real datasets from Johns Hopkins Covid-19 Data Center. The key elements of this chapter are:

- illustrating how statistical modeling, machine learning, and deep learning serve as the curve-fitting technique.
- determining which forecasting technique is the optimal model for predicting the number of confirmed cases.
- discussing potential solutions of the improvement, such as creating more complicated models, involving additional factors/features, and analyzing the data quality.

BACKGROUND

Several forecasting models have been applied by researchers aimed at predicting the outbreaks of many diseases such as HIV/AIDS, SARS, MERS, and swine flu. Yu et al. (2013) built an autoregressive integrated moving average (ARIMA) model to forecast the number of HIV infections in South Korea. Similar methods were applied in forecasting new HIV cases in the Ashanti region of Ghana (Aboagye-Sarfo et al., 2013). A machine learning based classification of SARS spatial distribution was done with a SVM algorithm (Hu and Gong, 2010). A similar study has been presented by using a maximum likelihood method with the construction phylogenetic tree to identify the spreading of SARS epidemic (Amiroch et al., 2018). H1N1, also known as the swine flu, ravaged the world in 2009. Relative studies have been done, for example, Sultana and Sharma (2018) used a neural forecasting model to predict swine flu in India. Similar models have been used in predicting the H1N1 cases (Khan et al., 2020). MERS was another coronavirus faced by the world. Kim et al. (2016) employed decision trees and apriori algorithm to depict the transmission route of the coronavirus. Moreover, the Naive Bayes and J48 decision tress classifiers were used to build predictive models for MERS-Cov (Al-Turaiki et al., 2016).

Despite limited data available in the early stage of the new virus outbreak, huge opportunities have been provided for the utilization of artificial intelligence and big data analytics in this research field. Masses of studies on prediction methods of the pandemic trend are presented using state-of-the-art techniques such as machine learning, deep learning, and time series forecasting. Sujatha et al. (2020) predicted the pandemic pattern and infected rate of Covid-19 using a set of machine learning models including linear regression, multilayer perceptron, and vector autoregression model. Similar models have been applied to forecasting confirmed cases over time with additional machine learning models such as XGBoost and multi-output regressor, decision tree (DT), support vector machine (SVM), and Naive Bayes (NB) models (Suzuki & Suzuki, 2020; Rustam et al., 2020; Singh et al., 2020; Muhammad et al., 2020). Moreover, different methods also take place on forecasting the trend of the pandemic in various nations and regions, for example, autoregression integrated moving average model (ARIMA) (Kumar et al., 2020; Chintalapudi et al., 2020) and exponential smoothing (Abebe, 2020). Lastly, deep learning algorithms such as long short-term memory

(LSTM) have been used to estimate the possible number of infected cases (Chimmula et al., 2020).

MAIN FOCUS OF THE CHAPTER

Issues, Controversies, Problems

The most significant issue of constructing a forecasting system for the number of confirmed COVID-19 cases is to determine the optimal model. However, the controversy is that most existing studies perform the predictive analysis relied on one or few models due to the main focus vary over research, e.g. some studies emphasize on deep learning approaches, therefore other methods will not be investigated. The model performance evaluation has not been provided in some studies. The input variables need to be pre-processed before the model fitting and testing because of the temporal component. However, the problem of stability and de-noising has not been considered in some studies. By summarizing various studies using daily reported data, the evaluation of the forecasting methods for the trend of COVID-19 can be classified as three categories including the family of machine learning models, the set of time series forecasting techniques, and the class of deep learning algorithms. This chapter proposes examining the utility of each model for forecasting the trend of COVID-19 in the U.S. based on the daily reported new cases. Empirical results by comparing multiple models in terms of their model performance and efficiency, appropriate models can be suggested to assess near futures of this pandemic, which can be readily applied by society and governments. Despite many factors that cannot be presently controlled, forecasting models are still effective for the pandemic trend analysis since the models are conductive with daily updated data.

Features of the Chapter

This chapter attempts to explain a broad range of forecasting techniques applied in the Covid-19 trend predictions, followed by case studies using the U.S. real-time daily report data with SVM, bayesian ridge regression, decision tree, XGBoost, ARIMA, and LSTM. A wavelet transformation has been applied in order to reduce the noise term and to make the input data stable. A comprehensive model evaluation and comparison will be presented with discussions of the model performance and efficiency for each model. This chapter also illustrates the most recent existing studies and the research trend in this field, along with the challenges that exist in the predictive models for Covid-19 with the possible research in the future. The main focus of the chapter is to present how artificial intelligence helps for forecasting the cases of Covid-19, and to further assist for constructing the smart healthcare system.

SOLUTIONS AND RECOMMENDATIONS

Forecasting Techniques

Machine Learning Models

To the predictive analysis perspectives, machine learning models can be used to solve classification and regression problems. Instead of constructing a machine learning classifier, forecasting the trend of the Covid-19 pandemic requires a regressor because it is a continuous variable based on the historical data. A broad range of machine learning models can be applied for the continuous cases, however, fitting all of them is time consuming. By reviewing the existing studies in this field, four machine learning algorithms, such as SVM, Bayesian ridge regression, decision tree, and XGBoost, have been chosen as representing the state-of-the-art approaches in terms of predicting the trend of the pandemic.

Officially introduced by Boser, Guyon, and Vapnik (1992), SVM has been widely applied for classifications with categorical variables, however, it may not be able to fit the model for a continuous case. In this chapter, support vector machines for regression (SVR) is applied for predicting the confirmed cases of Covid-19 infection as a continuous variable. The basic idea of running a SVM is to find a line (in lower dimensions) or a hyperplane (in higher dimensions) that separates different classes. Usually, the computational efficiency will be decreased with the increasing dimension of variables, which can be fixed by mapping the features to a higher dimensional space called kernels. SVR shares the same principle as SVM, but for regression problems that deal with finding a function of approximated mapping from inputs to real numbers on the basis of a training set. An optimization problem of the SVR is formulated as follows:

$$\min \frac{1}{2}\|w\|^2 \; subject\; to \left|y_i - \langle w, x_i \rangle - b\right| \leq \varepsilon$$

where y_i is the output with training sample x_i and the normal vector w. Bayesian ridge regression is the other option to predict future values based on estimating a probabilistic model of the regression problem with continuous variables. Similar models, such as the decision tree and the XGBoost, have been selected as candidates to build the forecasting models.

Bayesian regression methods can involve regularization parameters in the estimation procedure through performing uninformative priors. In ridge regression, the L_2 regularization finds a maximum a posteriori estimation based on a Gaussian prior over the coefficients w with the precision $\frac{1}{\lambda}$ from the data. Therefore, the bayesian ridge regression is to estimate a probabilistic model of the regression problems, as given as below:

$$p\left(y \mid X, w, \alpha\right) = N\left(y \mid Xw, \alpha\right)$$

Where α is a random variable from the data. This probabilistic model for y is assumed to be a Gaussian distribution based on Xw. The prior estimation for the coefficients w can be defined as a spherical Gaussian, as given as below:

$$p\left(w \mid \lambda\right) = N\left(w \mid 0, \frac{1}{\lambda} I_p\right)$$

Additionally, bayesian ridge regression chooses gamma distributions for the priors over α and λ, and these two parameters are estimated by maximizing the log marginal likelihood function.

Decision tree is a supervised learning method applied for classification and regression in machine learning. In classification trees, the target variable is usually a discrete set of values, while in regression trees the target variable can be continuous values. The goal of decision tree learning is to build a predictive model for the target variables through learning decision rules constricted by the data features. In a decision tree, each internal node is marked with an input feature, and the arcs that connect with the internal node are marked with corresponding possible values or lead to a subsequent decision node on a different input feature. Each leaf of the decision tree is marked with class label or a probability distribution.

XGBoost applies machine learning algorithms using the gradient boosting decision tree. Gradient boosting is an approach that minimizes the loss using a gradient descent algorithm. However, compared with the gradient boosting, XGBoost optimizes multiple aspects. XGBoost utilizes parallelized implementation to process the sequential tree building. This change can help XGBoost improve its computation performance through shorten the running time. Another influence factor that speed up the running time is that "depth-first" approach is used in XGBoost. Moreover, XGBoost has optimized hardware management to make the use of disk space more efficient. Considering model performance, XGBoost avoids overfitting or bias through both LASSO and ridge regularizations and adds built-in cross-validation method in each iteration.

Time Series Forecasting Model: ARIMA

ARIMA is one of the most classical time series forecasting models that can be used to predict future values of a series with its own lags and the lagged forecasting errors (Box & Jenkins, 2015). ARIMA models fit any non-seasonal time series data that is not a white noise, which are determined by three parameters such as the order of the autoregressive term (*p*), the number of differences required for stationarity (*d*), and the order of the moving average term (*q*). The general format of the ARIMA forecasting model is shown in the following equation:

$$\hat{y}_t = \mu + \phi_1 y_{t-1} + \ldots + \phi_p y_{t-p} - \theta_1 e_{t-1} - \ldots - \theta_q e_{t-q}$$

where ϕi and θi $_d$onate the autoregressive and moving average parameters, while ei $_i$s the white noise term. It starts at identifying the minimum differencing needed to make a stationary series. Autoregressive parameters are determined by inspecting the partial autocorrelation function (PACF) that describes the correlation between the data and its lags. Moving average terms can be identified by looking at the autocorrelation function (ACF) that indicates how many white noise terms are required to remove the autocorrelation pattern from the series. Determining the order and number of parameters for each term is necessary, otherwise the model will be failure.

Deep Learning Model: LSTM

LSTM is a typical recurrent neural network (RNN) that has been proposed to overcome the vanishing gradient problem by memorizing network parameters for a lone time (Hochreiter&Schmidhuber, 1997). Memory blocks, instead of neurons, are connected between layers in a LSTM network, which can be applied to create large RNNs. A block contains gates that process the flowing information and output using the sigmoid activation function. Only positive values can pass to the next gate in the sigmoid activation units, where output is in the range from 0 to 1. The gate of each unit has weights that can be trained by making the adjustment and adding information. Three types of gates are represented within a unit, shown as follows:

$$J_t = SD\left(w_J\left[h_{t-1}, k_t\right] + b_J\right)$$

$$G_t = SD\left(w_G\left[h_{t-1}, k_t\right] + b_G\right)$$

$$P_t = SD\left(w_P\left[h_{t-1}, k_t\right] + b_P\right)$$

where function of input gate Jt decides which values from the input to update the memory state; function of forget gate G_t controls what information to throw away from the block; function of output gate P_t determines what to output based on inner and the memory of the block. W_j, W_g, and W_p are recurrent weights for each gate that contains coefficients of neurons at that gate given result from previous time step and input to the function at the current time step. Bias of neurons at each gate are represented as b_j, b_g, and b_p, respectively.

Experimental Design and Results

Basic Statistics of the Training and Testing Sets

The experimental analysis has been proposed based on the data collection from the COVID-19 data repository by the center for systems science and engineering at the Johns Hopkins University, which is publicly accessible from a Github repository. The dataset contains the suspected, the confirmed, the recovered, and the dead cases by the end of each day, which is available in a time series format. The number of the confirmed patients in the United States has been collected from 01/22/2020 to 03/12/2021 as the input variable of each model. The dataset is divided into an 80% training and a 20% testing sets. Table 1 shows the basic statistics of the dataset.

Table 1. Basic statistics of the confirmed case data in the U.S.

Set	Count	Mean	St.d	Max	Min
Training	319	46185.1	46694.9	232785	0
Testing	80	169627.31	63775.42	300310	54186

Curve Fitting

Curve fitting is one of the most intuitional ways to predict the trend of the pandemic. Fitted curves can be applied for visual mining by constructing a function that has the best fit to a time series of data points. Figure 1 shows the fitted curve by each machine learning model, where the blue lines present the confirmed cases in the U.S. and the red lines indicate the predictive values. The decision tree algorithm and the XGBoost model are most likely fail to fit the curve because a huge gap between the predicted and the real cases after a certain time point. Bayesian ridge regression can fit the curve until January 2021, however, the prediction breaks away from the real data thereafter. Although gaps between the predicted and the real data exist from beginning to the end by fitting a SVM model, it is most likely that the model can fit the curve well. Decision tree, XGBoost, a Bayesian ridge regression yield almost zero on the goodness of fit (R2) measurement, indicating the performance of the model fitting is extremely ineffective, while the R2 score by SVM is 0.54.

Figure 1. U.S. confirmed case curve fitting over time by SVM

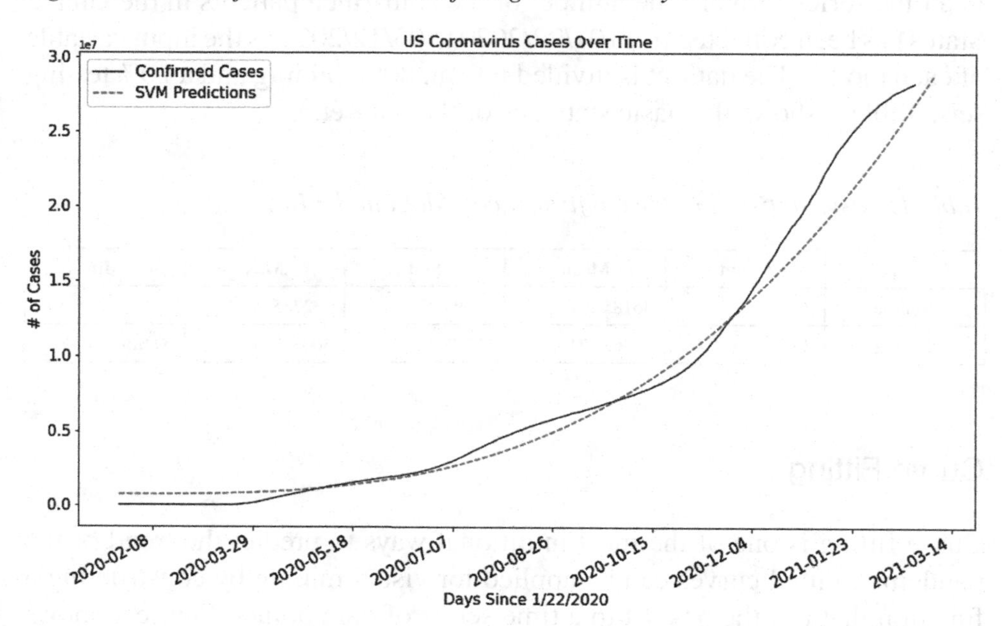

Figure 2 interprets the prediction (red line) comparing to the real data (blue line) by fitting the ARIMA model. Intuitively, ARIMA can forecast the future well because of the overlapped pattern between predictive and real values. The R2 score (0.76) for the ARIMA model confirms the visual result, which is the best goodness of fit measurement. For LSTM, the validation of the prediction result has been presented in Figure 3, where predicted values (orange line) have similar patterns comparing to the real data over time. The R2 score of the LSTM is about 0.66, indicating the model can somehow fit the curve.

Figure 2. U.S. confirmed case curve fitting over time by ARIMA

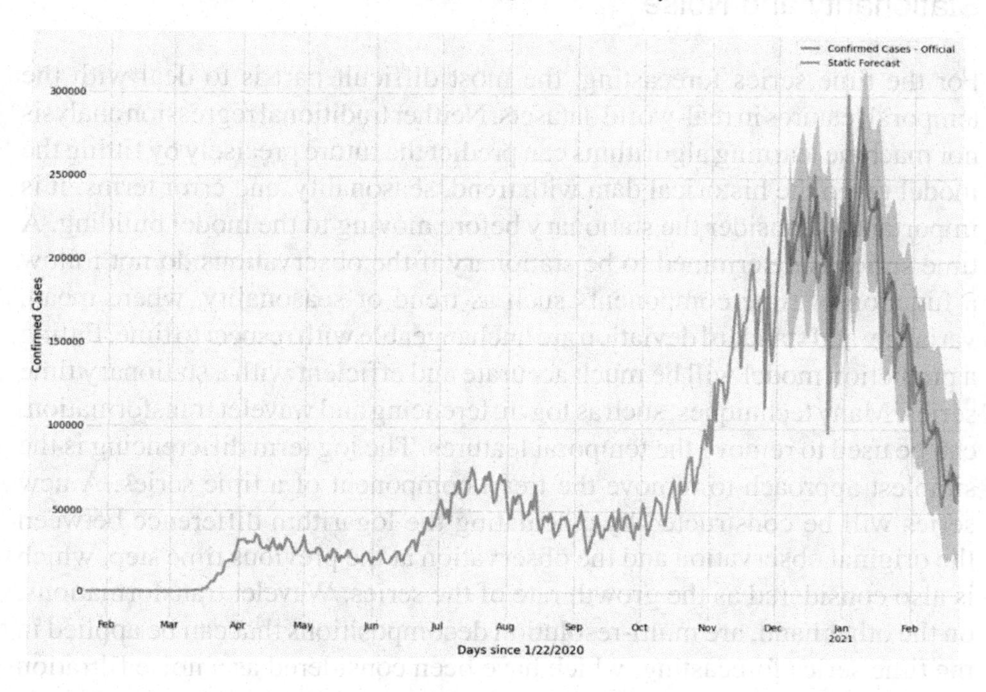

Figure 3. U.S. confirmed case curve fitting for predictions by LSTM

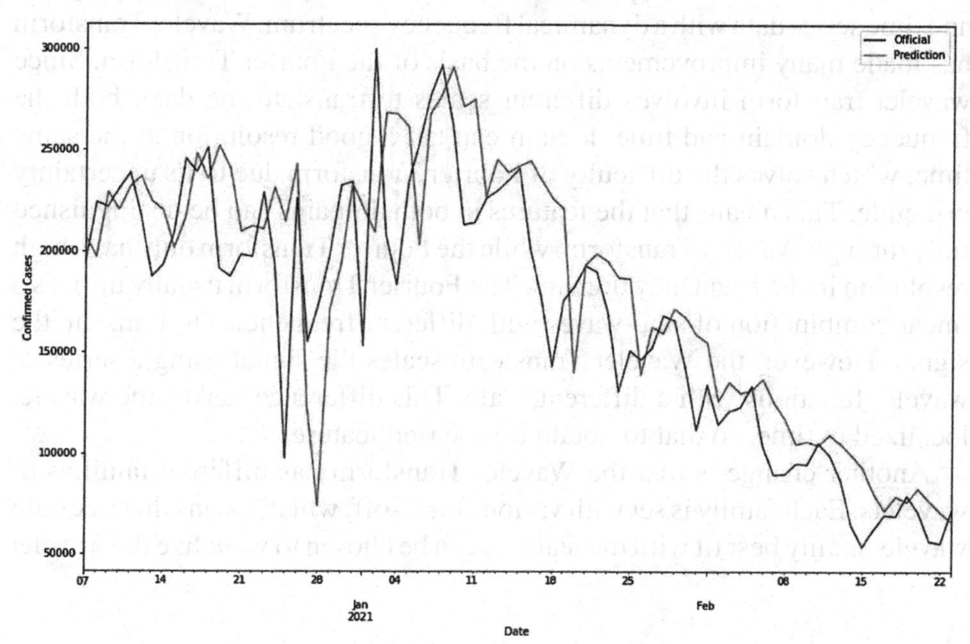

Stationarity and Noise

For the time series forecasting, the most difficult part is to deal with the temporal features in real-world datasets. Neither traditional regression analysis nor machine learning algorithms can predict the future precisely by fitting the model given the historical data with trend, seasonality, and error terms. It is important to consider the stationary before moving to the model building. A time series is determined to be stationary if the observations do not follow a function of time components such as trend or seasonality, where mean, variance, and standard deviation are unchangeable with respect to time. Fitting a prediction model will be much accurate and efficient with a stationary time series. Many techniques, such as log differencing and wavelet transformation, can be used to remove the temporal features. The log term differencing is the simplest approach to remove the trend component of a time series. A new series will be constructed by calculating the logarithm difference between the original observation and the observation at the previous time step, which is also considered as the growth rate of the series. Wavelet transformations, on the other hand, are multi-resolution decompositions that can be applied in the time series forecasting, which have been considered as a noise filtration technique to analyze signals. In this chapter, only wavelet transformation is applied to deal with the time-dependent components and the random noise in the dataset because of its effectiveness and popularity.

Wavelet Transform is a powerful approach to analyze and classify signals and time series data with a dynamical frequency spectrum. Wavelet Transform has made many improvements on the basis of the Fourier Transform. Since wavelet transform involves different scales to translate the data, both the frequency-domain and time-domain can get a good resolution at the same time, which solves the difficulty of Fourier Transform due to its uncertainty principle. This means that the features in both domains can be distinguished fully through Wavelet Transform, while the Fourier Transform only has a high resolution in the frequency domain. The Fourier Transform usually utilizes a linear combination of sine-waves with different frequencies to translate the signal. However, the Wavelet Transform scales the signal using a series of wavelet functions with a different scale. This difference makes the wavelet localized in time, so that to obtain time-based features.

Another change is that the Wavelet Transform has different families of wavelets. Each family is set with various trade-off, which means that a certain wavelet family best fit with the features can be chosen to visualize the wavelet in a close and smooth way. Additionally, in order to create a new type of wavelet, two mathematical constraints that must be satisfied are normalization and orthogonalization. In details, the wavelet should be localized in time and frequency and can be integrated and the inverse of the wavelet should also be computable. Moreover, every wavelet family still has many different wavelet subcategories, that can be discerned by the number of coefficients and the level of decomposition. It is possible to set the maximum level of decomposition for the input signal during the wavelet selection, since the maximum level is determined by the length of the input signal and the wavelet. The more the number of the coefficients is, the smoother the wavelet is. As the level of decomposition rises, the number of presented samples increases as well.

Wavelet Transform is categorized to the continuous and discrete wavelet transforms. Continuous wavelet transform uses arbitrary scales and wavelets. Mathematically, a continuous wavelet transform is defined in equation as follow:

$$X_w\left(a,b\right) = \frac{1}{\left|a\right|^{1/2}} \int_{-\infty}^{\infty} x\left(t\right) \overline{\Psi}\left(\frac{t-b}{a}\right) dt$$

Where (t) is a complex-valued function, which get scaled by a factor of a and translated by a factor of b. The overline of (t) means the conjugate complex. The equation for $X_w(a,b)$ is to convert $x(t)$ by the local oscillation in time b with scale a. The major difference between continuous and discrete wavelet transforms is that a discrete and orthogonal set of wavelets are performed by the discrete wavelet transform to resolve the signal.

Wavelet transform has been commonly applied as an ideal strategy for feature extraction to better understand the dynamical time series data in multiple research areas. In diagnosis analysis, wavelet transform plays an important in the ECG signal processing and denoising (Alfaouri and Daqrouq, 2008; Saritha et al., 2008). Besides, to avoid the high noise in financial data, wavelet threshold-denoising strategy is implemented in the data processing for historical stock price data and the transformed features can be trained for time series forecasting by machine learning and deep learning algorithms (Liang et al., 2019; Qiu et al., 2020). The study of Schlüter and Deuschele (2010) concluded that wavelet transform does benefit the performance of the time series forecasting. In urban-rural fringe analysis, spatial continuous wavelet transform, as a new reliable approach, is used to identify the urban-rural boundary more accurately than the subjective threshold method (Peng et al., 2018). In crack identification, continuous wavelet transform is involved for the model shapes of the cracked FGM beams and its coefficients are employed to estimate the crack depth (Zhu et al., 2019).

Model Evaluation and Comparison

As far as it has been shown in the case study, SVM, ARIMA, and LSTM models can fit the curve of confirmed cases in the United States. Instead of applying all models to forecast future infected case of COVID-19, a final forecasting model will be chosen based on its model performance matrix. Ideally, a predictive model should be more precise with less training time. Two major attributes, accuracy and root mean square error (RMSE), will be applied to measure the precision and error for fitting a model. The model efficiency can be described as the time of model fitting and prediction. The wavelet transformation has been applied per model before the model fitting process.

Table 2. Forecasting model evaluations by using wavelet transformation

Model	Accuracy	RMSE	Model Fitting Time (seconds)	Prediction Time (seconds)
ARIMA	98.94%	2.56	2.3773	0.1569
LSTM	96.85%	2.53	63.0175	2.6483
SVM	98.84%	2.56	3.0920	0.0013

Theoretically, SVM can handle the convex optimization problem and avoid overfitting, however, it is difficult to control the kernel with optimal size and the speed of training and testing sets (Jalalkamali et al., 2015). Although ARIMA works for seasonal and nonseasonal features with the noise from outliers reduced, the model stability cannot be held due to changes in sample observations and the model specification (Makridakis& Hibon, 1997). Despite results comparable to human expert performance, deep learning requires relatively large scales of data with the expensive training process (Krizhevsky, 2017). To determine the most accurate and efficient model, a model performance matrix has been compared shown in Table 2. Transformation methods for de-noising and stationary purposes have been applied separately to each model. Overall, predictive models with wavelet transformation yield the acceptable performance in terms of model accuracy measurements and error terms. LSTM as a deep learning method requires significantly longer time for the model fitting due to its complexity of the model structure. It is finally observed that the ARIMA model is more precise than any other approaches in terms of highest accuracy and smallest RMSE. Despite a longer prediction time comparing to SVM, ARIMA can fit the model in several seconds and predict the future in less than one second, which will be acceptable in any machine learning tasks. Fitting each model does not require much time. It only takes few seconds or a minute in a local computing environment. The model training speed will be increased in a high-performing computing platform, for example, a cloud-based computing infrastructure, which provides the feasible access and execution.

Predicting the dynamic trend of COVID-19 progression has been constantly demanding as the situation changing over time around the world. Many studies have been proposed to find and implementing different models with optimal parameters by applying ARIMA, LSTM and SVM. By summarizing a pool of existing studies, such models have been used in the early stage of the pandemic in various regions. Table 3 shows the model performance matrix with model training and predicting ranges and target regions per existing research. In Table 3, accuracy and RMSE are optimal results summarized from referenced studies as of January 2021. Accuracy for SVM is not available (NA) because it has not been presented in the study. It is not surprised that the model performance measurements changing given different input data. Overall, ARIMA and LSTM are providing better performance in terms of accuracy and RMSE, typically, LSTM has the least error term. It is no doubt that LSTM is an advanced model based on a deep learning, which can obtain better accuracy with less error. However, it also has some disadvantages, including a higher degree of the model complexity, a bigger risk of overfitting, and a significant sensitivity of random weight initializations.

Table 3. Model performance matrix per existing study

Model	Accuracy	RMSE	Data Range for Training	Data Range for Predicting	Target Region
ARIMA (Chintalapudi et al., 2020)	93.75%	514.74	15/02/20 - 31/03/20	01/04/20 - 01/06/20	Italy
LSTM (Chimmula&Zhang, 2020)	92.67%	34.83	21/01/20 - 31/03/20	01/04/20 - 15/04/20	Canada
SVM(Shahid et al., 2020)	NA	392.11	21/01/20 - 10/05/20	11/05/20 - 27/06/20	China

FUTURE RESEARCH DIRECTIONS

Existing studies have been presented towards forecasting the COVID-19 trendusing machine learning, deep learning, and statistical time series analysis. Training more complicated models is one of the most significant research trend in this field of studies. A bidirectional long short term memory (Bi-LSTM) has been applied to forecast the confirmed, the deaths, and the recovery cases for multiple countries, which outperforms other simpler models in terms of the lowest RMSE (Shahid et al., 2020). However, it does not consider the overfitting and the model efficiency. Saqib (2020) proposed a polynomial-Bayesian ridge regularization (PBRR) method that can be used to overcome the overfitting problem of the LSTM and ARIMA. However, the accuracy of PBRR is still lower than its by using LSTM and ARIMA. Plus the PBRR model has not been evaluated in terms of model efficiency and complexity, by which more time might be needed to train the model. The ARIMA forecasting method can grab better model performance in terms of higher accuracy and shorter fitting and predicting time, however, the method requires to consider many attributes such as the seasonal factors, the normality transformations, error terms, and trend components. A combination of trigonometric seasonal formulation, Box-Cox transformation, ARMA errors, and the trend component (TBAT) has been applied to forecast the trend of COVID-19 with a lower RMSE than ARIMA (Papastefanopoulos et al., 2020).

Another obvious trend of research is to use big data analytics for predicting patterns of the COVID-19 pandemic. Using a big data approach indicates that effectiveness of forecasting will be based upon a large volume of datasets with various types of attributes. Analyzing a large dataset has been done by accessing data sources from WHO, Johns Hopkins University, and national databases (Toda, 2020; Caccavo, 2020; Siwiak et al., 2020; Russo et al., 2020). Many attributes such as environmental features, disease control measures, socioeconomic variables, and clinical factors have been captured to describe the impacts on the COVID-19 pandemic (Bhattacharjee, 2020; Liu, et al., 2020; Sun et al., 2020; Bai et al., 2020). In the digital age, cyber-based information can be easily acquired through social media communication and internet searches that provide more and more information about the pandemic. Analysis of data from internet screech engines, mobile phones, social media channels, and newspapers has been proposed by cutting-edge techniques such has machine learning and big data analytics (Li et al., 2020; Lai et al., 2020; Anastassopoulou et al., 2020).

Although tremendous work has been presented in terms of targeting the predictive analysis of COVID-19, a gap between the research and the reality still exists. It provides an opportunity to fill up the gap, which will be a new research trend in this filed. By reviewing the existing work, many studies are relied on the Github data sources, which contain data from various data sources. Checking the reliability of repositories is necessary for the form of tracking the data sources and keeping them updated. Moreover, few studies have been presented to consider geo-spatial and demographical features into forecasting models, i.e. Zhu et al. (2020) and Dowd et al. (2020), hence, it may produce a better prediction of COVID-19 by involving such factors. Furthermore, modeling the different stages of the COVID-19 pandemic is necessary as the transmission rate and quarantine policies changing over time.

CONCLUSION

Precisely and efficiently forecasting the number of confirmed COVID-19 cases offers insightful information for health care professionals and managers to preview the future situation, thereby it is more effective to arrange the medical supplies. Moreover, forecasting techniques can assist governments to impose the disease prevention policies following a scientifically evidence-based decision-making process. The data-driven methods based on statistical analysis, machine learning, and deep learning have been used in constructing the real-time forecasting system. Several forecasting models, such as classical machine learning algorithms (i.e. SVM, Bayesian ridge regression, decision tree, and XGBoost), a time series forecasting model (ARIMA), and a deep learning model (LSTM) have been selected for a comprehensive study in terms of examining their predictive performances based on the model efficiency and complexity. Experimental results indicate SVM, ARIMA, and LSTM yield the acceptable and expected performance matrix with higher accuracy measurements and considerable training and prediction speeds. Determining the optimal model depends on input variables and parameters over time and across regions, thereby it is necessary to investigate the model performance matrix given different training sets. To improve the model fitting capability, wavelet transformation can be applied to reduce the noise. Additionally, potential solutions for improving the forecasting techniques, such as building more complex models, adding more factors and features, and considering data quality issues, have been derived from extra studies concentrated on this topic.

Although a comprehensive investigation for forecasting models has been proposed theoretically and experimentally in this chapter, building a forecasting model for COVID-19 is still challenging in the early stage of the pandemic and even throughout the whole spreading period. Data points are not enough due to the difficult facts of tracking the infected samples and a longer incubation period of COVID-19. Some data types are unstructured, which need the pre-processing and data cleaning to maintain quality and quantity of the data before it goes to modeling. In the early stage of the pandemic, data is just in approximate figures because of the limited testing capacity. In addition, the transparency of information sharing and data reporting may not be ensured given the chaotic situation at the beginning of the pandemic. Various variables are needed to be considered to the mathematical, statistical, and empirical perspective. Some attributes such as environmental variables (i.e. temperature, humidity, wind speed), availabilities of medical facilities (i.e. PPE, ICU, ventilator), underlying medical conditions (i.e. heart disease, diabetic), disease control policies (i.e. social distancing, quarantine, isolation), demographical information (i.e. population density, age and gender diversity, economical and social status), and human behaviors and sentiments (i.e. mobility, transportation pattens, awareness about the disease) can be captured from different data sources, which will be beneficial for forecasting the number of confirmed COVID-19 cases.

REFERENCES

Abebe, T. H. (2020). Forecasting the Number of Coronavirus (COVID-19) Cases in Ethiopia Using Exponential Smoothing Times Series Model. medRxiv. doi:10.1101/2020.06.29.20142489

Aboagye-Sarfo, P., Oduro, F., & Okyere, G. (2013). Time Series Forecast of New HIV Cases in the Ashanti Region of Ghana. *International Journal of Scientific and Engineering Research*, *4*, 546–549.

Al-Turaiki, I., Alshahrani, M., & Almutairi, T. (2016). Building predictive models for MERS-CoV infections using data mining techniques. *Journal of Infection and Public Health*, *9*(6), 744–748. doi:10.1016/j.jiph.2016.09.007 PMID:27641481

Alfaouri, M., & Daqrouq, K. (2008). ECG signal denoising by wavelet transform thresholding. *American Journal of Applied Sciences*, *5*(3), 276–281. doi:10.3844/ajassp.2008.276.281

Ali, F., El-Sappagh, S., Islam, S. R., Kwak, D., Ali, A., Imran, M., & Kwak, K. S. (2020). A smart healthcare monitoring system for heart disease prediction based on ensemble deep learning and feature fusion. *Information Fusion*, *63*, 208–222. doi:10.1016/j.inffus.2020.06.008

Amiroch, S., Pradana, M. S., Irawan, M. I., & Mukhlash, I. (2018, August). Maximum likelihood method on the construction of phylogenetic tree for identification the spreading of SARS epidemic. In *2018 International symposium on advanced intelligent informatics (SAIN)* (pp. 137-141). IEEE. 10.1109/SAIN.2018.8673334

Anastassopoulou, C., Russo, L., Tsakris, A., & Siettos, C. (2020). Data-based analysis, modelling and forecasting of the COVID-19 outbreak. *PLoS One*, *15*(3), e0230405. doi:10.1371/journal.pone.0230405 PMID:32231374

Arneson, D., Elliott, M., Mosenia, A., Oskotsky, B., Solodar, S., Vashisht, R., Zack, T., Bleicher, P., Butte, A. J., & Rudrapatna, V. A. (2020). CovidCounties is an interactive real time tracker of the COVID19 pandemic at the level of US counties. *Scientific Data*, *7*(1), 1–10. doi:10.103841597-020-00731-8 PMID:33199721

Bai, T., Tu, S., Wei, Y., Xiao, L., Jin, Y., Zhang, L., . . . Hou, X. (2020). Clinical and laboratory factors predicting the prognosis of patients with COVID-19: an analysis of 127 patients in Wuhan, China. China.

Bhattacharjee, S. (2020). *Statistical investigation of relationship between spread of coronavirus disease (COVID-19) and environmental factors based on study of four mostly affected places of China and five mostly affected places of Italy.* arXiv preprint arXiv:2003.11277.

Boser, B. E., Guyon, I. M., & Vapnik, V. N. (1992, July). A training algorithm for optimal margin classifiers. In *Proceedings of the fifth annual workshop on Computational learning theory* (pp. 144-152). 10.1145/130385.130401

Box, G. E., Jenkins, G. M., Reinsel, G. C., & Ljung, G. M. (2015). *Time series analysis: forecasting and control.* John Wiley & Sons.

Caccavo, D. (2020). Chinese and Italian COVID-19 outbreaks can be correctly described by a modified SIRD model. medRxiv. doi:10.1101/2020.03.19.20039388

Chen, M., Hao, Y., Hwang, K., Wang, L., & Wang, L. (2017). Disease prediction by machine learning over big data from healthcare communities. *IEEE Access: Practical Innovations, Open Solutions*, *5*, 8869–8879. doi:10.1109/ACCESS.2017.2694446

Chimmula, V. K. R., & Zhang, L. (2020). Time series forecasting of COVID-19 transmission in Canada using LSTM networks. *Chaos, Solitons, and Fractals*, *135*, 109864. doi:10.1016/j.chaos.2020.109864 PMID:32390691

Chintalapudi, N., Battineni, G., & Amenta, F. (2020). COVID-19 virus outbreak forecasting of registered and recovered cases after sixty day lockdown in Italy: A data driven model approach. *Journal of Microbiology, Immunology, and Infection*, *53*(3), 396–403. doi:10.1016/j.jmii.2020.04.004 PMID:32305271

Dowd, J. B., Andriano, L., Brazel, D. M., Rotondi, V., Block, P., Ding, X., Liu, Y., & Mills, M. C. (2020). Demographic science aids in understanding the spread and fatality rates of COVID-19. *Proceedings of the National Academy of Sciences of the United States of America*, *117*(18), 9696–9698. doi:10.1073/pnas.2004911117 PMID:32300018

Grasselli, G., Pesenti, A., & Cecconi, M. (2020). Critical care utilization for the COVID-19 outbreak in Lombardy, Italy: Early experience and forecast during an emergency response. *Journal of the American Medical Association*, *323*(16), 1545–1546. doi:10.1001/jama.2020.4031 PMID:32167538

Hamzah, F. B., Lau, C., Nazri, H., Ligot, D. V., Lee, G., Tan, C. L., ... Chung, M. H. (2020). CoronaTracker: Worldwide COVID-19 outbreak data analysis and prediction. *Bulletin of the World Health Organization*, *1*(32).

Hochreiter, S., & Schmidhuber, J. (1997). Long short-term memory. *Neural Computation*, *9*(8), 1735–1780. doi:10.1162/neco.1997.9.8.1735 PMID:9377276

Hu, B., & Gong, J. (2010, August). Support vector machine based classification analysis of SARS spatial distribution. In *2010 Sixth international conference on natural computation* (Vol. 2, pp. 924-927). IEEE. 10.1109/ICNC.2010.5583921

Jalalkamali, A., Moradi, M., & Moradi, N. (2015). Application of several artificial intelligence models and ARIMAX model for forecasting drought using the Standardized Precipitation Index. *International Journal of Environmental Science and Technology*, *12*(4), 1201–1210. doi:10.100713762-014-0717-6

Khan, M. A., Abidi, W. U. H., Al Ghamdi, M. A., Almotiri, S. H., Saqib, S., Alyas, T., ... Mahmood, N. (2021). Forecast the Influenza Pandemic Using Machine Learning. *CMC-Computers Materials & Continua*, *66*(1), 331–357. doi:10.32604/cmc.2020.012148

Kim, D., Hong, S., Choi, S., & Yoon, T. (2016, March). Analysis of transmission route of MERS coronavirus using decision tree and Apriori algorithm. In *2016 18th International conference on advanced communication technology (ICACT)* (pp. 559-565). IEEE.

Krizhevsky, A., Sutskever, I., & Hinton, G. E. (2017). ImageNet classification with deep convolutional neural networks. *Communications of the ACM*, *60*(6), 84–90. doi:10.1145/3065386

Kumar, P., Singh, R. K., Nanda, C., Kalita, H., Patairiya, S., Sharma, Y. D., & Bhagavathula, A. S. (2020). *Forecasting COVID-19 impact in India using pandemic waves Nonlinear Growth Models*. MedRxiv; doi:10.1101/2020.03.30.20047803

Lai, S., Bogoch, I. I., Ruktanonchai, N. W., Watts, A., Lu, X., Yang, W., & Tatem, A. J. (2020). *Assessing spread risk of Wuhan novel coronavirus within and beyond China, January-April 2020: A travel network-based modelling study*. MedRxiv.

Li, C., Chen, L. J., Chen, X., Zhang, M., Pang, C. P., & Chen, H. (2020). Retrospective analysis of the possibility of predicting the COVID-19 outbreak from Internet searches and social media data, China, 2020. *Eurosurveillance*, *25*(10), 2000199. doi:10.2807/1560-7917.ES.2020.25.10.2000199 PMID:32183935

Liang, X., Ge, Z., Sun, L., He, M., & Chen, H. (2019). LSTM with wavelet transform based data preprocessing for stock price prediction. *Mathematical Problems in Engineering*, *2019*, 2019. doi:10.1155/2019/1340174

Liu, P., Beeler, P., & Chakrabarty, R. K. (2020). COVID-19 progression timeline and effectiveness of response-to-spread interventions across the United States. medRxiv. doi:10.1101/2020.03.17.20037770

Los Alamos, N. L. (2020). *COVID-19 Confirmed and Forecasted Case Data*. Academic Press.

Luo, P., Li, Y., Tian, L. P., & Wu, F. X. (2019). Enhancing the prediction of disease–gene associations with multimodal deep learning. *Bioinformatics (Oxford, England)*, *35*(19), 3735–3742. doi:10.1093/bioinformatics/btz155 PMID:30825303

Makridakis, S., & Hibon, M. (1997). ARMA models and the Box–Jenkins methodology. *Journal of Forecasting*, *16*(3), 147–163. doi:10.1002/(SICI)1099-131X(199705)16:3<147::AID-FOR652>3.0.CO;2-X

Muhammad, L. J., Algehyne, E. A., Usman, S. S., Ahmad, A., Chakraborty, C., & Mohammed, I. A. (2021). Supervised Machine Learning Models for Prediction of COVID-19 Infection using Epidemiology Dataset. *SN Computer Science, 2*(1), 1-13.

Papastefanopoulos, V., Linardatos, P., & Kotsiantis, S. (2020). Covid-19: A comparison of time series methods to forecast percentage of active cases per population. *Applied Sciences (Basel, Switzerland)*, *10*(11), 3880. doi:10.3390/app10113880

Peng, J., Liu, Y., Ma, J., & Zhao, S. (2018). A new approach for urban-rural fringe identification: Integrating impervious surface area and spatial continuous wavelet transform. *Landscape and Urban Planning*, *175*, 72–79. doi:10.1016/j.landurbplan.2018.03.008

Petropoulos, F., & Makridakis, S. (2020). Forecasting the novel coronavirus COVID-19. *PLoS One*, *15*(3), e0231236. doi:10.1371/journal.pone.0231236 PMID:32231392

Qiu, J., Wang, B., & Zhou, C. (2020). Forecasting stock prices with long-short term memory neural network based on attention mechanism. *PLoS One*, *15*(1), e0227222. doi:10.1371/journal.pone.0227222 PMID:31899770

Russo, L., Anastassopoulou, C., Tsakris, A., Bifulco, G. N., Campana, E. F., Toraldo, G., & Siettos, C. (2020). Tracing day-zero and forecasting the COVID-19 outbreak in Lombardy, Italy: A compartmental modelling and numerical optimization approach. *PLoS One*, *15*(10), e0240649. doi:10.1371/journal.pone.0240649 PMID:33125393

Rustam, F., Reshi, A. A., Mehmood, A., Ullah, S., On, B. W., Aslam, W., & Choi, G. S. (2020). COVID-19 future forecasting using supervised machine learning models. *IEEE Access: Practical Innovations, Open Solutions*, *8*, 101489–101499. doi:10.1109/ACCESS.2020.2997311

Saqib, M. (2020). Forecasting COVID-19 outbreak progression using hybrid polynomial-Bayesian ridge regression model. *Applied Intelligence*, 1–11. PMID:34764555

Saritha, C., Sukanya, V., & Murthy, Y. N. (2008). ECG signal analysis using wavelet transforms. *Bulg. J. Phys*, *35*(1), 68–77.

Schlüter, S., & Deuschle, C. (2010). *Using wavelets for time series forecasting: Does it pay off?* (No. 04/2010). IWQW Discussion Papers.

Shahid, F., Zameer, A., & Muneeb, M. (2020). Predictions for COVID-19 with deep learning models of LSTM, GRU and Bi-LSTM. *Chaos, Solitons, and Fractals*, *140*, 110212. doi:10.1016/j.chaos.2020.110212 PMID:32839642

Singh, V., Poonia, R. C., Kumar, S., Dass, P., Agarwal, P., Bhatnagar, V., & Raja, L. (2020). Prediction of COVID-19 corona virus pandemic based on time series data using Support Vector Machine. *Journal of Discrete Mathematical Sciences and Cryptography*, 1-15.

Siwiak, M. M., Szczesny, P., & Siwiak, M. P. (2020). From a single host to global spread. The global mobility based modelling of the COVID-19 pandemic implies higher infection and lower detection rates than current estimates. *The Global Mobility Based Modelling of the COVID-19 Pandemic Implies Higher Infection and Lower Detection Rates than Current Estimates.*

Sujatha, R., & Chatterjee, J. (2020). *A machine learning methodology for forecasting of the COVID-19 cases in India*. Academic Press.

Sultana, N., & Sharma, N. (2018, December). Statistical Models for Predicting Swine F1u Incidences in India. In *2018 First international conference on secure cyber computing and communication (ICSCCC)* (pp. 134-138). IEEE. 10.1109/ICSCCC.2018.8703300

Sun, Y., Hu, X., & Xie, J. (2021). Spatial inequalities of COVID-19 mortality rate in relation to socioeconomic and environmental factors across England. *The Science of the Total Environment*, *758*, 143595. doi:10.1016/j.scitotenv.2020.143595 PMID:33218796

Suzuki, Y., & Suzuki, A. (2020). *Machine learning model estimating number of COVID-19 infection cases over coming 24 days in every province of South Korea (XGBoost and MultiOutputRegressor). medRxiv*. doi:10.1101/2020.05.10.20097527

Toda, A. A. (2020). *Susceptible-infected-recovered (sir) dynamics of covid-19 and economic impact.* arXiv preprint arXiv:2003.11221.

Yu, H. K., Kim, N. Y., Kim, S. S., Chu, C., & Kee, M. K. (2013). Forecasting the number of human immunodeficiency virus infections in the Korean population using the autoregressive integrated moving average model. *Osong Public Health and Research Perspectives*, *4*(6), 358–362. doi:10.1016/j.phrp.2013.10.009 PMID:24524025

Zhou, J., Theesfeld, C. L., Yao, K., Chen, K. M., Wong, A. K., & Troyanskaya, O. G. (2018). Deep learning sequence-based ab initio prediction of variant effects on expression and disease risk. *Nature Genetics*, *50*(8), 1171–1179. doi:10.103841588-018-0160-6 PMID:30013180

Zhu, L. F., Ke, L. L., Zhu, X. Q., Xiang, Y., & Wang, Y. S. (2019). Crack identification of functionally graded beams using continuous wavelet transform. *Composite Structures*, *210*, 473–485. doi:10.1016/j.compstruct.2018.11.042

Zhu, X., Zhang, A., Xu, S., Jia, P., Tan, X., Tian, J., & Yu, J. (2020). *Spatially explicit modeling of 2019-nCoV epidemic trend based on mobile phone data in mainland China*. MedRxiv; doi:10.1101/2020.02.09.20021360

ADDITIONAL READING

Adhikari, S. P., Meng, S., Wu, Y. J., Mao, Y. P., Ye, R. X., Wang, Q. Z., Sun, C., Sylvia, S., Rozelle, S., Raat, H., & Zhou, H. (2020). Epidemiology, causes, clinical manifestation and diagnosis, prevention and control of coronavirus disease (COVID-19) during the early outbreak period: A scoping review. *Infectious Diseases of Poverty*, *9*(1), 1–12. doi:10.118640249-020-00646-x PMID:32183901

Gamboa, J. C. B. (2017). Deep learning for time-series analysis. *arXiv preprint arXiv:1701.01887*.

Lalmuanawma, S., Hussain, J., & Chhakchhuak, L. (2020). Applications of machine learning and artificial intelligence for Covid-19 (SARS-CoV-2) pandemic: A review. *Chaos, Solitons, and Fractals*, *139*, 110059. doi:10.1016/j.chaos.2020.110059 PMID:32834612

Längkvist, M., Karlsson, L., & Loutfi, A. (2014). A review of unsupervised feature learning and deep learning for time-series modeling. *Pattern Recognition Letters*, *42*, 11–24. doi:10.1016/j.patrec.2014.01.008

Nielsen, A. (2019). *Practical time series analysis: Prediction with statistics and machine learning*. O'Reilly Media.

Pavlyshenko, B. M. (2016, August). Linear, machine learning and probabilistic approaches for time series analysis. In *2016 IEEE First International Conference on Data Stream Mining & Processing (DSMP)* (pp. 377-381). IEEE. 10.1109/DSMP.2016.7583582

Perc, M., Gorišek Miksić, N., Slavinec, M., & Stožer, A. (2020). Forecasting covid-19. *Frontiers in Physics*, *8*, 127. doi:10.3389/fphy.2020.00127

Rahimi, I., Chen, F., & Gandomi, A. H. (2021). A review on COVID-19 forecasting models. *Neural Computing & Applications*, 1–11. PMID:33564213

Zeroual, A., Harrou, F., Dairi, A., & Sun, Y. (2020). Deep learning methods for forecasting COVID-19 time-Series data: A Comparative study. *Chaos, Solitons, and Fractals*, *140*, 110121. doi:10.1016/j.chaos.2020.110121 PMID:32834633

KEY TERMS AND DEFINITIONS

ARIMA: A model that is used for time series analysis to characterize changes over time through statistical methods.

Curve-Fitting Technique: A method of constructing a mathematical function and adjusting it in order to achieve the best fit to all data points.

Deep Learning: A subset of machine learning based on artificial neural networks with more layers, which can improve the model performance significantly.

Early Disease Prevention: A series of measures that prevent an upcoming illness from widely spreading.

LSTM: An artificial recurrent neural network architecture that has advantages of processing with sequential data.

Machine Learning: A subject of artificial intelligence that aims at the task of computational algorithms, which allow machines to learning objects automatically through historical data.

SVM: A supervised learning method that analyze data for classification and regression problems by creating a line or a hyperplane which separate the data into classes.

Time-Series Forecasting: A prediction method for time-based historical data to build models to forecast future based on data characteristics.

Chapter 4
Mining Mobility Data in Response to COVID-19

ABSTRACT

Exploring human mobility changes and spatial dynamic patterns is crucial for assisting the policy-making process of non-pharmaceutical interventions. Examining the actual degree of practicing stay-at-home orders or travel restrictions becomes an underlying question that can be answered by tracking the human mobility within a target area over time. In this chapter, several visual mining tools have been performed with results of uncovering the reason why the United States fails the stay-at-home policy. The pandemic-mobility management system architecture has been illustrated with an example of its usage, which can be applied to monitor medical risks and pandemic-mobility indicators per region. Such a spatiotemporally hyperconnected resolution of human movements and pandemic information may assist public authorities to monitor the pandemic-mobility patterns, guide the health policymaking, and deepen the understanding of human behaviors in the context of COVID-19.

INTRODUCTION

As the new coronavirus spreading from coast to coast in the United States, over thirty million people have been confirmed a year after the COVID-19 outbreak. The federal government announced a national state of emergency in the middle of March followed by a major disaster in the New York metropolitan area. A month later, the COVID-19 pandemic has been determined as a disaster

DOI: 10.4018/978-1-7998-8793-5.ch004

in all states throughout the nation, announced by the Federal Emergency Management Agency (FEMA, 2020). Following an exponential growth in the number of infected cases with an increasing fatality rate, as of the end of April, state governments have been taking action to minimize the transmission rate. Almost all the state executives have issued the stay-at-home or shelter-in-place order by mid-April 2020 (Mervosh et al., 2020). The purpose of issuing travel restrictions or lockdown policies is to mitigate the spread of COVID-19 by reducing the human mobility concentrated on decreasing the probability of the contact. The impact of such policies on the disease control of COVID-19 has been explored by using epidemic models, such as the SIR model. Although some studies suggest that lockdowns or travel restrictions have different and limited effects across regions with significant opportunity costs (Bonardi et al., 2020; Chinazzi et al., 2020), such policies can reduce the daily growth rate of infected cases in the United States (Courtemanche et al., 2020). Similar suggestions have been provided based on the simulation in Chapter 2, which indicate that the lockdown policy can flat the curve effectively without new medicines or vaccines. The purpose of the lockdown policy is to reduce the reproduction rate that can be reflected by the effectiveness of the restricted mobility. Therefore, reducing the human mobility is an essential strategy at the early stage of the pandemic.

However, individuals in different areas may practice the stay-at-home orders or travel restrictions in different ways. Despite the importance of the quarantine enforcement and contact tracing suggested by the United States Centers for Disease Control and Prevention (CDC), their implementations vary widely by the communities across the nation. Therefore, measuring the actual reductions in social contacts and travels plays a critical role in terms of understanding how governments evaluate the effectiveness of disease control policies. It is important to follow the CDC public health guidance for COVID-19 because the coronavirus is primarily transmitted through person-to-person contact when an infected individual coughs, sneezes, or talks, hence, mobility data has been used as a powerful tool to identify potentially exposed people. A significant correlation between the growth rate of COVID-19 and the mobility pattern has been examined in the United States (Badr et al., 2020). The effectiveness of household quarantine and contact tracing has been evaluated by exploring mobility data on the second wave of COVID-19 (Aleta et al., 2020). Awareness of following the CDC guidance, on the other hand, can improve the effectiveness of COVID-19 preventions, which has been confirmed with a research poll indicating a large percentage of population is taking precautions to maintain distance from each other

(Saad, 2020). However, while research scientists can provide insights of the relationship between mobility and the coronavirus, it is still challenging to measure the efficacy of social distancing practices in terms of the ongoing pandemic with the difficulties of obtaining useful information.

Understanding human mobility has become a popular topic in urban science and smart city objects, as investigated by a wide range of research (Zhao et al., 2016; Wang et al., 2021). The data collection of the human mobility is the foundation of the whole development cycle of the decision-making process, which resorts to a set of social sensing and monitoring technologies that relevant data are collected from biological features, devices, and social media channels (Zhou et al., 2018; Wang & Taylor, 2016; Furini & Montangero, 2018). Visualizing and analyzing human mobility become a modern data-driven solution, as witnessed by many research topics via social sensing, ranging from visual mining of individual-level mobility patterns (Gonzalez et al., 2008), to urban planning (Yuan et al., 2012), including traffic congestions (Xu et al., 2016), sustainability problems (Prandi et al., 2017), and disease preventions (Charu et al., 2017). Such studies and applications provide rich experiences for exploring the human mobility against the COVID-19 pandemic. An early research is already started to examine how to track the COVID-19 infection pattern from mobility data with discussions on the effect of disease control policies (Kraemer et al., 2020).

Motivated by the current situation of COVID-19 and its challenges of the implementation of non-pharmaceutical disease control measures, and inspired by the previous experiences using human mobility, this chapter is proposed in the design of a visual mining approach applied to offer a comprehensive understanding of pandemic-mobility patterns. Several cutting-edge technologies for collecting mobility data will be introduced, along with the visual mining process for human mobility in the context of the COVID-19 pandemic. A novel pandemic-mobility dataset has been created in order to investigate the insights derived from mobility patterns, COVID-19 spreading features, and medical risk factors. The objectives of this chapter are listed as follows:

- introducing various human mobility measurements and indicators, such as mobility reports, GPS-based individual-level mobility data, and the social media-based mobility index.
- illustrating the data aggregation process for the novel dataset generation by combining selected mobility datasets, COVID-19 cases information, and medical risk indicators.

- presenting how visual mining evaluating the effect of the stay-at-home order by exploring the mobility patterns throughout nationwide human behaviors, seasonal mobility characteristics, and driven pattern segmentations among different cities.
- constructing a framework of the pandemic-mobility management system that aims on understanding the relationships between mobility patterns, virus spreading features, and medical risk factors in depth.

BACKGROUND

With the development of big data technologies over the last decade, analyzing mobile device location data becomes widely adopted to support policy-making objectives in terms of studying human mobility. Researchers are also examining location data in aggregate shape to better understand general patterns of human movements and behaviors based on the global positioning system (GPS) signals, cell site locations, and bluetooth beacons (Bachir et al., 2019; Song et al., 2010; Stange et al., 2011). Such analysis can also be applied to navigate the spreading of the COVID-19 outbreak and evaluate the effectiveness of public health interventions. COVID-19 mobility tracking programs have been proposed on several platforms by measuring human travel distance, i.e. Google's community mobility reports (Google, 2021), Apple's mobility trends reports (Apple, 2021), and Cuebiq's COVID-19 mobility insights (Cuebiq, 2021). Researchers have applied data from Facebook's Data for Good program to model the mobility patterns in Seattle by examining its effect on the COVID-19 outbreak (Burstein et al., 2020). Though such studies and reports are limited to the aggregated level of travel distance because of the lack of individual-level measurements. The limitation is understandable since data brokers need to protect personal privacy by law. However, peripheral analysis relied on such data sources may be useless by some degrees, for example, the mobility measurement created by aggregating general public mobility cannot determine whether an individual is staying at home or walking around; the percentage changes of the mobility index cannot guide governments to identify individuals who violate the restriction policies. Therefore, capturing the mobility of the individual response to the coronavirus, especially in advance of social control policies, is necessary and required.

Analyzing the individual-level mobility remains a challenge owing to the difficulties of data acquisitions. Although mobile devices, such as smartphones, have become popular and essential in daily lives, it is still challenging to obtain the personal location information due to the arguments of civil rights. The GPS capability of a mobile phone allows it to track the location within up to 10 feet, which its signals can be captured by smartphone apps such as maps, games, utility, and social media apps. The logged location datasets have been obtained by governments and researchers under the legally operational procedures that can protect citizens' privacy and personal information. Processing such a large-scale dataset, on the other hand, is another challenge that requires high-performing computing environments and efficient delivery tools. A cloud-based computing platform has been applied in terms of ingesting over 60 TB to identify the trips data on the individual level by a research group at University of Maryland (Xiong et al., 2020). The study relies on anonymized and privacy-protected location data from over 150 million monthly active mobile devices in the U.S. with COVID-19 data and census information, which uncovers the changes of mobility patterns and evaluations of the stay-at-home orders in the United States during the pandemic. Similar studies have also been proposed with limited data sharing to the public (Warren & Skillman, 2020; Ghader et al., 2020; Gao et al., 2020).

In the big data era, analyzing social media data becomes popular to support public health managements (Paul & Dredze, 2017). Many social media channels, such as Twitter, Facebook, Instagram, and Youtube, provide possible solutions in which users can indicate their geolocation information. Taking Twitter as an example, one can attach the exact coordinates to his or her tweets by tweeting from a GPS-enabled device. Such data can be displayed as the geolocation information corresponding to the coordinates to the user and also be obtained through Twitter's application programming interface (API). Research scientists have been applying the Twitter streaming API to download tweets associated with location datasets. A Twitter-based social mobility index has been proposed by three researchers from Johns Hopkins University and George Washington University (Xu et al., 2020). Over 0.4 billion tweets have been collected with geotags, concentrated on measuring the social distancing and travel patterns in the United States. Similar studies have also been presented in terms of tracking public transportation patterns, defining facets of social distancing, and exploring social distancing beliefs and human mobility by using geographical information derived from Twitter (Purnomo et al., 2020; Kwon et al., 2020; Porcher & Renault, 2021).

MAIN FOCUS OF THE CHAPTER

Issues, Controversies, Problems

Although existing studies have been presented in terms of mapping and measuring the human mobility in the context of COVID-19, exploring those large-scale datasets is limited to the data mining perspective. One of the most recent studies in this field has been done toward the course of the temporal anatomy for COVID-19 social distancing based on analyzing human mobility (Cot et al., 2021). However, its results are only relied on mining Google and Apple mobility reports. Despite various individual-level mobility indicators in the form of data collected from GPS devices or social media channels, neither the COVID-19 cases information nor the medical risk factors have been involved into the index. A comprehensive understanding of the mobility versus the pandemic is still a gap between various data sources and useful insights for the public health management. To fill up the gap, this chapter introduces a novel database, called COVID-MOBILITY-POOL, containing geolocation and pandemic information collected from six different open source repositories. A set of descriptive analysis has been presented in a visual mining format throughout the nation over time. The pandemic-mobility management system architecture will be illustrated in case of exploring insights from the dataset, which is helpful for navigating the pandemic control measures concentrated on tracking the human mobility.

Features of the Chapter

This chapter focuses on how big data analytics and data visualization work to explore the pandemic-mobility pattern that can be used for evaluating social distancing effects and implementing mobility restrictions during the COVID-19 pandemic. A data aggregation approach is introduced by illustrating how COVID-MOBILITY-POOL dataset is created in terms of its components, the structure of designs, and functionalities. A real-case study in the United State has been performed via visualizing mobility patterns for mapping nationwide human behaviors, exploring seasonal mobility features at the county-level, and segmenting driven patterns in major cities. Visual mining results indicate the overall mobility pattern overtime and across the nation, along with seeking the reasons why the stay-at-home order failed in the United States and the lessons from the past. The pandemic-mobility management system is proposed in terms of its architecture designs and functionalities by calculating the weights of involved factors, such as mobility measurements, COVID-19 cases information, and medical risk indicators, via variable importance based on machine learning algorithms. Experimentally, the pandemic-mobility management system can generate a new index that reflects the relationships between human mobility, pandemic spreading, and medical risk. Together with the discussion of potential elements that can be added to the architecture, such a management system designed for disease control by monitoring the human mobility will make significant contribution in the form of reducing the spread of COVID-19 via evidence-based policy-making processes of the mobility restriction.

SOLUTIONS AND RECOMMENDATIONS

Generating the COVID-MOBILITY-POOL Dataset

From the beginning of the COVID-19 outbreak in the Unite States, a large amount of mobility data has been collected through smartphones and social media channels. Relying on the single data source to analyze mobility patterns is bias, since geolocation datasets collected by different platforms only reflect partial information, for example, Google reports are relied on data collected from Android users, while Apple reports are based on data collected from Apple users. Additionally, no geographical information has been collected associated with the COVID-19 data such as the changes of confirmed cases over time. Therefore, it is essential to put all the available information together to reach a comprehensive understanding of mobility patterns with the pandemic. To fill up the gap between the research desire and the reality, a novel dataset, named as COVID-MOBILITY-POOL, has been proposed by gathering mobility information pools and pandemic datasets. The dataset is consisted of three major parts: mobility reports, individual-level mobility signals, and pandemic information.

Mobility reports can be acquired via different data sources, including Google reports, Apple reports, driving reports, and traffic reports. Human movement features have been detected by region, and across different categories of places, such as residential places, workplaces, transit stations, parks, grocery and pharmacy shops, retail and recreation sections. Such information can be download from the COVDI-19 Community Mobility Reports website that contains both state-level and county-level datasets for daily updates in the United States (Google, 2021). Similarly, the available mobility information, presented by Apple's COVID-19 Mobility Trends Reports, can be obtained by requests per region, sub-region, and city, which represent data on driving, transportation, and walking trends in the United States (Apple, 2021). Additional traffic and driving data can be extracted from information acquired from TomTom reports (TomTom, 2021) and Waze reports (Waze, 2021), of which traffic congestions in 80 cities and driving patters in over 40 cities have been provided from coast to coast in the United States.

However, above-mentioned datasets do not contain individual-level mobility signals that make sense for managing social distancing measures based on insights derived from such datasets. Tracking individual-level geolocation information may provide a micro vision, which can be formalized by generating indicators and indexes. Two open source datasets, such as the M50 index created by Descartes Labs (Warren & Skillman, 2020) and the Social Mobility Index developed by Mark Dredze's Research Group (Xu et al., 2020), have been providing since the early stage of the pandemic. Mobility statistics of the M50 index contains the daily distance of a certain population movements by tracking GPS signals at both state-level and county-level in the United States. The social mobility metric is generated by analyzing geolocation information from Twitter users with longitudinal data for nationwide cities and states.

Two aspects of data streaming, such as daily COVID-19 confirmed cases and medical risk indicators, forms the third part of the COVID-MOBILITY-POOL dataset. The New York Time has been releasing a set of data files containing cumulative numbers of COVID-19 cases in the U.S., at the state and county level, from late January 2020 until now (The New York Times, 2021). Data providers have been compelling the time series data from local governments and health departments in order to provide a complete record of the ongoing pandemic. City-level cumulative case counts have been collected from the Harvard Dataverse (China Data Lab, 2020). Additionally, the Index of Excess Risk (IER) can be applied to measure the substantial variation in medical risk due to predisposing factors throughout communities in the United States, which is available in the supplementary tables from a research paper concerning the individual-level risk for COVID-19 mortality (Jin et al., 2020). The IER value has been defined as a combination of different sociodemographic factors and predisposing health conditions weighted by the relative magnitude of the risk of fatality due to COVID-19, which can be considered, to the medical perspective, as a pandemic risk indicator. The medical risk dataset provides static IER values per state, county, and city, which has been merged into the COVID-MOBILITY-POOL dataset.

Although most mobility and pandemic-related datasets are publicly accessible, gathering such information is still challenging. The scope of measurements varies across different data sources, for example, pandemic-related datasets have data per state, county and city, while some mobility indicator measurements are only available at state or city levels. Therefore, the COVID-MOBILITY-POOL dataset has been divided into three aspects by geographical levels, including a state-level, a county-level, and a city-level data pool, respectively. Each pool has mobility information associated with pandemic data at its geographical level. Moreover, original datasets have various temporal scopes and frequencies of updates, i.e. social mobility index has a weekly updated reports, while others have the data stream in a daily format. Table 1 shows an overview of the data pools.

Table 1. The overview of the COVID-MOBILITY-POOL datasets

Geographical Level	Original Dataset	Dataset Type	Frequency of Updates
State level	Google reports	Mobility reports	Daily
	Apple reports	Mobility reports	Daily
	M50 index	Individual mobility	Daily
	Social mobility index	Individual mobility	Weekly
	COVID-19 confirmed case	Pandemic information	Daily
	IER	Pandemic information	Fix
County level	Google reports	Mobility reports	Daily
	Apple reports	Mobility reports	Daily
	M50 index	Individual mobility	Daily
	COVID-19 confirmed case	Pandemic information	Daily
	IER	Pandemic information	Fix
City level	Apple reports	Mobility reports	Daily
	TomTom reports	Mobility reports	Daily
	Waze reports	Mobility reports	Daily
	Social mobility index	Individual mobility	Weekly
	COVID-19 confirmed case	Pandemic information	Daily
	IER	Pandemic information	Fix

Visual Mining for Mobility Patterns

Mapping Nationwide Human Behaviors

Mapping human behaviors describes the general movement patterns from the beginning of COVID-19 to the most recent. Insights of the mobility dynamic can be depicted by visualizing the time series data from the Google reports data that provides such visual mining for the United States shown in Figure 1. Human movements have been spliced into six different categories, including retail and recreation, grocery and pharmacy, parks, transit stations, workplaces, and residential. As shown in Figure 1, all categorized movements, except the residential movement, drop significantly during the stay-at-home period (from middle March to early April, 2020). The grocery and pharmacy movement increases slightly after the lockdown announcement and has three spikes during the holiday season because of the panic buying behavior in the early stage of the pandemic and holiday consumptions, respectively. Despite a tiny increase in the residential movement after the stay-at-home order, it remains stable all the time in the whole year. Other movements remain stable and periodical because of the social distancing policy and the economic activity restriction. However, the parks movement changes frequently with higher volatility and somehow shows a strong seasonality.

The general movement pattern explains the reason for the failure of the stay-at-home policy in the United States. The residential movement should overwhelmingly take over the total mobility patterns as governments impose the stay-at-home order, however, its changes are not significant. The visual mining result confirms the conclusion that the daily percentage of residents staying at home in the majority of states never exceeds 50% (Xiong et al., 2020). Despite a strong restriction policy set, such as economy reopening delays, working-from-home options, school closing regulations, and customer limits regulations in stores, it is hard for people to follow the stay-at-home order because of more flexible times for individuals walking in parks. Although an individual has been required to perform 6-feet social distancing and wearing a mask in parks, some of them violate the rule or even ignore the regulation, hence such phenomena increases the risk of the infection (Taylor & Asmundson, 2021).

Visualizing the general human mobility pattern provides insights for governments in terms of policy-making purposes. For instance, state governments may put more efforts on how to educate and regulate people who enter to parks by practicing social distancing apart from each other and monitoring the face-covering order. On the other hand, local governments are desired to monitor the mobility pattern in a lower bound of the data aggregation. As collecting and mining detailed data, it is possible to explore insightful data visualization at different geographical levels, such as a county level.

Figure 1. Mobility changes by categories of places in the United States based on Google reports

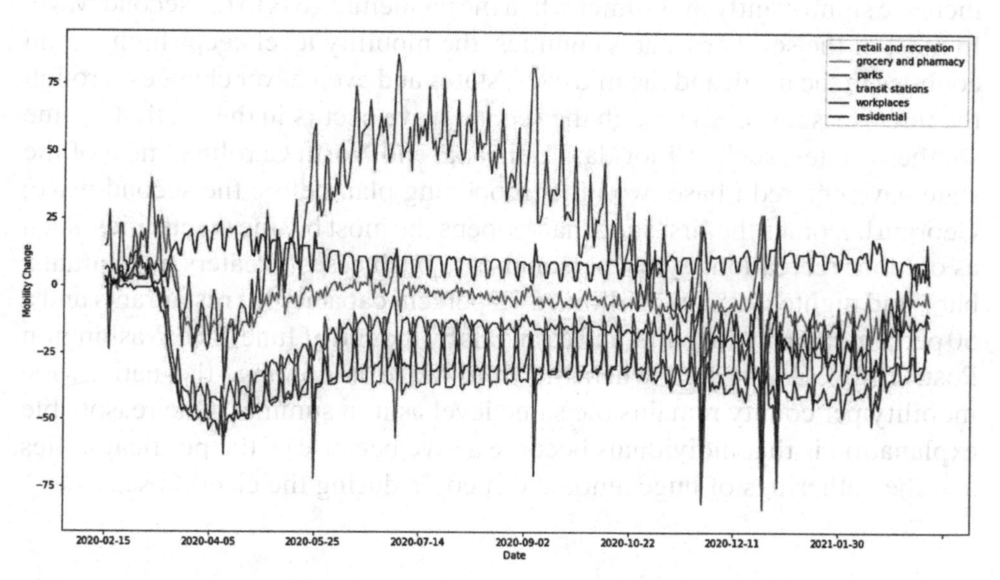

Exploring Seasonal Mobility Patterns per County

The COVID-MOBILITY-POOL dataset also provides the county-level mobility data that can be used to visualizing changes over time. Instead of a time series plot, a choropleth map provides an effectively visual mining method that can be applied for overlooking the mobility pattern per county, which is adopted for either a real-time steaming data flow or a periodically static measurement. The seasonal movement indicator per county has been derived from the individual-level mobility index (m50) by calculating the mean of mobility measurements given a certain period, as shown in Figure 2. Three choropleth maps describe the storyline of human movement patterns through spring, summer, and fall in 2020. The mobility has been stagnated because the stay-at-home order was issued nationwide in the spring when mobility measurements reached the lowest level during the whole year, especially in most northeastern states and the west coast. The Map is lighted up as the summer comes, which indicates a higher degree of the mobility as the weather turns warmer. The mobility level keeps the same level thereafter in the fall, which is similar to that of the summer time.

The visual mining for seasonal mobility patterns per county provides another explanation of why the stay-at-home order fails in the United States. Despite a lower degree of the mobility in the spring, individual movements increase significantly in summer when the pandemic goes to the second wave. In most of the southern states counties, the mobility level keeps higher than counties in the north and the mid-west states and even never changes through the three seasons, even though the second wave occurs in the south. In some southern states, such as Florida, Louisiana, and North Carolina, most of the state have entered Phase two of its reopening plan before the second wave; Georgia becomes the first state that reopens the most business activities, such as outdoor recreation, gyms, personal-care businesses, theaters, restaurants, bars, and nightclubs; Texas allows a 75 percent capacity for restaurants and a 50 percent capacity for parks and carnivals by the end of June (The Washington Post, 2021). As of fall, when the third wave is coming, the nationwide mobility per county remains the same level as it in summer. One reasonable explanation is that individuals become active because of the political rallies and the gatherings of huge amount of people during the election season.

Figure 2. Individual-level mobility changes (m50 index) per county in 2020 by seasons. (A). Spring mobility choropleth map based on data from March 1 to April 30; (B). Summer mobility choropleth map based on data from May 1 to July 31; (C). Fall mobility choropleth map based on data from Aug 1 to Oct 31. A blank area means no data available in the corresponding county.

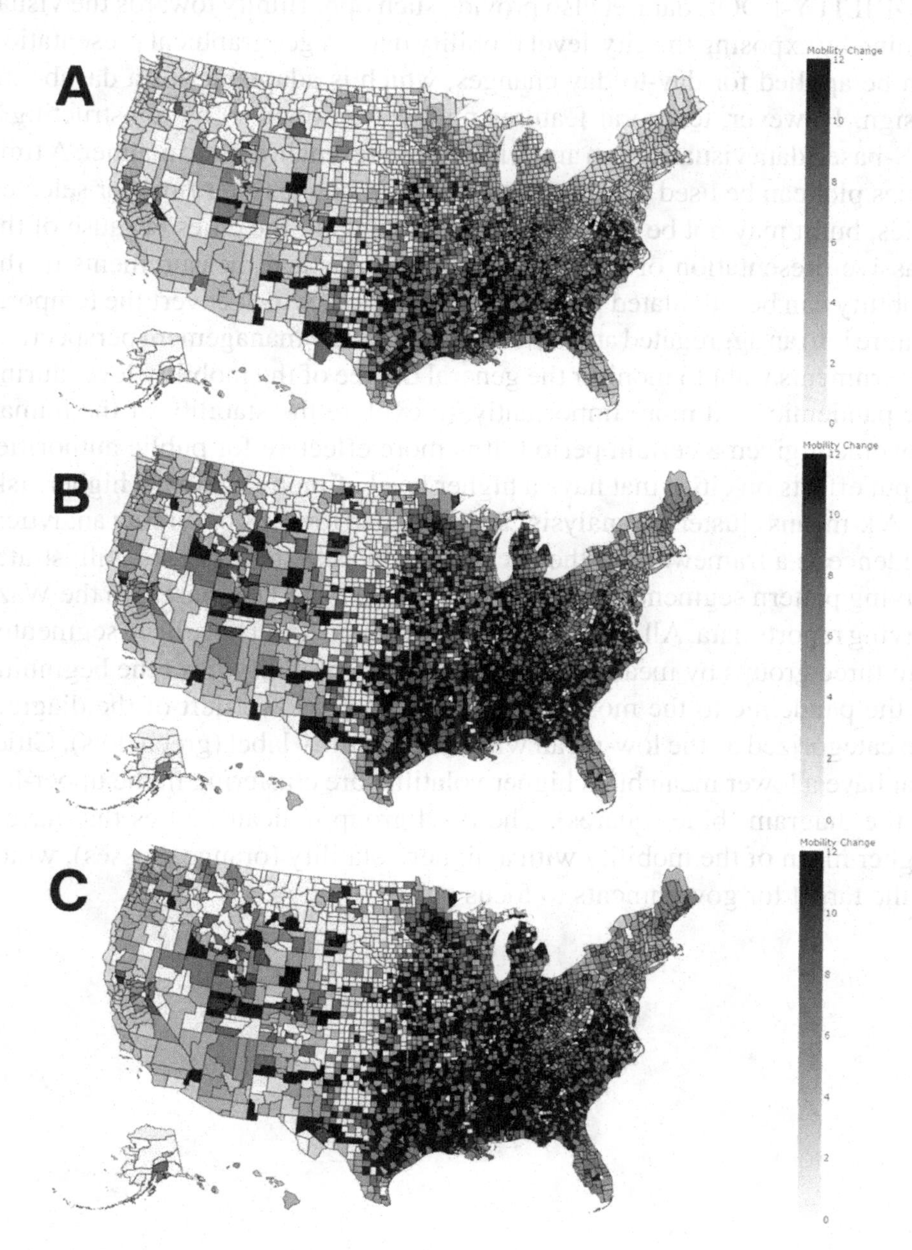

Segmenting City-level Driving Patterns

Visualizing the city-level mobility pattern is important in the context of the pandemic due to the higher population density in cities. The COVID-MOBILITY-POOL dataset also provide such opportunity towards the visual mining by exposing the city-level mobility data. A geographical presentation can be applied for day-to-day changes, which is adaptable for a dashboard design. However, temporal features may not be obvious by constructing a GIS-based data visualization and it is hard to present in a static manner. A time series plot can be used to depict the dynamic changes over time for selected cities, but it may not be effective for visualizing all the cities because of the massive presentation of the visual output. Statistical measurements of the mobility can be calculated given a certain period, which covert the temporal feature into an aggregated attribute. To the mobility management perspective, governments want to monitor the general degree of the mobility level during the pandemic, and more importantly, to explore the stability of the human movement given a certain period. It is more effective for public authorities to put efforts on cities that have a higher level of mobility with a higher risk.

A k-means clustering analysis can be applied towards providing analytical evidence in a framework of the decision support system. Figure 3 illustrates driving pattern segmentations for major cities in the U.S. based on the Waze driving reports data. All cities in the Waze report dataset have been segmented into three groups by mean and volatility of the mobility from the beginning of the pandemic to the most recent. Cities in the lower-left of the diagram are categorized as the low-mean with low volatility label (green dots). Cities that have a lower mean but a higher volatility are clustering in the upper-left of the diagram (blue squares). The third group indicates cities that have a higher mean of the mobility with a higher volatility (orange crosses), which is the target for governments to focus on.

Figure 3. Driving pattern segmentations in the United States based on Waze driving reports

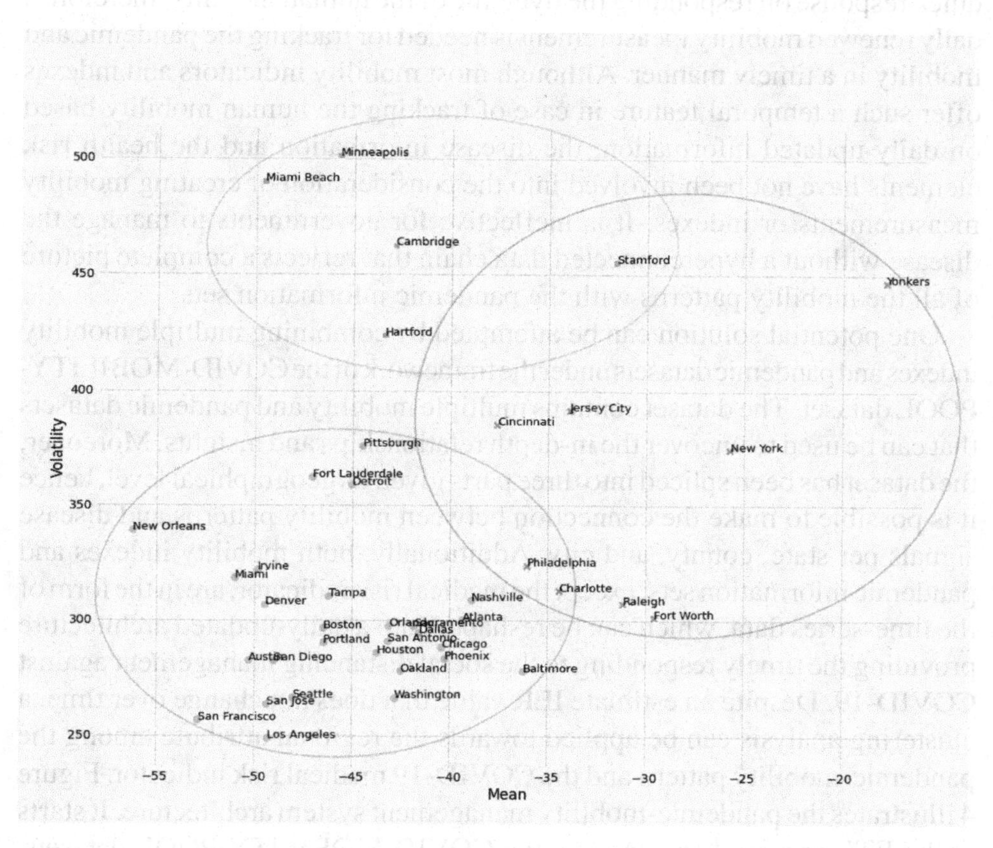

Architecture of the Pandemic-Mobility Management System

Regarding studies on the mobility during the pandemic, analytical results vary with different mobility measurements. For instance, some researchers reveal a positive relationship between human mobility and COVID-19 infections based on analyzing mobile device datasets (Xiong et al., 2020), while others, according to the analytical results from mining Google and Apple reports, imply that mobility patterns have very limited impact on the spreading speed of the coronavirus (Cot et al., 2020). A disconnected mobility dataset only provides an incomplete vision of the situation towards deploying the social distancing supervisions, thus merging multiple mobility datasets as one is necessary and important. Additionally, managing the social distancing measures requires a quick response on responding the dynamic of the human mobility, therefore a daily renewed mobility measurement is needed for tracking the pandemic and mobility in a timely manner. Although most mobility indicators and indexes offer such a temporal feature in case of tracking the human mobility based on daily updated information, the disease information and the health risk elements have not been involved into the consideration of creating mobility measurements or indexes. It is ineffective for governments to manage the disease without a hyperconnected data chain that reflects a complete picture of all the mobility patterns with the pandemic information sets.

One potential solution can be attempted by combining multiple mobility indexes and pandemic datasets under the framework of the COVID-MOBILITY-POOL dataset. The dataset contains multiple mobility and pandemic datasets that can be used to uncover the in-depth relationships and insights. Moreover, the dataset has been spliced into three parts given its geographical level, hence it is possible to make the connection between mobility patterns and disease signals per state, county, and city. Additionally, both mobility indexes and pandemic information sets, except the medical risk indicator, are in the form of the time-series data, which can be reshaped into a daily-updated architecture providing the timely responding to the social distancing management against COVID-19. Despite an estimate IER value that does not change over time, a clustering analysis can be applied towards the regional attribute among the pandemic-mobility pattern and the COVID-19 medical risk indicator. Figure 4 illustrates the pandemic-mobility management system architecture. It starts at the ETL process for data from the COVID-MOBILITY-POOL datasets, followed by generating the new mobility index that is a linear combination of weighted mobility components. The IER value can be extracted directed from the dataset with respect to different geographical levels, which is one of the variables in the clustering analysis that makes the segmentation upon medical risks versus mobility patterns.

Figure 4. The data flow char for the pandemic-mobility management system

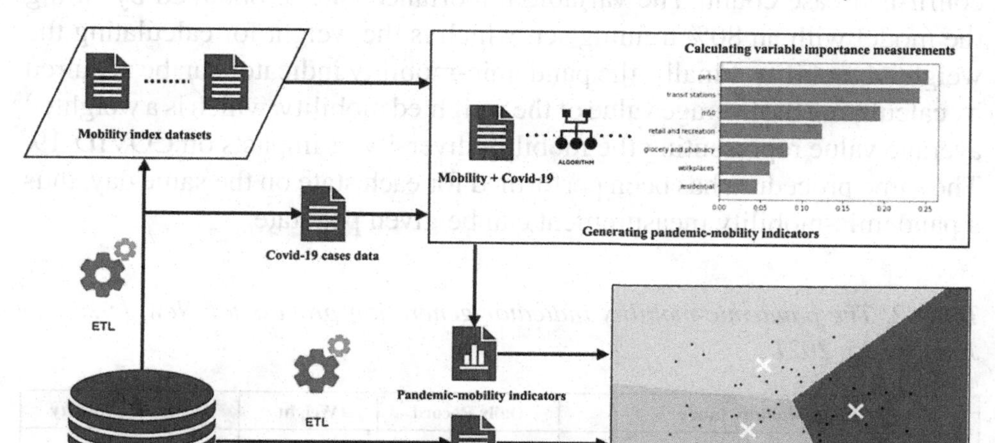

The most challenging part of establishing the architecture is to generate the new measurement representing all mobility signals derived from the above-mentioned mobility reports and indexes with respect to the human dynamic feature and its connection to the coronavirus. Taking the average number for mobility indexes per day at each geographical level can be considered as a simple way of aggregating multiple mobility measurements into one. However, each mobility indicator may make different impacts on the number of infected cases. The idea of the pandemic-mobility indicator is to calculate a weighted average number for multiple mobility indexes, where the weight can be determined by analyzing the variable importance using a linear model, such as a random forest algorithm. An example of the pandemic-mobility indicator generation for a certain state in the United States on a single day has been illustrated in Tabel 2. To calculate the pandemic-mobility indicator, first of all, all mobility indicators and indexes are needed to be extracted per state. In this case, only Google reports and one of the individual-level mobility indexes (m50) can be used for further processing. Google reports have been spliced into six sub-sets by category. Secondly, the daily record for each state has been collected per mobility index with the number of the state-level COVID-19 confirmed cases on the same day. The table contains seven mobility attributes that are input variables in a fitted random forest model, where the output variable is the pandemic feature indicating the daily confirmed case count. The variable importance can be obtained by fitting the model with an 80% training set, which is the weight for calculating the weighted mobility. Finally, the pandemic-mobility indicator can be acquired by calculating the average value of the weighted mobility, which is a weighted average value representing the mobility diversity of impacts on COVID-19. The same procedure has been performed for each state on the same day, thus a pandemic-mobility measurement can be given per state.

Table 2. The pandemic-mobility indicator generation process for New Jersey on January 15, 2021

Input Mobility Index	Daily Record	Weight	Weighted Mobility
Google reports - parks	-20	0.25	-5
Google reports - transit stations	-49	0.24	-11.76
Individual-level mobility - m50	3.76	0.19	0.71
Google reports - retail and recreation	-32	0.12	-3.84
Google reports - workplaces	-39	0.08	-3.12
Google reports - grocery and pharmacy	-13	0.06	-0.78
Google reports - residential	16	0.05	0.8
The pandemic-mobility indicator	**-3.28**		

On the policy-making perspective, it is more insightful to explore the different attributes among regional mobility patterns and pandemic features along with risk levels. With the limited resources and urgent situation, governments need to allocate the enforcement more accurately and efficiently. Regions with higher medical risk levels and more active pandemic-mobility dynamics should be taken into the top priority. The scientific evidence must be provided before the policy-making process, which is possible under the framework of the pandemic-mobility indicator with the COVID-MOBILITY-POOL dataset. The IER value, i.e. a fixed value indicating the medical risk level within an area, has been given, which can be applied to segment states by clustering their COVID-19 medical risk levels and pandemic-mobility dynamics. Figure 5 presents an example of the analytical evidence providing process concentrated on determining which states have the highest degree of medical risk and the most dynamic mobility associated with COVID-19. Blue squares are grouped as states that have lower medical risks with lower degrees of the pandemic-mobility; green dots are labeled as the high pandemic-mobility degree but low medical risk states; orange crosses are states that have higher medical risks with higher pandemic-mobility measurements. Governments need to pay more attention on states with higher IER values and pandemic-mobility indicators.

Figure 5. Medical risk levels and pandemic-mobility indicators segmentations for states

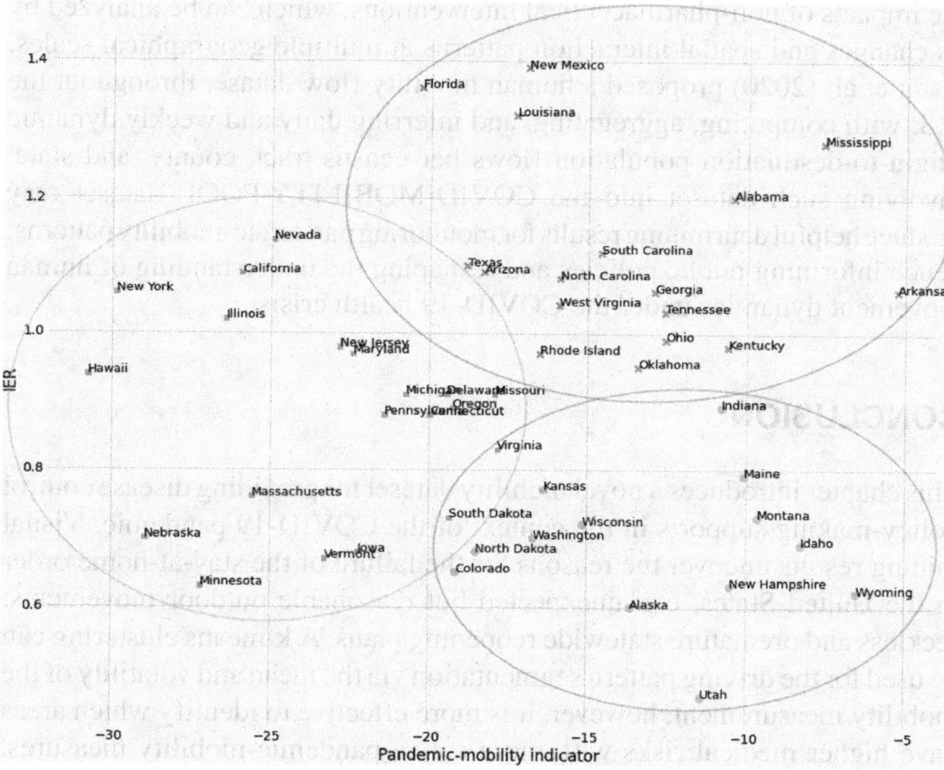

FUTURE RESEARCH DIRECTIONS

Supplementing the COVID-MOBILITY-POOL dataset will enlighten the research extension by adding related mobility indicators based on data collected from multiple data transaction media, such as Bluetooth, WiFi, and QR code. Several COVID-19 tracking Apps have been widely used in different counties or regions. For instance, the Stopp Corona used in Austria has been developed on a public key and private key rolling basis using signal-distance relationships detected from the Bluetooth; the StayHomeSafe is a Bluetooth-GPS-WiFi based decentralized contact-tracing App that has been applied to self-quarantined individuals staying at home in Hong Kong; the health code on WeChat and Alipay has been applied nationwide with over 60% of the population coverage in China (Li & Guo, 2020). Generating additional mobility measurements relied on such data may complete the dataset by covering human movements via multiple dimensions as much as possible. Dynamic human mobility flow, on the other hand, is crucial for assessing the impacts of non-pharmaceutical interventions, which can be analyzed by its changes and spatial interaction patterns at multiple geographical scales. Kang et al. (2020) proposed a human mobility flow dataset throughout the U.S. with computing, aggregating, and inferring daily and weekly dynamic origin-to-destination population flows per census tract, county, and state. Involving such dataset into the COVID-MOBILITY-POOL dataset may produce helpful data mining results for monitoring pandemic-mobility patterns, hence informing public policies and reshaping the understanding of human movement dynamics under the COVID-19 health crisis.

CONCLUSION

This chapter introduces a novel mobility dataset for providing disease control policy-making supports in the context of the COVID-19 pandemic. Visual mining results uncover the reasons for the failure of the stay-at-home order in the United States, e.g. unexpected but reasonable outdoor movements; reckless and premature statewide reopening plans. A k-means clustering can be used for the driving pattern segmentation via the mean and volatility of the mobility measurement, however, it is more effective to identify which areas have higher medical risks with more active pandemic-mobility measures. The pandemic-mobility management system architecture has been illustrated with an example of its usage, which can be applied to monitor medical risks and pandemic-mobility indicators per region.

Several potential problems remain in the process of generating the pandemic-mobility indicator. Firstly, the pandemic-mobility indicator is calculated based on the mobility and pandemic measurements and records for a single day, which cannot reflect the dynamic features over time during a certain period. To involve the temporal components into the index generation process, calculating a 7-day or a 14-day moving average can be applied instead of the daily measurement for each mobility index and the pandemic record.

Secondly, the weight is calculated by fitting a linear model, however, it may follow a non-linear relationship between the mobility index and the number of COVID-19 confirmed cases. Weights may vary by fitting the different models, hence exploring the in-depth relationship between mobility patterns and the pandemic features makes the estimation of the weight more accurate. Thirdly, despite the fitted model based on a partially random selected training set that represents the general attributes as a whole, a weighted average number of the mobility assumes that all regions follow the same attributes. For instance, in the above-mentioned case, the mobility indicator for parks has been added more weight according to the variable importance for all states. However, it does not imply that every state has the same mobility patterns that individuals' movements to parks will make the most significant impact to COVID-19. Therefore, weights for each state may be required for calculating individually by fitting multiple models with the temporal components. Finally, different types of mobility indexes may reflect an overlapped information retrieval that indicate the same mobility patterns at the same time per region. For example, in the above-mentioned case, mobility patterns derived from Google reports may contain the same attributes comparing information collected from the individual-level mobility index (m50) because of the similar data collection process from the same data providers. For instance, both Google researchers and m50 modelers collect GPS data from Android users, thus mobility patterns derived from Google reports and m50 may be overlapped as they aim at the same individual who provides exactly the same geolocation but has been double-counted. Therefore, a similarity analysis across the different mobility indexes may solve the problem by analyzing the root of data collection before generating the pandemic-mobility index.

REFERENCES

Aleta, A., Martin-Corral, D., Piontti, A. P., Ajelli, M., Litvinova, M., Chinazzi, M., ... Moreno, Y. (2020). Modelling the impact of testing, contact tracing and household quarantine on second waves of COVID-19. *Nature Human Behaviour*, *4*(9), 964–971. doi:10.103841562-020-0931-9 PMID:32759985

Apple. (2021). *Mobility trends reports*. https://www.apple.com/covid19/mobility

Bachir, D., Khodabandelou, G., Gauthier, V., El Yacoubi, M., & Puchinger, J. (2019). Inferring dynamic origin-destination flows by transport mode using mobile phone data. *Transportation Research Part C, Emerging Technologies*, *101*, 254–275. doi:10.1016/j.trc.2019.02.013

Badr, H. S., Du, H., Marshall, M., Dong, E., Squire, M. M., & Gardner, L. M. (2020). Association between mobility patterns and COVID-19 transmission in the USA: A mathematical modelling study. *The Lancet. Infectious Diseases*, *20*(11), 1247–1254. doi:10.1016/S1473-3099(20)30553-3 PMID:32621869

Bonardi, J. P., Gallea, Q., Kalanoski, D., & Lalive, R. (2020). Fast and local: How did lockdown policies affect the spread and severity of the covid-19. *Covid Economics*, *23*, 325–351.

Burstein, R., Hu, H., Thakkar, N., Schroeder, A., Famulare, M., & Klein, D. (2020). *Understanding the impact of COVID-19 policy change in the greater Seattle area using mobility data*. Institute for Disease Modeling.

Charu, V., Zeger, S., Gog, J., Bjørnstad, O. N., Kissler, S., Simonsen, L., Grenfell, B. T., & Viboud, C. (2017). Human mobility and the spatial transmission of influenza in the United States. *PLoS Computational Biology*, *13*(2), e1005382. doi:10.1371/journal.pcbi.1005382 PMID:28187123

China Data Lab. (2020). *US Metropolitan Daily Cases with Basemap*. doi:10.7910/DVN/5B8YM8

Chinazzi, M., Davis, J. T., Ajelli, M., Gioannini, C., Litvinova, M., Merler, S., Pastore y Piontti, A., Mu, K., Rossi, L., Sun, K., Viboud, C., Xiong, X., Yu, H., Halloran, M. E., Longini, I. M. Jr, & Vespignani, A. (2020). The effect of travel restrictions on the spread of the 2019 novel coronavirus (COVID-19) outbreak. *Science*, *368*(6489), 395–400. doi:10.1126cience.aba9757 PMID:32144116

Cot, C., Cacciapaglia, G., & Sannino, F. (2021). Mining Google and Apple mobility data: Temporal anatomy for COVID-19 social distancing. *Scientific Reports*, *11*(1), 1–8. doi:10.103841598-021-83441-4 PMID:33602967

Courtemanche, C., Garuccio, J., Le, A., Pinkston, J., & Yelowitz, A. (2020). Strong Social Distancing Measures In The United States Reduced The COVID-19 Growth Rate: Study evaluates the impact of social distancing measures on the growth rate of confirmed COVID-19 cases across the United States. *Health Affairs*, *39*(7), 1237–1246. doi:10.1377/hlthaff.2020.00608 PMID:32407171

Cuebiq. (2021). *COVID-19 mobility insights*. https://www.cuebiq.com/visitation-insights-covid19/

FEMA. (2020, April 11). *Disaster Information*. https://www.fema.gov/disasters

Furini, M., & Montangero, M. (2018). Sentiment analysis and twitter: A game proposal. *Personal and Ubiquitous Computing*, *22*(4), 771–785. doi:10.100700779-018-1142-5

Gao, S., Rao, J., Kang, Y., Liang, Y., & Kruse, J. (2020). Mapping county-level mobility pattern changes in the United States in response to COVID-19. *SIGSpatial Special*, *12*(1), 16–26. doi:10.1145/3404820.3404824

Ghader, S., Zhao, J., Lee, M., Zhou, W., Zhao, G., & Zhang, L. (2020). *Observed mobility behavior data reveal social distancing inertia*. arXiv preprint arXiv:2004.14748.

Gonzalez, M. C., Hidalgo, C. A., & Barabasi, A. L. (2008). Understanding individual human mobility patterns. *Nature, 453*(7196), 779-782.

Google. (2021). *COVID-19 community mobility reports*. https://www.google.com/covid19/mobility/

Jin, J., Agarwala, N., Kundu, P., Harvey, B., Zhang, Y., Wallace, E., & Chatterjee, N. (2020). Individual and community-level risk for COVID-19 mortality in the United States. *Nature Medicine*, 1–6. PMID:33311702

Kang, Y., Gao, S., Liang, Y., Li, M., Rao, J., & Kruse, J. (2020). Multiscale dynamic human mobility flow dataset in the US during the COVID-19 epidemic. *Scientific Data*, *7*(1), 1–13. doi:10.103841597-020-00734-5 PMID:31896794

Kraemer, M. U., Yang, C. H., Gutierrez, B., Wu, C. H., Klein, B., Pigott, D. M., du Plessis, L., Faria, N. R., Li, R., Hanage, W. P., Brownstein, J. S., Layan, M., Vespignani, A., Tian, H., Dye, C., Pybus, O. G., & Scarpino, S. V. (2020). The effect of human mobility and control measures on the COVID-19 epidemic in China. *Science*, *368*(6490), 493–497. doi:10.1126cience.abb4218 PMID:32213647

Kwon, J., Grady, C., Feliciano, J. T., & Fodeh, S. J. (2020). Defining facets of social distancing during the COVID-19 pandemic: Twitter analysis. *Journal of Biomedical Informatics*, *111*, 103601. doi:10.1016/j.jbi.2020.103601 PMID:33065264

LiJ.GuoX. (2020). Global deployment mappings and challenges of contact-tracing apps for COVID-19. *Available at* SSRN 3609516. doi:10.2139/ssrn.3609516

Mervosh, S., Lu, D., & Swales, V. (2020). See which states and cities have told residents to stay at home. *The New York Times, 3.*

Paul, M. J., & Dredze, M. (2017). Social monitoring for public health. *Synthesis Lectures on Information Concepts, Retrieval, and Services*, *9*(5), 1–183. doi:10.2200/S00791ED1V01Y201707ICR060

Perra, N. (2021). Non-pharmaceutical interventions during the COVID-19 pandemic: A review. *Physics Reports*, *913*, 1–52. doi:10.1016/j.physrep.2021.02.001 PMID:33612922

Porcher, S., & Renault, T. (2021). Social distancing beliefs and human mobility: Evidence from Twitter. *PLoS One*, *16*(3), e0246949. doi:10.1371/journal.pone.0246949 PMID:33657145

Prandi, C., Nunes, N., Ribeiro, M., & Nisi, V. (2017, December). *Enhancing sustainable mobility awareness by exploiting multi-sourced data: The case study of the madeira islands. In 2017 Sustainable Internet and ICT for Sustainability (SustainIT).* IEEE.

Purnomo, E. P., Loilatu, M. J., Nurmandi, A., Qodir, Z., Sihidi, I. T., & Lutfi, M. (2021). How Public Transportation Use Social Media Platform during Covid-19: Study on Jakarta Public Transportations' Twitter Accounts? *Webology, 18*(1), 1-19.

Saad, L. (2020). Americans step up their social distancing even further. Academic Press.

Song, C., Koren, T., Wang, P., & Barabási, A. L. (2010). Modelling the scaling properties of human mobility. *Nature Physics*, *6*(10), 818–823. doi:10.1038/nphys1760

Stange, H., Liebig, T., Hecker, D., Andrienko, G., & Andrienko, N. (2011, November). Analytical workflow of monitoring human mobility in big event settings using bluetooth. In *Proceedings of the 3rd ACM SIGSPATIAL international workshop on indoor spatial awareness* (pp. 51-58). 10.1145/2077357.2077368

Taylor, S., & Asmundson, G. J. (2021). Negative attitudes about facemasks during the COVID-19 pandemic: The dual importance of perceived ineffectiveness and psychological reactance. *PLoS One*, *16*(2), e0246317. doi:10.1371/journal.pone.0246317 PMID:33596207

The New York Time. (2021). *Coronavirus (Covid-19) Data in the United States*. https://github.com/nytimes/covid-19-data

The Washington Post. (2021). *Where states reopened and cases spiked after the U.S. shutdown*. https://www.washingtonpost.com/graphics/2020/national/states-reopening-coronavirus-map/

TomTom. (2021). *TomTom Traffic Index*. https://www.tomtom.com/en_gb/traffic-index/

Wang, A., Zhang, A., Chan, E. H., Shi, W., Zhou, X., & Liu, Z. (2021). A Review of Human Mobility Research Based on Big Data and Its Implication for Smart City Development. *ISPRS International Journal of Geo-Information*, *10*(1), 13. doi:10.3390/ijgi10010013

Wang, Q., & Taylor, J. E. (2016). Process map for urban-human mobility and civil infrastructure data collection using geosocial networking platforms. *Journal of Computing in Civil Engineering*, *30*(2), 04015004. doi:10.1061/(ASCE)CP.1943-5487.0000469

Warren, M. S., & Skillman, S. W. (2020). *Mobility changes in response to COVID-19*. arXiv preprint arXiv:2003.14228.

Waze. (2021). *Waze COVID-19 Impact Dashboard*. https://www.waze.com/covid19

Xiong, C., Hu, S., Yang, M., Luo, W., & Zhang, L. (2020). Mobile device data reveal the dynamics in a positive relationship between human mobility and COVID-19 infections. *Proceedings of the National Academy of Sciences of the United States of America*, *117*(44), 27087–27089. doi:10.1073/pnas.2010836117 PMID:33060300

Xiong, C., Hu, S., Yang, M., Younes, H., Luo, W., Ghader, S., & Zhang, L. (2020). Mobile device location data reveal human mobility response to state-level stay-at-home orders during the COVID-19 pandemic in the USA. *Journal of the Royal Society, Interface*, *17*(173), 20200344. doi:10.1098/rsif.2020.0344 PMID:33323055

Xu, F., Lin, Y., Huang, J., Wu, D., Shi, H., Song, J., & Li, Y. (2016). Big data driven mobile traffic understanding and forecasting: A time series approach. *IEEE Transactions on Services Computing*, *9*(5), 796–805. doi:10.1109/TSC.2016.2599878

Xu, P., Dredze, M., & Broniatowski, D. A. (2020). The Twitter Social Mobility Index: Measuring Social Distancing Practices With Geolocated Tweets. *Journal of Medical Internet Research*, *22*(12), e21499. doi:10.2196/21499 PMID:33048823

Yuan, J., Zheng, Y., & Xie, X. (2012, August). Discovering regions of different functions in a city using human mobility and POIs. In *Proceedings of the 18th ACM SIGKDD international conference on Knowledge discovery and data mining* (pp. 186-194). 10.1145/2339530.2339561

Zhao, K., Tarkoma, S., Liu, S., & Vo, H. (2016, December). Urban human mobility data mining: An overview. In *2016 IEEE International Conference on Big Data (Big Data)* (pp. 1911-1920). IEEE. 10.1109/BigData.2016.7840811

Zhou, Y., Lau, B. P. L., Yuen, C., Tunçer, B., & Wilhelm, E. (2018). Understanding urban human mobility through crowdsensed data. *IEEE Communications Magazine*, *56*(11), 52–59. doi:10.1109/MCOM.2018.1700569

ADDITIONAL READING

Bonczek, R. H., Holsapple, C. W., & Whinston, A. B. (2014). *Foundations of decision support systems*. Academic Press.

Hu, S., Xiong, C., Yang, M., Younes, H., Luo, W., & Zhang, L. (2021). A big-data driven approach to analyzing and modeling human mobility trend under non-pharmaceutical interventions during COVID-19 pandemic. *Transportation Research Part C, Emerging Technologies*, *124*, 102955. doi:10.1016/j.trc.2020.102955 PMID:33456212

Huang, Q., & Wong, D. W. (2015). Modeling and visualizing regular human mobility patterns with uncertainty: An example using Twitter data. *Annals of the Association of American Geographers*, *105*(6), 1179–1197. doi:10.1080/00045608.2015.1081120

Kang, Y., Gao, S., Liang, Y., Li, M., Rao, J., & Kruse, J. (2020). Multiscale dynamic human mobility flow dataset in the US during the COVID-19 epidemic. *Scientific Data*, *7*(1), 1–13. doi:10.103841597-020-00734-5 PMID:31896794

Kaufman, L., & Rousseeuw, P. J. (2009). *Finding groups in data: an introduction to cluster analysis* (Vol. 344). John Wiley & Sons.

Longley, P., & Batty, M. (2003). *Advanced spatial analysis: the CASA book of GIS*. ESRI, Inc.

Loo, B. P., Tsoi, K. H., Wong, P. P., & Lai, P. C. (2021). Identification of superspreading environment under COVID-19 through human mobility data. *Scientific Reports*, *11*(1), 1–9. doi:10.103841598-021-84089-w PMID:33633273

Perra, N. (2021). Non-pharmaceutical interventions during the COVID-19 pandemic: A review. *Physics Reports*, *913*, 1–52. doi:10.1016/j.physrep.2021.02.001 PMID:33612922

Wang, A., Zhang, A., Chan, E. H., Shi, W., Zhou, X., & Liu, Z. (2021). A review of human mobility research based on big data and its implication for smart city development. *ISPRS International Journal of Geo-Information*, *10*(1), 13. doi:10.3390/ijgi10010013

Williams, N. E., Thomas, T. A., Dunbar, M., Eagle, N., & Dobra, A. (2015). Measures of human mobility using mobile phone records enhanced with GIS data. *PLoS One*, *10*(7), e0133630. doi:10.1371/journal.pone.0133630 PMID:26192322

KEY TERMS AND DEFINITIONS

Data Aggregation: A process of collecting data from different data sources and organizing it in a unified format for data processing.

Decision Support System: A computer-based framework that can process and analyze the large scale of data for extracting useful knowledges and information, which can be applied to solve problems in decision-making.

GIS: A system that can convert all types of data into a map to present all descriptive information and understand geographic context.

Human Mobility: A measurement that describes how people move within a network or system through tracking and analyzing human behavior patterns demographically and geographically over time.

K-Means Clustering: A clustering method that aims to separate observations into k clusters through minimizing squared Euclidean distances in each cluster.

Non-Pharmaceutical Disease Control: A set of actions, apart from medicine and vaccination, that communities can slow down the spread of a disease, a.k.a. non-pharmaceutical interventions (NPIs).

Stay-at-Home Order: An order forced by a government authority to isolate people at home for mitigating the pandemic spreading.

Visual Mining: An interactive graphical method that can manage performance and allow the users to analyze and explore the insight of data.

Chapter 5
Monitoring Social Distancing With Real-Time Detection and Tracking

ABSTRACT

Social distancing is one of the suggested solutions by the health authorities to reduce the spreading speed of the COVID-19 in public areas. A six-foot physical distancing has been set by the majority of public governors as a mandatory social regulation. However, it is difficult to monitor whether individuals practice the social distancing regulation or not. State-of-the-art technologies, such as computer visions, artificial intelligence, and big data analytics, can help for automated people detection and tracking in the crowd for indoor and outdoor environments using surveillance cameras. In this chapter, several types of popular object detection and tracking schemes in monitoring social distancing are illustrated with implementations of a cutting-edge human detection model by testing its reliability using a sample video. A real-world case study for social control management system is also introduced with its architecture designs and implementations in the context of the COVID-19 pandemic.

INTRODUCTION

The COVID-19 pandemic has completely changed the world with dire consequences to the global healthcare system. It is urgent to control the spread

DOI: 10.4018/978-1-7998-8793-5.ch005

of the disease, and reduce its negative impacts. Developing medications and vaccines is the most effective way to limit the speed of infection, confirmed by epidemiological studies in Chapter 2. Despite the efficiency of pharmaceutical solutions, no appropriate cure or available treatment has been developed until the last quarter of 2020. While medical research scientists keep working on producing effective medications for the new coronavirus, before October, no antiviral drug has been approved for use in the treatment of COVID-19 requiring hospitalization. Veklury is the first anti-COVID treatment approved by the United States Food and Drug Administration (FDA), however, FDA also declares that it does not include the entire population (FDA, 2020). Despite an ongoing research on the vaccine development, no certain vaccine has been reported as of November when Pfizer and BioNTech announced their vaccine against COVID-19 has been achieved a success in the Phase 3 clinical trial (Pfizer, 2020). Given such harsh conditions from March to November, non-pharmaceutical measures, such as stay-at-home orders and social distancing regulations, have been executed as an alternative way to reduce the spread of COVID-19. However, the stay-at-home policy, as concluded in Chapter 4, has been determined as a failure because of an unexpected movement pattern in the United States, thereby, it is more crucial to monitor the social distancing in the public.

Practicing social distancing is one of the most effective precautions to prevent the spread of an infectious disease. It has been referred by the majority of the global communities since the COVID-19 pandemic. The purpose of practicing social distancing is to minimize the proximity of individuals physical contacts in crowded spaces, thus reducing the accumulative infection risk. Initially, the coronavirus is believed to be transmittable through the air only if people sneeze or cough, however, the World Health Organization (WHO) denies this inference in July 2020, and further announces that the new coronavirus can be spread by tiny particles suspended in the air after individuals talk or even breathe in crowded, closed, or poorly ventilated environments (WHO, 2020). Medical research has confirmed that people with mild or even no symptoms can also carry the novel virus, which implies that it is essential for the entire population to practice social distancing (Kim et al., 2020). Suggested by the epidemiologists, individuals must maintain at least 6 feet as a minimum distance between each other during a pandemic (Olsen et al., 2003). As one of the non-pharmacological preventions, practicing social distancing has been proved to be effective and necessary for controlling the transmission of contagious virus, including SARS, H1N1, and COVID-19 (Ferguson et al., 2006; Thu et al., 2020). The objective of practicing social distancing is to reduce transmission rate, thereby flattening the Gaussian curve and spreading cases over a longer time to relieve pressure on the health care capacity (Fong et al., 2020). A sharp spike of cumulative cases will happen without social distancing, and consequently, will lead to a disease control failure with an exponential growth of the death number.

However, monitoring and alerting the society to maintain an adequate distance from each other is a difficult task, especially when people walk on the street or gather in crowd. The human-based tracking and monitoring way requires huge amounts of social resources with less efficiency. In the digital age, technology-based approaches, such as human detections using wireless signals and tracking technologies of big data analytics, have been proposed to step in to assist the healthcare systems and social managements in coping with COVID-19 challenges (Nguyen et al., 2020). The India government has applied GPS and Bluetooth data to monitor the movements of infected patients or suspected individuals by assisting healthy people to maintain a safe distance from the risky groups (Jhunjhunwala, 2020). A state-of-the-art solution, called MySD, has been attempted for public to observe social distance advice by leveraging smart phone hardware facilities to determine safe distance (Rusli et al., 2020). A cross-national study on the efficacy of social distancing during the COVID-19 pandemic has been presented by integrating multiple transactional datasets with big data analytics (Delen et al., 2020). These studies vary from the individual localization and tracking to the human movement analysis, however, all of them are relied on data derived from the signal front-end, i.e. a smartphone. Despite the feasibility and operability of the above-mentioned technologies, the assumption that all individuals carry a smartphone when they go outside may not be practical in the reality. Therefore, alternative ways of performing social distancing surveillance and measurements are essential and important in the context of implementing disease control policies during the COVID-19 pandemic.

Motivated by the challenges from the current situation in regard to the implications of non-pharmacological disease control measures, this chapter is proposed to investigate the social distancing monitor system by introducing different types of popular object detection and tracking schemes in monitoring the social distancing as well as the implementations of the system by testing its reliability using a well-designed human detector. A real-world case presentation for social control management system is also introduced with its architecture designs and implementations in the form of future research directions. The objectives of the chapter are listed as follows:

- reviewing most recent studies in object detections and social distancing monitoring systems, along with discussing limitations and challenges among the existed researches in this field.
- illustrating most popular object detection and tracking techniques, such as anchor boxes, faster RCNN, Single Shot Detector, You Only Look Once, and Deepsort.
- investigating the full development stage of implementing a social distancing monitor using a pre-trained algorithm with a new surveillance video for the testing process.
- introducing and discussing an innovated social control system for non-pharmacological disease control measures during the COVID-19 pandemic.

BACKGROUND

With the development of pattern recognition and deep learning during the last decade, AI-based computer vision objectives have been widely used in interpreting and exploring the visual data from images and videos. Cutting-edge technologies, such as machine learning, deep learning, image recognition, and motion detection, have been applied in smart transportation architectures, public surveillance systems, and intelligent safety monitors in the form of constructing a smart city (Zhao et al., 2019; Idrees et al., 2018; Hipps et al., 2017). The utilization of computer vision with artificial intelligence can enable researchers to extract pedestrian features by analyzing spatial-temporal data of the image sequences. Such technologies may further turn CCTV cameras in a real-time infrastructure capacity into a computer-based intelligent monitor system that can determine whether individuals follow the social distancing policy or not. Advanced human detection algorithms are required at the top priority for building the intelligent social distancing monitor, which is one of the most crucial parts in object detections and smart objectives. Despite the adequate research in people detection and tracking, existing studies are suffering from precision problems because of the challenges under various conditions, e.g. lighting conditions, on the street or in other public places (Nguyen et al., 2016). Although a manual tuning methodology can be employed to classify human activities, the functionality has its limitation that always brings underlying issues (Gawande et al., 2020).

As the computer hardware capability strengthens rapidly, deep learning algorithms allow research scientists to create more accurate models and faster monitors compared with regular machine learning models. Deep learning models, such as convolutional neural networks (CNNs) and deep neural networks (DNNs) have been applied in feature extraction and complex object classification (Gidaris & Komodakis, 2015; Szegedy et al., 2013), which can contribute to social distancing monitoring as well. Punn et al. (2020) developed a DNNs-based monitor for human detection to measure the number of individuals who violated the social distancing. Similar studies have been proposed given various research objectives, i.e. detecting the people distancing in a certain manufactory (Khandelwal et al., 2020), monitoring social distancing constrains in crowded scenarios (Sathyamoorthy et al., 2020), and assessing infection risk measurements on the street (Rezaei & Azarmi, 2020). Such studies focus on applying common types of object detection models such as fast RCNN, Single Shot MultiBox Detector (SSD), and You Only Look Once (YOLO) in terms of establishing social distancing monitor systems in the context of COVID-19. However, limitations vary from research to research, e.g. insufficient statistical analysis of the output results is provided; system accuracy measurements are estimated based on different datasets without comparable ground truths; no further individual-level information is available for the tracking and management system.

MAIN FOCUS OF THE CHAPTER

Issues, Controversies, Problems

The object detection models, such as SSD, YOLO family, and RCNN family, have been applied as the major component in measuring the social distancing. Moreover, several pre-trained object detection algorithms, such as MobileNet, ResNet, Xception, VGG, and Inception, have also been investigated and used to perform the feature extraction and object detection tasks. However, such studies are relied on a common-used video, e.g. the sample CCTV output of Oxford Town Center, for the testing process. Despite well-performed sample outputs of the proposed models for monitoring social distancing on surveillance footage, it is still difficult to draw any conclusion that their proposed algorithms can be applied in other videos for the same purpose. The controversy behind research in this filed is the lack of extensive model testing. Besides, most existing studies focus on fine-tuned models with a comparative analysis concentrated with model performance measures, however, the model applicability and feasibility has not yet been examined. Significant challenges that have not been investigated in most existing studies include unknown functionalities in complicate environments and unstable lighting conditions, less computational efficiency, complicated and non-transparent model structures, and lack of real-world representations.

Features of the Chapter

To deal with the challenges from the current studies, this chapter is designed for explaining how artificial intelligence assists to monitor social distancing violations using a pre-trained detector, named as DeepSOCIAL. Such an algorithm is a DNN-based human detector that has been proposed to detect and track static and dynamic individuals in public places for monitoring social distancing metrics under multiple visual conditions. The system is able to operate in various challenging conditions with the outperforming model efficiency and complexity. DeepSOCIAL can be formalized upon a three-stage module that contains human detection, people tracking and identification, and inter-distance estimation. Such a framework has been integrated and used on CCTV surveillance cameras with the real-time performance. In this chapter, such modules are illustrated with a sample test procedure in terms of implementing a social distancing monitor under the framework of the DeepSOCIAL model. Instead of the common-used video, a new dataset will be applied in the testing process, along with examples throughout the whole stages of the development of the monitor. An innovated social control management system, called DeepCovidPark, will be introduced and discussed in the form of future research direction, which is a real-world representation of successfully applied social distancing monitor in the context of the COVID-19 pandemic.

SOLUTIONS AND RECOMMENDATIONS

Object Detection and Tracking Algorithms

Anchor Boxes

To detect multiple objects, object detection algorithms usually sample many regions in the input image to further determine the targeted objects in the certain region. Therefore, anchor boxes are introduced, which can build multiple bounding boxes with various sizes and aspect ratios. As presented in the hyperparameter summary for anchor boxes generation (Punn et al., 2020), different detection model may generate different anchor boxes, based on different region sampling method. The size of anchor box can be constructed with the width $kd\sqrt{r}$ and the height $hd\sqrt{r}$, respectively, where w and h represent width and height parameters. Assume size parameter is $d\epsilon(0,1]$ and the aspect ratio is r>0. In the training process, the anchor box is set to two types of labels: the class that the anchor box belongs to and the offset of the ground-truth bounding box based on the classification and regression loss. The anchor box is labeled as negative (0) and positive (1), according to the similarity between the anchor box (i) and the ground-truth bounding box (g). Class label vector assigned to the positive anchor box is $l_0\epsilon\left(k_1,k_2,\ldots,k_n\right)$, where k_n symbolizes the category of the nth object. For negative anchor box, l_0 is equal to 0. The encoding vector for ground-truth bounding box can be described as $g\left(f_i|i\right)$. Assume the image is I, then the loss of each anchor box prediction can be calculated as follow,

$$Loss\left(i \mid I;w\right) = \alpha 1_i^{obj} L_{reg}\left(g\left(f_i \mid i\right) - Y_{reg}\left(I \mid i;w\right)\right) + \beta L_{cls}(l_i, Y_{cls}(I \mid i;w))$$

where α and β are the trained parameters corresponding to the regression and classification loss, 1_i^{obj} represents i^{th} anchor box as a positive anchor. $Y_{reg}\left(I \mid i;w\right)$ and $Y_{cls}(I \mid i;w)$ symbolize the predicted object class and the bounding box offset, respectively.

Faster RCNN

Different from Fast RCNN, Faster RCNN replaces selective search (SS) with region proposal network (RPN), which can speed up the computation time of the algorithm by sharing convolutional features with the down-stream detection network. RPN is a fully convolutional network that can efficiently predict object bounds and scores at each position. The classification and regression loss functions for the faster RCNN can be defined as follows,

$$L_{cls}\left(p_k, p_k^*\right) = -p_k^* \log\left(p_k\right) - \left(1 - p_k^*\right)\log\left(1 - p_k\right)$$

$$L_{reg}\left(t^{u},v\right)=\sum_{k\epsilon\left(x,y,w,h\right)}L_{1}^{smooth}\left(t_{k}^{u}-v_{k}\right)$$

$$L_{1}^{smooth}\left(m\right)=\begin{cases}0.5m^{2}, if\left|m\right|<1\\ \left|m\right|-0.5, otherwise\end{cases}$$

where p_k is the predicted probability for anchor k being an object and p_k^* is the ground-truth label of anchor k. $t^u = \left(t_x^u, t_y^u, t_w^u, t_h^u\right)$ is the predicted bounding-box regression offsets and indicates the top-left coordinates, height and width of the uth object. $v = \left(v_x, v_y, v_w, v_h\right)$ is the true bounding-box regression targets . L_1 loss, which is involved in the calculation of regression loss, can avoid exploding gradients by tuning of the learning rates (Girshick, 2015).

Single Shot Detector (SSD)

Compared with faster RCNN, SSD is more efficient because it can find all objects within an image in one shot by using a fully convolutional network. Its convolutional network can create bounding boxes in fixed sizes and estimate a score for each of them based on whether objects appear in these boxes. In order to realize the final detections, SSD usually extract feature maps firstly using a base pre-trained network, and then perform convolution filters to recognize objects by a multi-scale feature layers, in which multi-class classification and bounding box regression are involved. Besides, in this process non-max suppression (NMS) unit can avoid overlapping boxes and guarantee each object only has one box by calculating intersection over union (IoU) parameter. The overall loss function of the SSD model also consists of localization loss and confidence loss as defined in the equations:

$$L_{loc}\left(x,b,g\right)=\sum_{i\epsilon\, pos}^{n}\sum_{t\epsilon\, c_x,c_y,w,h}x_{ij}^{k}smooth_{L_1}\left(b_{i}^{t}-\hat{g}_{j}^{t}\right)$$

$$\hat{g}_{j}^{c_x}=\frac{\left(g_{j}^{c_x}-e_{i}^{c_x}\right)}{e_{i}^{w}}$$

$$\hat{g}_j^{c_y} = \frac{\left(g_j^{c_y} - e_i^{c_y}\right)}{e_i^h}$$

$$\hat{g}_j^w = \log\left(\frac{g_j^w}{e_i^w}\right)$$

$$\hat{g}_j^h = \log\left(\frac{g_j^h}{e_i^h}\right)$$

$$x_{ij}^p = \begin{cases} 1, if\ IoU > 0.5 \\ 0, otherwise \end{cases}$$

Where *b* represents the predicted box, and *g* is the ground truth box. A Smooth L_1 loss is utilized in the localization loss between predicted box and the ground truth box parameters. (c_x, c_y) are the offsets for the anchor box *e*. If i^{th} anchor box matches with the jth ground truth box, IoU will be greater than 0.5 and then x_{ij}^p will be equal to 1, and vice versa.

The confidence loss is a softmax loss over multiple class confidences to make a class prediction.

$$L_{conf}\left(x, q\right) = -\sum_{i\in Pos}^{N} x_{ij}^p \log\left(\hat{q}_i^p\right) - \sum_{i\in Neg} \log\left(\hat{q}_i^0\right)$$

$$\hat{q}_i^p = \frac{exp\left(q_i^p\right)}{\sum_p exp\left(q_i^p\right)}$$

where *N* is the number of matched anchor boxes and *q* is the class confidence score. The overall loss for SSD model can be calculated by a weighted summation shown as below:

$$L\left(x, q, b, g\right) = \frac{1}{N}\left(L_{conf}\left(x, q\right) + \alpha L_{loc}\left(x, b, g\right)\right)$$

You Only Look Once (YOLO)

YOLO is a popular real-time object recognition algorithm because of its advantage in terms of the model efficiency. YOLO can take the entire image in a single instance and predict the coordinates of the bounding box and the corresponding class probabilities for the boxes. YOLO has different versions, including YOLO v1-4. YOLO v1 usually divides the input image into S´S grid. The grid cell can detect the assigned object. Different from the Inception modules in GoogleNet, YOLO v1 uses 1´1 reduction layer connected with 3´3 convolutional layers. YOLO v3 has 24 convolutional layers connected with 2 fully connected layers. Based on YOLO v1, YOLO v2 attempted to improve recall and reduce the localization errors, by batch normalization, high resolution classifier, and multi-object prediction per grid cell. Darknet-19, a classification model containing 19 convolutional layers and 5 max-pooling layers, is used as a backbone for YOLO v2 to guarantee the classification accuracy and release the complexity of the problem. As an upgraded version, YOLO v3 directly utilizes the confidence score of a box, predicted by logistic regression, as its probability of objectness. The loss function of YOLO v3 is described through the formula below:

$$
{}_{coord}\sum_{i=0}^{G^2}\sum_{j=0}^{B}1_{i,j}^{obj}\left(\left(t_x-\hat{t}_x\right)^2+\left(t_y-\hat{t}_y\right)^2+\left(t_w-\hat{t}_w\right)^2+\left(t_h-\hat{t}_h\right)^2\right)
$$

$$
+\sum_{i=0}^{G^2}\sum_{j=0}^{B}1_{i,j}^{obj}\left(-\log\left(\left(t_o\right)\right)+\sum_{k=1}^{C}BCE\left(\hat{y}_k,\left(s_k\right)\right)\right)+{}_{noobj}\sum_{i=0}^{G^2}\sum_{j=0}^{B}1_{i,j}^{noobj}\left(-\log\left(1-\left(t_o\right)\right)\right)
$$

where ${}_{coord}$ represents the weight of the error of coordinate and ${}_{noobj}$ represents the weight of the case for cell without object. $\left(t_x,t_y,t_w,t_h,t_{o,}G_1,\ldots,G_N\right)$ is the prediction vector and $\left(\hat{t}_x,\hat{t}_y,\hat{t}_w,\hat{t}_h,\hat{y}_1,\ldots,\hat{y}_N\right)$ is the vector of ground truth labels, in which *N* symbolizes the number of total classes. $1_{i,j}^{obj}$ denotes if j^{th} bounding box in grid cell *i* controls the object. Moreover, binary cross-entropy is indicated as *BCE*.

Deepsort

Deepsort framework is a pragmatic approach to track custom objects in a video. In multi-object tracking of Deepsort framework, Kalman filter is used to track and to find out missing tracks. Besides, he Hungarian algorithm is also used to associate detections to the tracked objects. Thus, the state can be described with 8 spatial variables as follow,

$$x = \left[u, v,, h, u', v',' , h'\right]^T$$

where (u,v) is the centroid of the bounding box, λ is the aspect ratio, and h is the height of the image. $u', v','$ and h' are the respective velocities of u,v and h. After tracking through the Kalman filter, Mahalanobis distance, as a distance metric, is performed with the Hungarian algorithm to quantify the association between new detections and new predictions and to incorporate the uncertainties,

$$d^{(1)}\left(i,j\right) = \left(d_j - y_i\right)^T S_i^{-1}\left(d_j - y_i\right)$$

where d_j is j^{th} bounding box and (y_i, S_i) is the projection of the ith track distribution into measurement space. The Mahalanobis distance can measure the standard deviations of the detection from the mean track location, in order to estimate uncertainty. When the threshold of Mahalanobis distance is set at a 95% confidence interval by the inverse $\chi 2$ distribution, unlikely associations can be avoided. This decision can be made through the indicator shown as below:

$$y_{i,j}^{(1)} = 1\left[d^{(1)}\left(i,j\right) \le t^{(1)}\right]$$

When the association between the ith track and j^{th} detection is admission, the indicator $y_{i,j}^{(1)}$ is equal to 1. However, the Mahalanobis distance may have disadvantages for tracking through occlusions when rapid displacements happen in the image plane. Therefore, a second metric is used to measure the smallest cosine distance between i^{th} track and j^{th} detection, as indicated as follow:

$$d^{(2)}\left(i,j\right) = min\{1 - r_j^T r_k^{(i)} \mid r_k^{(i)} \epsilon R_i$$

Its association indicator can be described as a binary variable as follow:

$$y_{i,j}^{(2)} = 1\left[d^{(2)}\left(i,j\right) \le t^{(2)}\right]$$

Furthermore, a weighted summation is applied to combine these two metrics as follow,

$$x_{i,j} = d^{(1)}\left(i,j\right) + \left(1 - \right)d^{(2)}\left(i,j\right)$$

where $x_{i,j}$ is the combined indicator and is the weight variable that can assign the weight of each metric on the combined association cost. To resolve the deficiency in tracking through occlusions, different viewpoints, non-stationary cameras, appearance descriptor, which is a trained single feature vector, is utilized.

Implementing a Social Distancing Monitor by DeepSOCIAL

Human Detection

The overall structure of the human detection stage starts at inputting image sequences collected from a surveillance camera that generate a sample video file, followed by passing images to the DNNs model. The output will be the classified objects in the scene with their unique localization bounding boxes. Typical challenges such as differences in clothes, variations in actions, multiple distances measurements, and various lighting conditions can be solved by developing a robust human detection model. A DNN-based object detection framework consists of three major components, including inputs such as augmentations and activation functions, detection cores such as backbone architectures and neck modules, and outputs such as the head classification.

A pool of rich datasets will be needed for obtaining a robust human detector, which should contain people in different gender and age groups with a huge number of predefined annotations and labeled objects. Large image datasets such as Google Open Image and Common Objects in Context can be used to satisfy the training demand by providing over three million labeled human beings. Two types of training options, Bag of Freebies (BoF) and Bag of Specials (BoS), are included in YOLOv4, which have been applied to different parts of the network. BoF deals with the increasing generalization of the model's training strategy, while BoS can significantly improve the object detection precision. Mosaic data augmentations have been employed for integrating multiple images into one in order to reduce the batch size. A cross-iteration batch normalization technique has been applied to address the batch size reduction issue that leads to noises from the estimated mean and variance (Yao et al., 2020). Activation functions for BoF such as ReLU, SELU, Swish, and Mish have been investigated and evaluated based on the model performance results for human detection applications, where the Mish function has the minimum loss, less training time, and higher accuracy (Misra, 2019).

Instead of using extra layers to improve the model accuracy, a skip-connection technique has been suggested for reducing the difficulty of the model training process (Chen & Qi, 2018). Several models such as Cross-Stage-Partial and DenseNet have been applied based on the same logic to connect layers, which are used in the design of many backbone architectures, including CSPResNeXt50, CSPDarknet53, and EfficinetNet-B3 (Wang et al., 2020; Huang et al., 2017; Alhichri et al., 2021). The CSPDarknet53 is chosen as the optimal backbone model for the human detection task because of its multiple objects detecting capability, which has been proved by theoretical justifications and experimental results under the YOLOv4 framework (Bochkovskiy et al., 2020; Hu et al., 2021). Extra layers between the backbone and the head have been implemented in most modern deep learning models, which is considered as the neck module that composes the feature collection from multiple stages of the backbone network (Rodríguez et al., 2019). To improve the accuracy of image recognitions, some deep learning models have been designed with fully-connected layers for the image classification task, however, only fixed dimensions of images can be accepted. Neither images with low resolution nor classifications of the small objects may not be able to deal with, and consequently, applications towards image recognitions in surveillance cameras with different sizes of input objects and various resolutions of image frames. Images with different sizes can be addressed directly in YOLOv4 because it has no fully convolutional layers; dealing with small objects can be done by performing a pyramid technique to extract multiple scales of input images from the backbone with the enhancement of the receptive field (Jia et al., 2012). The YOLO feature pyramid network (FPN) module has been proposed for enhancing the receptive fields and achieving better performance on the small objects by extracting features in multi-scale detections from the backbone (Lin et al., 2017). Alternative configurations for the neck module of object detections have been presented and examined in the form of various options, such as the spatial pyramid pooling (SPP) layer and the spatial attention module (SAM), which together make contributions to the effective and robust human detection models concentrated on optimizing the parameters (He & Wang, 2014; Woo et al., 2018).

The final stage of the DNN-based object detection is the head module that is designed for classifying the objects with calculations of the object's size and coordinates of the correspondent bounding boxes. The region proposal is applied before conducting the classification in a two-stage detector that is formed as dense and sparse stages. A set of candidate bounding boxes is extracted by a selective search, and then the detector resizes the object proposals to be a fixed size, followed by feeding them into the deep learning models, which is similar to R-CNN models (Ren et al., 2016). However, such a method is not adaptable for the scenarios with restricted computational resources (Sharifi et al., 2020). Other detectors, such as Single Shot Multibox Detector (SSD) and YOLO, calculate the dimensions of object proposals by using regression analysis with interpreting class probabilities, which can be considered as a significant improvement in terms of the model efficiency and the training cost (Liu et al., 2020; Bochkovskiy et al., 2020). Similar to the configuration in YOLOv3, YOLOv4 applies predefined boxes that are used to detect different objects, followed by a trained object detection model that can predict object classes from anchor boxes. To match the ground-truth data, the dimensions of the actor box will be adjusted by using an offset based on the prediction and regression loss.

This chapter applies a pre-trained model, called DeepSOCIAL (Rezaei & Azarmi, 2020), with two different CCTV surveillance video files to test the human detection module. The DeepSOCIAL model investigates common multi-object annotated datasets, such as PASCAL VOC, COCO, Image Net, and Google Open Images, with a collection of almost 20,000 classes that is suitable for the people detection task. A fine-tuning optimization of the YOLO-based model has been used with the CSPDarkNet53-PANet-SPP-SAM as the backbone architecture that forms the DeepSOCIAL model. Figure 1 presents the robustness of the human detector in two challenging open source datasets, i.e. the VIRAT video dataset (Oh et al., 2011) and the Oxford Town Center dataset (Harvey & LaPlace, 2019). Visual outputs indicate that the DeepSOCIAL model enables the human detector given different scenarios, including partial visibility and small objects. The overall testing results are superior and capable for the next step of constructing a social distancing monitor.

Figure 1. Human detection performances of the DeepSOCIAL model in two different datasets. (A). the VIRAT video dataset; (B). the Oxford Town Center dataset.

People Tracking and Identification

After initializing the human detection stage, the people tracking and identification phase will be the intermediate stage among the social distancing monitor model. The Simple Online and Real-time (SORT) tracking technology can be used as the framework for the Kalman filter along with the Hungarian optimization to track the people (Bewley et al., 2016; Rezaei & Klette, 2017). The position of a person can be predicted based on the current measurement by performing the Kalman filter with the mathematical modeling of the human movement. Unique ID numbers can be assigned to identify a certain object in a set of image frames by using a combination of optimization algorithms (i.e. Hungarian algorithm) to determine whether an individual in the current frame is the same detected object in any other frames or not. The bounding box will be refreshed with the incoming observed status once a detected object connects with a new observation. This procedure is computed based on the velocity and acceleration components estimated by the Kalman filter. The information association issue can be addressed by calculating the intersection over union and the difference between the actual input values and the estimated values by using the Hungarian algorithm and the Kalman filter.

One of the potential questions raised from the reality is that public authorities and some epidemiologists argue the way of dealing with couples who have close relationships, for example, family members, in social distancing monitoring (Greenstone & Nigam, 2020). Some studies indicate that every single person should practice social distancing without any exceptions, while others suggest couples and family members can walk closer in public areas without being counted as violations of the social distancing regulation (Okabe-Miyamoto et al., 2021). Whether the social distancing should be applied to family members is still a controversial issue in most regions across the United States and around the world. In common sense, it is quite difficult to control the social distancing of family members or couples, even though a strict policy is proposed to restrict individuals' distancing. Therefore, solutions regarding activations of couple detections must be included in the social distancing monitoring algorithm. This challenge can be addressed by applying a temporal data analysis method that determines two persons as a couple if they keep less than a certain distance measurement apart in an adjacency for more than a given duration, i.e. a few seconds.

Figure 2 demonstrates the people tracking and identification performances of the DeepSOCIAL model in the test procedure using the VIRAT video dataset. Every single individual has been denoted and detected in the green bounding box with a unique ID given for each one (shown in Figure 2A). An example of the representation for detected couples in a sample image frame has been presented in Figure 2B, where the orange bounding boxes indicate the coupled people with their unique IDs for the group.

Figure 2. People tracking and identification performances of the DeepSOCIAL model. (A). Examples of single individuals tracking and identification; (B). Examples of coupled persons tracking and identification.

Calculating the Inter-distance Measurement

The final stage of implementing a social distancing monitor is to estimate the inter-distance between individuals among each image frame of a sample video. Saleem et al. (2019) proposed a stereo-vision technique that has been widely used for distance estimation, however, such a technology cannot be a feasible solution for applying the social distancing monitor algorithm in any given public spaces using only a basic CCTV surveillance camera. A monocular approach can be deployed by using a single camera projection of a 3D scene into a 2D perspective photo plane (Lemaire et al., 2007). However, such a solution may lead to unrealistic pixel-distances among objects, thereby causing the perspective effect that it is difficult to perceive uniform distribution of distances in the entire image. The distance between farther persons seems much shorter than those who are closer to the camera, therefore, measuring the Euclidean distance criterion may be erroneous in terms of the inter-persons distance estimation. In the image frame extracted from the video, the original 3D space has been reduced to a 2D panel that the depth parameter is not available, therefore, a camera calibration by setting the depth parameter to be zero is the first step towards eliminating the perspective effect as performing a calibrated inverse perspective mapping (IPM) transition (Zhu & Feng, 2019). Obviously, the camera setting information, such as its location, height, angle of view, and the intrinsic parameters, are required to be known (Rezaei & Klette, 2017).

By using the IPM, the 2-dimensional pixels can be mapped to the corresponding world coordinates with the rotation matrix, the translation matrix, and the intrinsic parameters. As a transformation matrix, the rotation matrix is applied to perform a rotation in Euclidean space. In order to keep the object's size, shape and orientation as the same, the translation matrix is required as a type of transformation when the figure moves from one location to another on the coordinate plane. The intrinsic parameters are formed as a matrix having focal length measurements, calibration coefficients in horizontal and vertical pixel units, and the principal point shifts that adjust the optical axis of the 2D plane. The relationship between 3D coordinate points and the 2D projection points can be represented by using homogeneous coordinates that map the world coordinate points into the image points based on the camera location and the reference frame. Such parameters are provided by the rotation matrix, the translation matrix, and the intrinsic parameters matrix. The dimensions of the above-mentioned matrixes can be reduced and transferred from the perspective space to inverse perspective space, considering the image plane perpendicular to the depth access in the world coordinate system.

Figure 3 illustrates the final output of a social distancing monitor using DeepSOCIAL model. Figure 3A shows a sample of representation for detected individuals in an image frame with two types of social distancing detections. People who keep safe distance are denoted as green bars with a 6-feet circle, while the bars will turn to red when social distancing rules are violated. The yellow bars in Figure 3B indicate people detected as couples have a radios of a given number of distance that ensures a minimum safety distance of 6 feet for each of them to the neighbors around.

Figure 3. Social distancing monitors using the DeepSOCIAL model. (A). the sample of single individuals; (B). the sample of coupled persons. Three types of detections are safe (green bars), violations (red bars), and couples (yellow bars).

FUTURE RESEARCH DIRECTIONS

Despite the expected testing outputs using DeepSOCIAL model, several limitations still remain in terms of establishing the social control system during the pandemic. The DeepSOCIAL model can only detect social distancing violations based on the fixed CCTV cameras, which needs multiple monitors for covering the whole area, thereby a huge number of investments on the deployment. Moreover, a warning system that provide real-time alert signals may improve the social control effects by notifying those who violate the social distancing regulation immediately, however, it is difficult to be set up by using the DeepSOCIAL monitoring system. Additionally, although couples or single individuals can be detected and tracked in the DeepSOCIAL model, it cannot provide more detailed information, such as body temperatures, risks of exposure with the coronavirus carriers, and pedestrians' face covering conditions, for each individual captured on the screen. Motivated by the challenges from the current existing social distancing monitors, an innovated social control system, called DeepCovidPark, has been proposed in the form of sandbox testing procedures by aggregating social distancing monitoring algorithms and big data analytics, along with the real-world applications in social control managements during the pandemic.

Developed by the global research group for youth at IR big data and AI Lab, the DeepCovidPark is designed for monitoring and managing people who enter into a park (Li et al., 2020). The system testing procedure has been done in June 2020 at Quanshan Forest Park in the city of Xuzhou, China. Instead of the human detection based on the CCTV cameras, DeepCovidPark uses a drone to collect surveillance videos, thereby monitored areas will not be restricted and the cost of its deployment can be effectively reduced. The collected video files will be uploaded into a cloud-based data center that provides the storage of the dataset, followed by the automatic data processing for the social distancing detection model. The system is designed as a two-stage architecture that can be used to monitor whether individuals apply the social distancing regulation and the face covering requirement or not. Its real-time alarm sensors will provide the suggestion to the central control system of the park by generating risk assessments for detected areas. Each individual will be captured a close-up image when people are less than a 6-feet safe distance, followed by a face mask detection by the human-based identification process. Finally, the central control system will notify those who violate the pandemic regulations through the alarm system covering the whole area of the park.

The social distancing monitoring module in DeepCovidPark applies Andrew Ng's detection model using artificial intelligence technology to analyze real-time video streams and monitor whether people are applying social distancing regulations in the public places or not (Landing AI, 2020). Such the algorithm is realized through three major steps: calibration, detection, and measurement. In the stage of calibration, the monocular (single-camera) video stream can be transformed to a bird's-eye view. The calibration method takes four points in the frame and maps them to the corners of a rectangle in the bird's-eye view. In the detection stage, the system draws a bounding box around each pedestrian using an open-source R-CNN architecture based pedestrian detection network. The minimal post-processing is also applied to clean up the output bounding boxes in this stage. In the measurement stage, the coordinate for each individual position is estimated in the bird's-eye view plane, followed by calculating the bird's-eye view distance between every instance. When the distance is below the minimum acceptable distance, the system highlights these people in red and draws up a line between this pair of people to emphasize and issue a reminder to keep social distancing.

Before entering to the park, each visitor will be assigned a unique ID that is used to track his or her movements and conditions during the visiting time. The park's authority will check the body temperatures and scan the health codes for all visitors in order to take records associated with their IDs. The health code, initiated by Alibaba and Tencent, is an innovative app that tracks a person's travel and contact history, along with biometric information (i.e.

body temperature) via smartphones, which has been adopted by the Chinese government (Tan, 2020). The individual level of health risks can be identified by scanning the QR code generated from the app that provide three different colors associated with imposed disease control regulations, such as green (travel allowed after temperature taken), yellow (a 14-day quarantine at home), and red (suspected or confirmed patients who need to be hospitalized or isolated immediately). The health code scanning becomes an ideal social management tool for the disease control system because of its high usage coverage and the risk assessment, therefore it is essential to record such information for visitors at each gate of the park. Figure 4 provides a sample screenshot of the system control dashboard from the real-time DeepCovidPark. The whole park has been divided by five areas based on the visitor distribution. The drone will fly around each area collecting real-time videos that will be uploaded into the social distancing monitoring module for calculating the number of violations and risk levels for the area.

Figure 4. The real-time dashboard based on the DeepCovidPark

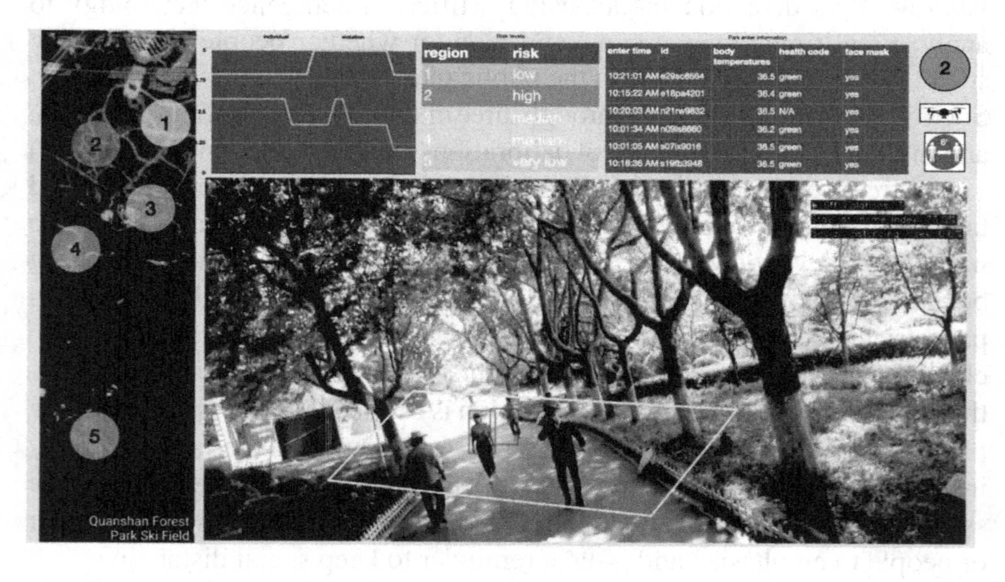

CONCLUSION

This chapter illustrates several popular object detection and tracking schemes by implementing a cutting-edge human detector model, called DeepSOCIAL, to detect and track individuals in a sample public place by monitoring social distancing measurements in the context of COVID-19. Various types of state-of-the-art modules, such as backbones, necks, and heads have been introduced and examined given a real case study using the DeepSOICAL model that offers a viewpoint independent human detection algorithm. The testing outputs indicate that the DeepSOCIAL model can be applicable for a wider range of social monitoring tasks, no matter how the camera is setup in terms of angles and positions. However, the DeepSOCIAL model cannot uncover the detected individual's personal information associated with the COVID-19 pandemic for the disease control management, thereby in-depth research and innovations are necessary for the real world applications in COVID-19.

The DeepCovidPark is one of the most successful attempts by combining the computer-vision based technology and big data analytics, which has been proved to be feasible with the applicable testing procedure for the social control management in parks during the pandemic. DeepCovidPark system offers a comprehensive individual-level detection and tracking architecture that allows parks' authorities to monitor the visiting regulations during the pandemic. Instead of the CCTV cameras deployed under the DeepSOCIAL framework, the DeepCovidPark applies a drone to collect monitoring videos. The alarm system has been designed based on the social distancing detection algorithm that is formalized in a cloud-computing environment, providing high-performance computing capabilities for the real-time data stream. Its risk assessment provides not only the violation measurement for practicing social distancing regulations, but also personal information derived from the health code. Such an innovated system design can minimize the pandemic risk with the evidence for further references of tacking visitors who exposure to the coronavirus during the visiting time.

Despite the advancement and superiority of the DeepCovidPark system, its level of automation and intelligence is relatively low, i.e. many decision-making processes, such as providing warning messages, checking body temperatures, and identifying face covering conditions, are still relied on personals who works at the gates of the park or in the central control center. Therefore, further studies may focus on how to improve the system in terms of its automatic intelligence by transforming human-based tasks to AI-based procedures. For example, identifying whether visitors wear face masks or not can be completely implemented by using deep learning algorithms, which will be investigated in the next chapter.

REFERENCES

Alhichri, H., Alswayed, A. S., Bazi, Y., Ammour, N., & Alajlan, N. A. (2021). Classification of Remote Sensing Images Using EfficientNet-B3 CNN Model With Attention. *IEEE Access: Practical Innovations, Open Solutions*, *9*, 14078–14094. doi:10.1109/ACCESS.2021.3051085

Bewley, A., Ge, Z., Ott, L., Ramos, F., & Upcroft, B. (2016, September). Simple online and realtime tracking. In 2016 IEEE international conference on image processing (ICIP) (pp. 3464-3468). IEEE. doi:10.1109/ICIP.2016.7533003

Bochkovskiy, A., Wang, C. Y., & Liao, H. Y. M. (2020). *Yolov4: Optimal speed and accuracy of object detection.* arXiv preprint arXiv:2004.10934.

Chen, C., & Qi, F. (2018, October). Single image super-resolution using deep CNN with dense skip connections and inception-resnet. In *2018 9th International Conference on Information Technology in Medicine and Education (ITME)* (pp. 999-1003). IEEE. 10.1109/ITME.2018.00222

Delen, D., Eryarsoy, E., & Davazdahemami, B. (2020). No place like home: Cross-national data analysis of the efficacy of social distancing during the COVID-19 pandemic. *JMIR Public Health and Surveillance*, *6*(2), e19862. doi:10.2196/19862 PMID:32434145

Ferguson, N. M., Cummings, D. A., Fraser, C., Cajka, J. C., Cooley, P. C., & Burke, D. S. (2006). Strategies for mitigating an influenza pandemic. *Nature*, *442*(7101), 448–452. doi:10.1038/nature04795 PMID:16642006

Fong, M. W., Gao, H., Wong, J. Y., Xiao, J., Shiu, E. Y., Ryu, S., & Cowling, B. J. (2020). Nonpharmaceutical measures for pandemic influenza in nonhealthcare settings—Social distancing measures. *Emerging Infectious Diseases*, *26*(5), 976–984. doi:10.3201/eid2605.190995 PMID:32027585

Gawande, U., Hajari, K., & Golhar, Y. (2020). *Pedestrian Detection and Tracking in Video Surveillance System: Issues, Comprehensive Review, and Challenges*. Recent Trends in Computational Intelligence.

Generals, W. H. O., & Speeches, D. (2020). *Opening remarks at the media briefing on COVID-19.* WHO Generals and Directors Speeches.

Gidaris, S., & Komodakis, N. (2015). Object detection via a multi-region and semantic segmentation-aware cnn model. In *Proceedings of the IEEE international conference on computer vision* (pp. 1134-1142). 10.1109/ICCV.2015.135

Girshick, R. (2015). Fast r-cnn. In *Proceedings of the IEEE international conference on computer vision* (pp. 1440-1448). IEEE.

Greenstone, M., & Nigam, V. (2020). *Does social distancing matter?* University of Chicago, Becker Friedman Institute for Economics Working Paper, (2020-26).

Harvey, A., & LaPlace, J. (2019). *MegaPixels: Origins, ethics, and privacy implications of publicly available face recognition image datasets*. Megapixels.

He, S. K., & Wang, X. (2014). Spatial pyramid pooling in deep convolutional neural network for visual recognition. In CVPR (Vol. 2). Academic Press.

Hipps, R., Chopra, T., Zhao, P., Kwartler, E., & Jaume, S. (2017). Geospatial Analytics to Improve the Safety of Autonomous Vehicles. *International Journal of Knowledge-Based Organizations*, *7*(3), 40–51. doi:10.4018/IJKBO.2017070104

Hu, X., Wei, Z., & Zhou, W. (2021, March). A video streaming vehicle detection algorithm based on YOLOv4. In *2021 IEEE 5th Advanced Information Technology, Electronic and Automation Control Conference (IAEAC)* (*Vol. 5*, pp. 2081-2086). IEEE. 10.1109/IAEAC50856.2021.9390613

Huang, G., Liu, Z., Van Der Maaten, L., & Weinberger, K. Q. (2017). Densely connected convolutional networks. In *Proceedings of the IEEE conference on computer vision and pattern recognition* (pp. 4700-4708). IEEE.

Idrees, H., Shah, M., & Surette, R. (2018). Enhancing camera surveillance using computer vision: A research note. *Policing*, *41*(2), 292–307. doi:10.1108/PIJPSM-11-2016-0158

Jhunjhunwala, A. (2020). Role of telecom network to manage Covid-19 in India: Aarogya setu. *Transactions of the Indian National Academy of Engineering*, *5*(2), 157–161. doi:10.100741403-020-00109-7

Jia, Y., Huang, C., & Darrell, T. (2012, June). Beyond spatial pyramids: Receptive field learning for pooled image features. In *2012 IEEE Conference on Computer Vision and Pattern Recognition* (pp. 3370-3377). IEEE.

Khandelwal, P., Khandelwal, A., & Agarwal, S. (2020). *Using computer vision to enhance safety of workforce in manufacturing in a post covid world.* arXiv preprint arXiv:2005.05287.

Kim, G. U., Kim, M. J., Ra, S. H., Lee, J., Bae, S., Jung, J., & Kim, S. H. (2020). Clinical characteristics of asymptomatic and symptomatic patients with mild COVID-19. *Clinical Microbiology and Infection*, *26*(7), 948–e1. doi:10.1016/j.cmi.2020.04.040 PMID:32360780

Landing, A. I. (2020). *Landing AI Creates an AI Tool to Help Customers Monitor Social Distancing in the Workplace.* https://landing.ai/landing-ai-creates-an-ai-tool-to-help-customers-monitor-social-distancing-in-the-workplace/

Lemaire, T., Berger, C., Jung, I. K., & Lacroix, S. (2007). Vision-based slam: Stereo and monocular approaches. *International Journal of Computer Vision*, *74*(3), 343–364. doi:10.100711263-007-0042-3

Li, B., Huang, T., & Wang, X. (2020). *A New Solution of The Social Distancing and Face Mask Monitor Using Deep Learning Algorithms*. https://github.com/pzhaoir/Social-Distancing-and-Face-Mask-Monitor/blob/main/SocialDis_FaceMask_Final.pdf

Lin, T. Y., Dollár, P., Girshick, R., He, K., Hariharan, B., & Belongie, S. (2017). Feature pyramid networks for object detection. In *Proceedings of the IEEE conference on computer vision and pattern recognition* (pp. 2117-2125). IEEE.

Liu, L., Ouyang, W., Wang, X., Fieguth, P., Chen, J., Liu, X., & Pietikäinen, M. (2020). Deep learning for generic object detection: A survey. *International Journal of Computer Vision*, *128*(2), 261–318. doi:10.100711263-019-01247-4

Misra, D. (2019). *Mish: A self regularized non-monotonic neural activation function.* arXiv preprint arXiv:1908.08681, 4.

Nguyen, C. T., Saputra, Y. M., Van Huynh, N., Nguyen, N. T., Khoa, T. V., Tuan, B. M., . . . Ottersten, B. (2020). *Enabling and emerging technologies for social distancing: A comprehensive survey.* arXiv preprint arXiv:2005.02816.

Nguyen, D. T., Li, W., & Ogunbona, P. O. (2016). Human detection from images and videos: A survey. *Pattern Recognition*, *51*, 148–175. doi:10.1016/j.patcog.2015.08.027

Oh, S., Hoogs, A., Perera, A., Cuntoor, N., Chen, C. C., Lee, J. T., & Desai, M. (2011, June). A large-scale benchmark dataset for event recognition in surveillance video. In *CVPR 2011* (pp. 3153–3160). IEEE. doi:10.1109/CVPR.2011.5995586

Okabe-Miyamoto, K., Folk, D., Lyubomirsky, S., & Dunn, E. W. (2021). Changes in social connection during COVID-19 social distancing: It's not (household) size that matters, it's who you're with. *PLoS One*, *16*(1), e0245009. doi:10.1371/journal.pone.0245009 PMID:33471811

Olsen, S. J., Chang, H. L., Cheung, T. Y. Y., Tang, A. F. Y., Fisk, T. L., Ooi, S. P. L., Kuo, H.-W., Jiang, D. D.-S., Chen, K.-T., Lando, J., Hsu, K.-H., Chen, T.-J., & Dowell, S. F. (2003). Transmission of the severe acute respiratory syndrome on aircraft. *The New England Journal of Medicine*, *349*(25), 2416–2422. doi:10.1056/NEJMoa031349 PMID:14681507

Pfizer. (2020). *Pfizer and BioNTech Announce Vaccine Candidate Against COVID-19 Achieved Success in First Interim Analysis from Phase 3 Study*. https://www.pfizer.com/news/press-release/press-release-detail/pfizer-and-biontech-announce-vaccine-candidate-against

Punn, N. S., Sonbhadra, S. K., & Agarwal, S. (2020). *Monitoring COVID-19 social distancing with person detection and tracking via fine-tuned YOLO v3 and Deepsort techniques*. arXiv preprint arXiv:2005.01385.

Ren, S., He, K., Girshick, R., & Sun, J. (2016). Faster R-CNN: Towards real-time object detection with region proposal networks. *IEEE Transactions on Pattern Analysis and Machine Intelligence*, *39*(6), 1137–1149. doi:10.1109/TPAMI.2016.2577031 PMID:27295650

Rezaei, M., & Azarmi, M. (2020). Deepsocial: Social distancing monitoring and infection risk assessment in covid-19 pandemic. *Applied Sciences (Basel, Switzerland)*, *10*(21), 7514. doi:10.3390/app10217514

Rezaei, M., & Klette, R. (2017). Computer vision for driver assistance. Cham: Springer International Publishing.

Rezaei, M., Sarshar, M., & Sanaatiyan, M. M. (2010, February). Toward next generation of driver assistance systems: A multimodal sensor-based platform. In *2010 The 2nd International Conference on Computer and Automation Engineering (ICCAE)* (Vol. 4, pp. 62-67). IEEE. 10.1109/ICCAE.2010.5451782

Rodríguez, P., Velazquez, D., Cucurull, G., Gonfaus, J. M., Roca, F. X., & Gonzàlez, J. (2019). Pay attention to the activations: A modular attention mechanism for fine-grained image recognition. *IEEE Transactions on Multimedia*, *22*(2), 502–514. doi:10.1109/TMM.2019.2928494

Rusli, M. E., Yussof, S., Ali, M., & Hassan, A. A. A. (2020, August). MySD: A Smart Social Distancing Monitoring System. In *2020 8th International Conference on Information Technology and Multimedia (ICIMU)* (pp. 399-403). IEEE. 10.1109/ICIMU49871.2020.9243569

Saleem, N. H., Chien, H. J., Rezaei, M., & Klette, R. (2019). Effects of ground manifold modeling on the accuracy of stixel calculations. *IEEE Transactions on Intelligent Transportation Systems*, *20*(10), 3675–3687. doi:10.1109/TITS.2018.2879429

Sathyamoorthy, A. J., Patel, U., Savle, Y. A., Paul, M., & Manocha, D. (2020). *COVID-robot: Monitoring social distancing constraints in crowded scenarios.* arXiv preprint arXiv:2008.06585.

SharifiA.ZibaeiA.RezaeiM. (2020). DeepHAZMAT: Hazardous materials sign detection and segmentation with restricted computational resources. Available at SSRN 3649600. doi:10.2139/ssrn.3649600

Szegedy, C., Toshev, A., & Erhan, D. (2013). *Deep neural networks for object detection.* Academic Press.

Tan, S. (2020). *China's Novel Health Tracker: Green on Public Health, Red on Data Surveillance.* https://www.csis.org/blogs/trustee-china-hand/chinas-novel-health-tracker-green-public-health-red-data-surveillance

Thu, T. P. B., Ngoc, P. N. H., Hai, N. M., & Tuan, L. A. (2020). Effect of the social distancing measures on the spread of COVID-19 in 10 highly infected countries. *The Science of the Total Environment*, *742*, 140430. doi:10.1016/j.scitotenv.2020.140430 PMID:32623158

U.S. Food and Drug Administration. (2020). *FDA Approves First Treatment for COVID-19.* https://www.fda.gov/news-events/press-announcements/fda-approves-first-treatment-covid-19#:~:text=Veklury%20is%20the%20first%20treatment,to%20receive%20FDA%20approval

Wang, C. Y., Liao, H. Y. M., Wu, Y. H., Chen, P. Y., Hsieh, J. W., & Yeh, I. H. (2020). CSPNet: A new backbone that can enhance learning capability of CNN. In *Proceedings of the IEEE/CVF conference on computer vision and pattern recognition workshops* (pp. 390-391). 10.1109/CVPRW50498.2020.00203

Woo, S., Park, J., Lee, J. Y., & Kweon, I. S. (2018). *CBAM: Convolutional block attention module.* arXiv preprint arxiv:1807.06521.

Yao, Z., Cao, Y., Zheng, S., Huang, G., & Lin, S. (2020). *Cross-iteration batch normalization.* arXiv preprint arXiv:2002.05712.

Zhao, P., Xiao, H., & Wang, X. (2019). Satellite Image Recognition for Smart Ships Using A Convolutional Neural Networks Algorithm. *International Journal of Decision Science, 10*(2), 85–91.

Zhu, J., & Fang, Y. (2019). Learning object-specific distance from a monocular image. In *Proceedings of the IEEE/CVF International Conference on Computer Vision* (pp. 3839-3848). 10.1109/ICCV.2019.00394

ADDITIONAL READING

Bian, S., Zhou, B., Bello, H., & Lukowicz, P. (2020, September). A wearable magnetic field based proximity sensing system for monitoring COVID-19 social distancing. In *Proceedings of the 2020 International Symposium on Wearable Computers* (pp. 22-26). 10.1145/3410531.3414313

Chen, C., Liu, M. Y., Tuzel, O., & Xiao, J. (2016, November). R-CNN for small object detection. In *Asian conference on computer vision* (pp. 214-230). Springer.

Cyganek, B. (2013). *Object detection and recognition in digital images: theory and practice*. John Wiley & Sons. doi:10.1002/9781118618387

Hassaballah, M., & Awad, A. I. (Eds.). (2020). *Deep learning in computer vision: principles and applications*. CRC Press. doi:10.1201/9781351003827

Liu, L., Ouyang, W., Wang, X., Fieguth, P., Chen, J., Liu, X., & Pietikäinen, M. (2020). Deep learning for generic object detection: A survey. *International Journal of Computer Vision, 128*(2), 261–318. doi:10.100711263-019-01247-4

Ngan, K. N., & Li, H. (Eds.). (2011). *Video segmentation and its applications*. Springer Science & Business Media. doi:10.1007/978-1-4419-9482-0

Sharma, L., & Lohan, N. (2019). Internet of Things with Object detection: Challenges, Applications, and Solutions. In *Handbook of Research on Big Data and the IoT* (pp. 89–100). IGI Global. doi:10.4018/978-1-5225-7432-3.ch006

Szeliski, R. (2010). *Computer vision: algorithms and applications*. Springer Science & Business Media.

Yang, D., Yurtsever, E., Renganathan, V., Redmill, K. A., & Özgüner, Ü. (2021). A vision-based social distancing and critical density detection system for COVID-19. *Sensors (Basel)*, *21*(13), 4608. doi:10.339021134608 PMID:34283141

Zou, Z., Shi, Z., Guo, Y., & Ye, J. (2019). Object detection in 20 years: A survey. *arXiv preprint arXiv:1905.05055*.

KEY TERMS AND DEFINITIONS

Computer Vision: An automation technology that makes computers to gain high-level understanding from images and videos throughout acquiring, processing, analyzing, and recognizing digital data by transforming visual images into numerical or symbolic information.

Deep Learning: A broad family of machine learning models based on neural networks. Typical deep learning models are deep neural networks, convolutional neural networks, recurrent neural networks, deep belief networks, and deep reinforcement learning.

Deepsort: A machine learning model that can detect and track targeted objects in a video.

Non-Pharmaceutical Disease Control: A set of actions, apart from medicine and vaccination, that communities can slow down the spread of a disease, a.k.a. non-pharmaceutical interventions (NPIs).

Object Detection: A computer vision technique that can recognize objects from image or video.

RCNN: A machine learning model for computer vision that can apply selective search to extract features from images.

Social Distancing Monitor: A technology that is designed to warn individuals when they get too close to each other, particularly relying on communications or contacts in short distances.

SSD: A detector built through a single convolutional neural network to identify object in images.

YOLO: A real-time object detection system that can directly train on complete images to achieve the best detection performance.

Chapter 6
Artificial Intelligence Methods for Face Covering Detections

ABSTRACT

With the global spreading of the COVID-19 pandemic, the face-covering regulation has become a common disease control policy issued by the public authorities nationwide in the United States and the rest of the world. Many public and private service providers require people to wear an appropriate mask correctly; however, it is still challenging to monitor whether individuals practice such a policy or not. In the artificial intelligence era, deep learning models with computer vision technologies can be applied as an efficient solution for this challenge in the context of the COVID-19 pandemic. In this chapter, a deep transfer learning-based face mask detection model has been proposed with the testing procedure for multiple purposes, such as identifying whether individuals wear face masks or not, recognizing which type of the face-covering a person wears, and detecting whether individuals wear masks correctly or not, with a real-time human-computer interaction representation.

INTRODUCTION

In today's age, COVID-19 becomes a crucial public health issue that affects people's quality of life, leading to huge impacts to the social and economic system. The new coronavirus spreads easily and widely in crowd and close contact due to its high transmission rate, and becomes one of the most significant challenges around the world. Advised by health organizations

DOI: 10.4018/978-1-7998-8793-5.ch006

and epidemiologists, the usage of personal protective equipment (PPE) has been recommended in crowded and closed environments (Shen et al., 2021; Garrigou et al., 2020). The common types of PPE, such as gloves, eye protections, gowns, and face masks, can minimize exposure to the virus, thereby reducing the risk of the COVID-19 infection (Howell et al., 2021). Public authorities have begun working on making new rules by forcing individuals to follow the disease control policies, including practicing social distancing and facial covering. Wearing a mask becomes the requirement in public area in order to reduce the infection and spreading rate of COVID-19 since it can be transmitted via airdrops. As normal people are forced by the face covering regulations across the globe, it is challenging to monitor whether individuals follow such rules or not, especially in the high density of the population. In order to win the battle against the pandemic, guidance and surveillance in the crowd have been deployed by government officials to ensure that face covering regulations are applied with computer vision technologies and object detection algorithms (Wu et al., 2020). Such attempted applications can be implemented by integrating surveillance systems and artificial intelligence (AI) techniques.

Despite the huge impact on the society, the COVID-19 pandemic has given rise to opportunities in terms of research cooperations and industrial applications. Artificial intelligence along with computer vision technologies provide cutting-edge solutions to fight against the virus in many forms. As introduced in previous chapters, machine learning and deep learning allow research scientists and policymakers to build the early warning mechanism for the pandemic outbreaks and to monitor human behaviors in regards of disease control measures and regulations. The computer-based face covering detection, as one of the innovated approaches in the social control system during the pandemic, has been applied in the process of monitoring large groups of people. A novel AI-based prototype has been applied in the surveillance cameras of the Paris Metro System in order to monitor whether riders wear face masks or not (Fouquet, 2020). A face mask detection system, developed by LeewayHertz, has been proposed by using existing IP cameras and CCTV surveillance cameras combined with computer vision to monitor people without face masks (Pispati et al., 2020). Executed by the National Institute of Standard and Technology (NIST), a study has been proposed under the Ongoing Face Recognition Vendor Test (FRVT) documents accuracy of algorithms to detect masked individuals in the United States, which is the first report on evaluating the performance of face recognition models on protective face masks in COVID-19 (Ngan et al., 2020). The national first facial recognition tool that can identify citizens when they are wearing a face mask, developed by a Chinese facial recognition company, has been developed for assisting disease control purposes in China, which is more technologically advanced because it can extract personal information from mask-wearers (Pollard, 2020). Although ethical issues of the usage of facial recognition for surveillance systems in public have been argued for a lone time (Brey, 2004), its application focusing on the face mask detection is significant and demanded. Reported by DatakaLab, the purpose of using face mask detection algorithms is not to arrest individuals who violate the rule but to provide scientific suggestions that can help governments to reduce the risk from COVID-19 (Faizah, et al., 2021).

As covering faces becomes a new normal, several new issues arise. People are in dispute over how to choose the right way of the facial covering because of the variety of filtration efficiency and prevention effectiveness of different types of face masks. Although most public and private providers do not require customers to wear a medical-used face mask, wearing a high-protective mask can reduce the infection risk significantly (Dugdale & Walensky, 2020). N95 is such a face mask that has been recommended for health care professionals and essential workers when caring for patients or exploring to the high risk of the infection (CDC, 2020). Therefore, it makes sense to monitor whether people wear medical-used masks or not in dangerous zones. Moreover, regular individuals are encouraged to wear surgical masks rather than homemade face shields because the homemade masks are less effective than surgical masks (Eikenberry et al., 2020). Hence, identifying mask types is important for the disease control policy in public spaces. Besides, some individuals are wearing their masks incorrectly because of bad practices and behaviors. In this sense, a mask wearing detector application, called CheckYourMask, has been developed towards checking people whether they wear masks correctly or not (Hammoudi et al., 2020). However, such an app does not include a detection module for identifying mask types.

Motivated by challenges of the disease control measures, this chapter illustrates how artificial intelligence serves as the solution toward monitoring the face covering conditions. By examining most recent studies in this field, several typical face mask detection algorithms will be illustrated with an experimental testing procedure using the selected model and a new image set of training and testing algorithms for identifying multiple face covering conditions on both static images and real-time video streams. The objectives of this chapter are:

- introducing typical face mask detection approaches, such as the CNN-based architecture, hybrid models, and deep transfer learning-based methods.
- illustrating how to train a face mask detector using the deep transfer learning-based approach with fine-tuned MobileNetV2.
- investigating and evaluating the model performance per detection task, including the with or without masks detection, the face covering types identification, and the correctly/incorrectly masked faces detection, along with implementing such models in real-time video streams.
- discussing limitations and future research directions, i.e. fine-tuning methods and tiny face techniques, in the form of possible solutions for the current problems and challenges in face mask detectors.

BACKGROUND

Before the COVID-19 outbreak, most studies in masked face detections are concentrated on face instruction and facial recognition using traditional machine learning approaches. Ejaz et al. (2019) proposed a machine learning-based algorithm for recognizing masked and unmasked faces through principal component analysis. A novel GAN-based network has been used for removing mask from the face based on trained models using paired synthetic datasets (Din et al., 2020). To the disease control perspective, such studies may not be suitable since the main focus is on identifying individuals who are not applying face covering, thereby it may help to control the further risk of COVID-19 transmission. With the pandemic spreading globally, more and more research studies have been presented concentrating on detecting whether a person is wearing a mask or not by using a broad range of cutting-edge technologies. Loey et al. (2021a) proposed a deep transfer learning method with multiple machine learning models for recognizing the people who are not wearing face masks. Similarly, Oumina et al. (2020) applied various machine learning classifiers to process face covering detections based on feature exertions by using multiple deep learning methods. A super-resolution and classification networks (SRCNet) has been applied in developing a novel face mask wearing detection model (Qin & Li, 2020). A new artificial intelligence prototype for the face mask detection has been proposed by using hand-crafted and deep learning features with machine learning classifiers (Liu & Agaian, 2021).

Most existing studies on this topic aim on applying deep learning approaches directly to build state-of-the-art face mask detectors. Jiang et al. (2020) proposed a one-stage detector, called the RetinalFaceMask algorithm, which is high-performed face mask detection procedure consisted of a feature pyramid network with the convolutional neural network (CNN) and the transfer learning method. MobileNetV2, a CNN-based image classification method, has been developed with the high accuracy and efficiency (Sanjaya & Rakhmawan, 2020). Das et al. (2020) presented a simplified method to detect face mask wearing conditions using a set of deep learning packages, including TensorFlow, Keras, and OpenCV. A transfer learning model has been proposed by fine-tuning the InceptionV3 that is a pre-trained deep learning algorithm in order to automate the procedure of face mask detections (Chowdary et al., 2020). The YOLOv2-based medical face masks detection has been developed with a ResNet-50 deep transfer learning model for the feature extraction (Loey et al., 2021b). Based on the CNN model, Fasfous et al. (2021) introduced a binary neural network-based face mask wearing and positioning detector (BinaryCoP) that can be deployed at entrances of indoor locations to mitigate the spread of COVID-19.

Several studies are concentrated on implementing deep learning-based face mask detection algorithms into a real-time architecture. A real-time face mask recognition with alarm system has been developed by using deep learning algorithms (Militante & Dionisio, 2020). Yadav (2020) proposed a real-time monitor to detect safe social distancing and face masks by implementing deep learning models on raspberry pi4. An edge computing-based deep learning (ECMask) framework has been applied to build a real-time face mask identification for COVID-19 (Kong et al., 2021). Nagrath et al. (2021) proposed a real-time face mask detection system using single shot multi-box detector and MobileNetV2 (SSDMNV2) based on the deep neural network (DNN) model. A hybrid mask detection and social distancing monitoring system has been presented for complex pictures processing using R-CNN, Fast R-CNN, and Faster R-CNN algorithms (Meivel et al., 2021). The web-based efficient AI recognition of masks (WearMask) has been deployed on any common devices that connect to the internet, designed by a server-less edge-computing architecture with integrating YOLO, NCNN, and WebAssembly (Wang et al., 2021).

MAIN FOCUS OF THE CHAPTER

Issues, Controversies, Problems

To the technical perspective, face mask detection algorithms are based on object detection from images, which is probably the most essential aspect of computer vision. Both supervised and unsupervised learning can be applied in this field for outfitting the image recognition tasks based on object detection. Common object detection methods are categorized as machine learning detectors, such as Scale-invariant feature transform (Lowe, 1999) and Histogram of oriented gradients (Lowe, 2004), and deep learning detectors, such as region convolution neural network (R-CNN (Girshick et al., 2014)), Fast R-CNN (Girshick, 2015), Faster R-CNN (Ren et al., 2016), Single shot multi box detector (SSD (Liu et al., 2016), and YOLO v1-4 (Redmon et al., 2016; Redmon & Farhadi, 2017; Redmon & Farhadi, 2018; Bochkovskiy et al., 2020). As discussed in the previous chapter, YOLO family models are faster and more accurate than R-CNN family models, which have been widely used in many object detection cases. However, the overfitting problem is still one of the major concerns in applying YOLO for the real-time object detection (Güven, 2019). In general, the object detection model can be integrated between the feature extraction process using deep learning and the classification process using machine learning. A deep transfer learning for feature extractions combined with multiple classical machine learning algorithms has been applied in many object detection tasks in terms of handling the overfitting problem in the image processing for computer vision by using deep learning algorithms (Nixon & Aguado, 2019).

Features of the Chapter

Existing studies concerning face mask detectors have been reviewed and summarized, followed by three types of deep learning-based face mask detection models, such as the CNN-based architectures, the hybrid deep learning models as the feature extractor with the classical machine learning-baed classifiers, and the deep transfer learning-based methods. A deep transfer learning-based approach with fine-tuning the MobileNetV2 architecture has been selected as the proposed face mask detector, which is applied to solve the real-world problems, including a basic prototype of the face mask detection, a face covering types detector, and a detector of the masked face conditions.

Each face mask detector has been implemented in the form of the human-computer interaction with a real-time representation. The proposed face mask detector can identify whether individuals wear a face mask or not, recognize which types of the face covering a person wears, and monitor whether people wear face masks correctly or not, which will benefit to the current situation by implementing the social management and control regulations during the COVID-19 pandemic.

SOLUTIONS AND RECOMMENDATIONS

Typical Face Mask Detection Approaches

The CNN-based Architecture

Deep learning has been witnessing a monumental growth in filling the gap between the humans' expectations and computational capacities. Research scientists and data professionals have been working on various aspects of studies in this field to benefit the applied computer vision. The purpose is to enable computers to recognize the real world as humans do, using the knowledge set for multiple tasks such as image recognition and video analysis. The Convolutional Neural Network (CNN) model becomes popular in numerous computer vision tasks (Khan et al., 2018), which is a deep learning algorithm that can be applied in many image classification and recognition tasks. The architecture of a CNN model is similar to the connected neurons in the human brain, in which individual neurons only reflect to stimulations in restrictions per visual field. The entire visual region can be further covered by a collection of such fields. The spatial and temporal dependencies in images can be captured through the relevant filters in a CNN architecture that performs a better fitting to the image dataset. Its reduction in the number of parameters and reusability of weights allows the network to be trained to recognize the complexity of the image. To obtain a better prediction, features extracted from the image will not be lost with a reduced form of the image processing in CNN.

The kernel, usually a matrix, is the first component of the convolution operation in the convolutional layer. The kernel shifts multiple times based on the stride length, operating a matrix multiplication process among filters and portions of the image over where the kernel is scanning. The filter moves with a certain stride value from the beginning of the image and then hops down with the same stride value. The process will repeat until the entire image is scanned. The goal of the convolution operation is to extract features from the image. A CNN architecture has multiple convolutional layers that are responsible for different operational tasks. The first convolutional layer is to extract the low-level features such as color, edges, and gradient orientation, while the extra layers deal with high level features that are similar to how human would understand the image. The valid padding technique convolved addresses feature reduced in dimensionality, while the same padding technique can be applied if the dimensionality either increases or remains the same as compared with the input image. To decrease the computational requirement for processing the image through dimensional reduction, the pooling layer is added in order to reduce the spatial size of the convoluted feature. It is helpful for extracting dominant features that are unchanged in rotation and position, thereby maintaining the efficiency of the model training process.

Two types of pooling methods, max pooling and average pooling, perform as the noise suppressant. Average polling plays dimensionality reduction as the de-noise mechanism, while max pooling completely discards the noise along with dimensionality reduction. Therefore, max pooling performs much better than average pooling. The extra layers of a CNN model are formed by putting the convolutional layer and the max pooling layer together. The number of such layers may be increased, depending on the complexity of the input image, for the feature extraction enables the model to recognize the image. Finally, a regular neural network can be applied for the classification process by adding a fully connected layer. The image will be flattened into a column vector that is fed to a feed-forward neural network with back-forward propagation applied to each training iteration with the classification process using the softmax function. Commonly-used CNN architectures, such as LeNet, VGGNet, ResNet, and AlexNet, are available in building deep learning algorithms that power computer vision techniques as the innovated solution for various real-world problems, i.e. face mask detections.

A typical CNN-based face mask detector consists of an image pre-processing and a pre-trained CNN model that contains multiple 2-D convolution layers connected with dense neurons. For each labeled image in the dataset that includes faces with and without masks, the RGB image needs to convert to the Gray-scale type, followed by resizing and normalizing it into a multi-dimensional array. Regularly, a modern image recognition process relies on grayscale images without emphasizing the method applied to convert from color-to-grayscale because of its limited effect when using robust descriptors. The size of the training dataset that is required to obtain goodness of performance will be increased by importing nonessential information, therefore, utilizing the color-to-grayscale method can extract essential features as it may diminish the computational requirement (Ahmad et al., 2018). Most CNN architectures only accept fine-tuned images, causing many problems throughout the data collection process and the model implementation. Reconfiguring and reshaping the input image before augmentation can solve the problem with the normalization of the pixel range (Ghosh et al., 2019). To the CNN model building side, it starts at adding a convolution layer of *C* filters and the second convolution layer of *S* filters, followed by inserting a flatten layer to the network. A dense layer with *N* neurons is needed along with the final dense layer for two categories, i.e. with mask or without mask. Figure 1 illustrates the whole picture of the CNN-based face mask detection. The first convolution layer is followed by the rectified linear unit (ReLU) and MaxPooling layers with a kernel size that specifies the height and width of the 2-D convolution window. The second convolution layer has less filters with the same kernel size, which is also followed by ReLU and MaxPooling layers. Then a dense layer containing *N* neurons is added with a ReLU activation function. The second dense layer with two categories is added as the final layer using the Softmax activation function.

Figure 1. Face mask detection based on CNN architecture with the image pre-processing

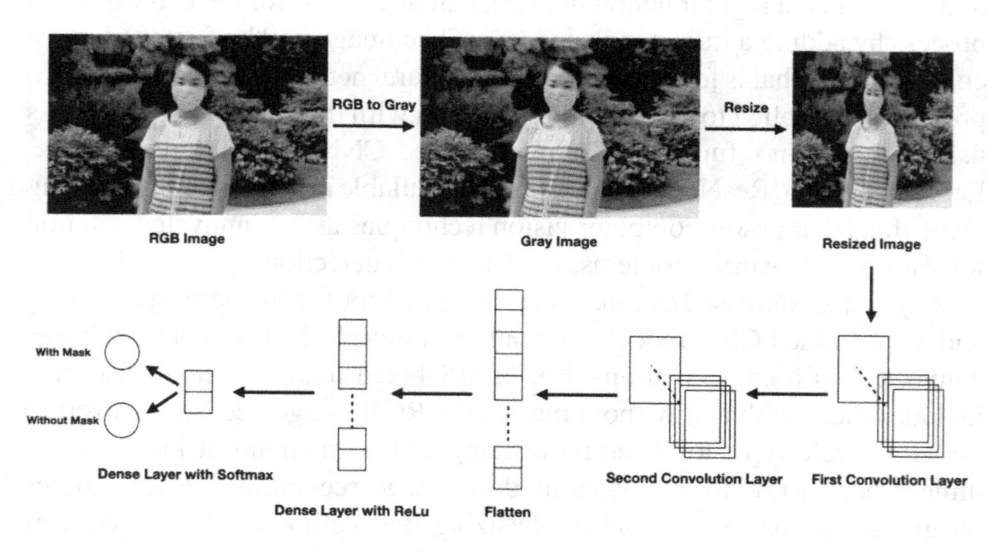

The Hybrid Method

Most existing studies based on deep learning methods do not provide the loss verse epoch plot for the validation set, thereby it is difficult to investigate whether the overfitting exists or not. Despite the superior of the CNN-based architecture, the overfitting problem may remain in the model training and evaluation process due to the number of parameters set in the network. Setting more parameters indicates a higher capacity for the model as it can approximate complex functions with more complicated decision boundaries. Ideally, the model will perform better if the right capacity is given by all useful aspects of the data, while it is difficult to have enough capacity to model the noise in the dataset, particularly in case of datasets with limited sample size. A fully-connected neural network as a classifier is more likely to overfit the model due to its more parameters. In general, deep learning models learn every single one of training samples without having the ability to generalize. Therefore, it is easier to overfit the model in a high-sufficient network that just memorizes samples, especially with small datasets. One potential solution will be increasing the size of the training set, however, in most cases, it is time-consuming and less efficient for the purpose of the project.

The other method that can be applied to overcome the overfitting problem is to change the fully-connected deep learning-based classification layers into classical machine learning classifiers. This hybrid approach contains two major components: the first one is the feature extractor that is usually based on a deep learning algorithm; the second one is a set of traditional machine learning classification models, such as SVM, decision tress, random forrest, logistics regression, etc. CNN-based architectures, such as VGGNet and ResNet, can be used for the feature extraction purpose. Most recent studies, i.e. Khojasteh et al. (2019) and Wen et al. (2019), indicate that ResNet-50 can achieve better performance in terms of classification results when it is applied as a feature extractor. The ResNet-50 algorithm is a residual neural network that has 50 layers under the convolutional neural network architecture based on residual learning method. Other versions of the ResNet model, such as ResNet-18 and ResNet-101, along with ResNet-50, can be used to drive out the problem of vanishing gradients that have their specific residual block. ResNet-50 begins with a convolution layer, followed by 16 residual blocks that have three layers of convolution layer with kernels, and ends with a full-connected layer. Such an architecture has been widely used on the feature extraction tasks with the benefit of depth and reduced computational expense (Varshni et al., 2019). The last layer in ResNet-50 has been removed and replaced by a set of classical machine learning classifiers that will not overfit the training process, comparing with that in the CNN-based classifier. A hybrid architecture using ResNet-50 as the feature extractor with machine learning classifies has been presented in Figure 2. To the classification side, performance metrics are essential to be investigated in order to evaluate the overall performance of the set of machine learning models. Consequently, the optimal model can be obtained based on the common performance measures, including accuracy, AUCs, and F1 score.

Figure 2. The hybrid architecture for face mask detection proposed ResNet-50 as the feature extractor with classical machine learning algorithms as the classifier

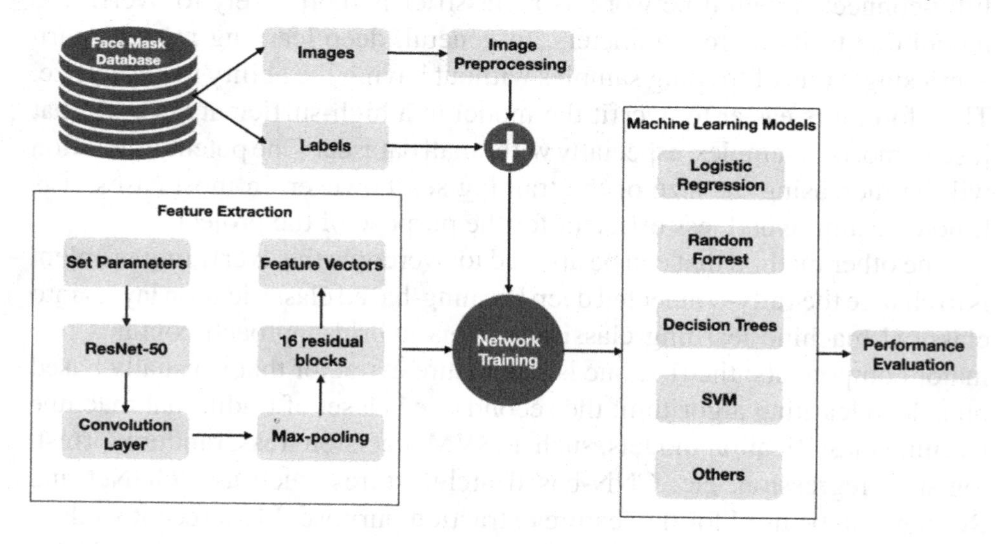

Although the overfitting problem is less likely to happen by using machine learning models comparing with the deep learning classification process, the probability of such a problem is not zero. One potential solution is to perform a n-fold cross validation, however, it may affect the learning process because of the small dataset. The other solution can be applied by validating trained model with different datasets. For instance, Loey et al. (2021a) conducted the experimental design on three different datasets, including the Real-World Masked Face Dataset (RMFD), the Simulated Masked Face Dataset (SMFD), and the Labeled Faces in the Wild (LFW) dataset. RMFD has both masked and unmasked face images in reality, while masked faces are simulated in

SMFD and LFM. To evaluate the performance of each machine learning algorithm, a training-testing strategy can be operated by training over RMFD and testing over all available datasets, and repeat the same procedure for the rest datasets. In the above-mentioned study, authors made the experiments trails even more complicated by combining and splitting those three original datasets. The main purpose of the experimental design is to reduce the risk of overfitting.

The Pre-trained Deep Transfer Learning Approach

Despite the effective problem-solving attempt by using a hybrid method, it seems more complicated as multiple stages are needed in splitting the face mask detector into two components: one is the feature extraction by deep learning; the other is the classification by machine learning. Therefore, the hybrid method may require higher computational capacity for the model building process because of its algorithm complexity. The CNN-based method provides a simplified way for the task, however, its overfitting problem should be addressed before the model training and testing procedure. Similarly, deep neural networks can be applied for the image recognition due to the better performance than classical algorithms. However, besides the overfitting problem, training a deep neural network is much more expensive because of its extra requirements in terms of the computational power and the time-consuming on the model training. The common assumption on machine learning and deep learning allows the trained model to be reused if the training and testing data are drawn from the same feature space and the same distribution, however, the model should be rebuilt from scratch when the feature and distribution change. It is quite expensive and time-consuming for the data recollection and the model rebuilding.

To develop an efficient model that makes the network to be trained faster and cost-effective, a deep transfer learning method can be used throughout the whole process of the image classification. Similar to the human's knowledge learning process, the idea of using transfer learning is based on the fact that it can intelligently apply information acquired previously from multiple tasks or domains that can be used to figure out new problems. From the algorithm perspective, transfer learning allows the trained data of the neural network to be transformed to the new model in terms of parametric weights. Such a process can boost the performance of the new network training even when it is trained with a small sample size. Several pre-trained models, such as ResNet,

Xception, InceptionV3, VGG-16, VGG-19, and MobileNet, are extensively used for computer vision tasks. Table 1 provides the general model efficiency and complexity matrix for available deep transfer learning algorithms. The model performance evaluation is based on training with 14 million images from the ImageNet dataset (*Keras Applications,* n.d.). The accuracy is the top-5 accuracy that refers to the model performance, whereas depth refers to the topological depth of the network, which induces activation layers, batch normalization layers, etc.

Table 1. Efficiency and complexity matrix for typical pre-trained deep transfer learning models

Model	Size	Accuracy	Parameters	Depth
MobileNet	16 MB	0.895	4,253,864	88
MobileNetV2	14 MB	0.901	3,538,984	88
VGG-16	528 MB	0.901	138,357,544	23
VGG-19	549MB	0.900	143,667,240	26
ResNet-50	98 MB	0.921	25,636,712	N/A
ResNet-101	171 MB	0.928	44,707,176	N/A
ResNet-152	232 MB	0.931	60,419,944	N/A
InceptionV3	92 MB	0.937	23,851,784	159
Xception	88 MB	0.945	22,910,480	126

Source: (Keras Applications, n.d.)

Obviously, the objective of the face mask detection is an ad-hoc task that requires datasets designed for the model training. Although Table 1 offers the preview of the candidate models that can be applied in this project, the optimal algorithm selection is still essential by evaluating the model performance matrix with training and testing multiple models. Chowdary et al. (2020) provided such a model comparison with experimental trials in the Google Colab environment. The experimental results indicate that the proposed InceptionV3-based face mask detector achieves a higher accuracy than other pre-trained models as the feature extractor.

The typical strategy for image recognition using deep transfer learning is to apply the pre-trained model as the feature extractor, followed by fine-tuning the selected pre-trained models. For the InceptionV3-based face mask detection project, the last layer of the pre-trained network has been removed and replaced by fine-tuning with adding five extra layers, including an average pooling layer with a 5 by 5 pool size, a flattening layer, a dense layer containing 128 neurons with the ReLU activation function as well as a 0.5 dropout rate, and a decision layer with 2 neurons and softmax activation function. The deep transfer learning-based feature extractor has been trained from 80 epochs with 42 steps for each epoch. One of the potential problem in this project is that limited samples are available because of the security and privacy issues, which may cause the model training process to learn. To solve the problem, authors performed image augmentation before the model training process. Input images have been augmented with several artificial modification operations, such as resizing, rotating, zooming, flipping horizontally, blurring, etc., followed by rescaling the whole dataset to a certain pixels and transforming augmented images to a greyscale representation. Figure 3 illustrates the architecture representation of the InceptionV3-based face mask detector. To evaluate the model performance of the candidate feature extractors, several measurements, such as accuracy, precision, sensitivity, specificity, Intersection over Union (IoU), and Matthews Correlation Coefficient (MCC) can be applied to select the optimal transfer learning-based network as the feature extractor.

Figure 3. Schematic representation of the InceptionV3-based face mask detector

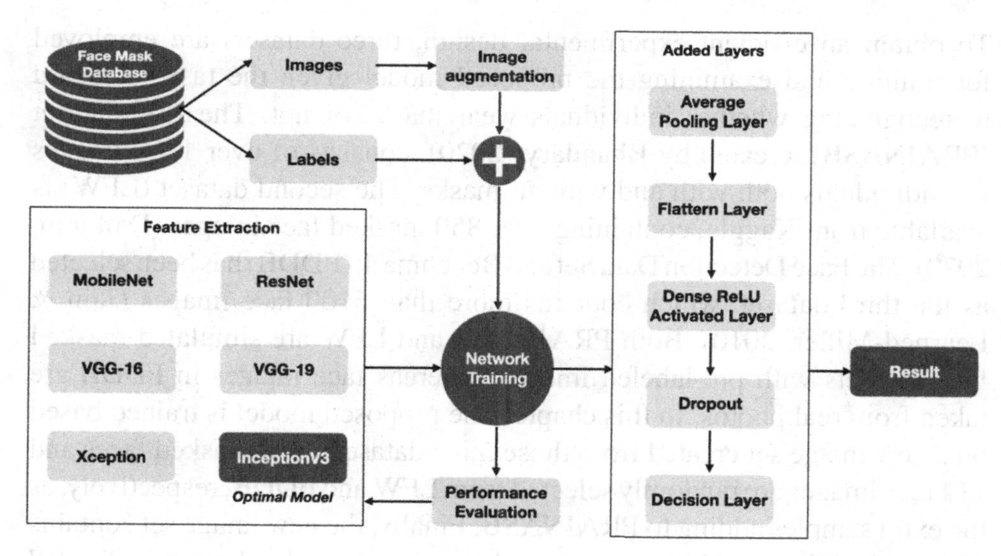

Model Training and Testing for Static Face Mask Detectors

To the project deployment perspective, testing the face mask detection algorithm is the most important part in implementing the proposed face mask detector into the real-world applications. Despite various methods that can be applied to build the image recognition models, it is critical to consider the model efficiency and complexity by leveraging the accuracy, overfitting problems, difficulties of the network training process, and the cost efficiency issue. By summarizing the existing studies in this field, a deep transfer learning-based approach by fine-tuning the MobileNetV2 architecture has been selected as face mask detector applied to the following tasks. The proposed model is based on a high-efficient network that can be used on embedded devices with the limited requirement for computational capacity. The whole network has been trained by using TensorFlow and Keras, along with the image augmentation process and the conversion of the class vector to the binary class matrix in data processing. Finally, testing and implementing the face mask detection for images can be finished by using OpenCV.

Recognizing Whether Face Masked or Not

To obtain an efficient experimental design, three datasets are employed for training and examining the proposed model given the task in regard to recognizing whether individuals wear masks or not. The first dataset (PRAJNASB), created by Bhandary (2020), consists of over 1300 images for individuals both with and without masks. The second dataset (LFW) is available from Kaggle, containing over 850 masked face images (Dalkiran, 2020). The Face Detection Data Set and Benchmark (FDDB) has been selected as the third dataset, which contains more than 5100 face images (Jain & Learned-Miller, 2010). Both PRAJNASB and LFW are simulated masked face datasets with pre-labeled images, whereas face images in FDDB are taken from real photos. In this chapter, the proposed model is trained based on a new image set created from those three datasets. 110 masked faces and 114 face images are randomly selected from LFW and FDDB, respectively, as the extra samples adding to PRAJNASB. Finally, the new image set contains 800 masked faces and 800 no-masked face images. The dataset is divided into an 80% training set and a 20% set for validating. The trained model obtains a high accuracy (nearly 99%) on both training and test sets. Figure 4 indicates the training and testing accuracy with loss curves of the proposed face mask detection model.

Figure 4. Training and testing epochs verse loss and accuracy curves for the face mask detection model

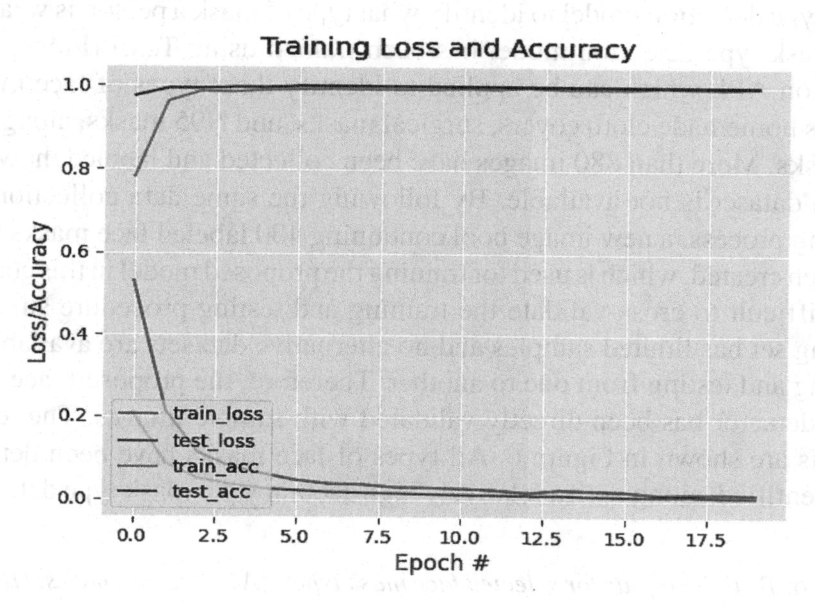

The proposed face mask detector is then validated by testing its effectiveness given two sample images collected by authors and from Google. The results are shown in Figure 5. The proposed face mask detector works well in the single individual case, whereas some people cannot be identified in the multiple individual masks detection. The reason will be discussed in the next section of the chapter.

Figure 5. Testing outputs for selected masked faces. (A). detecting single individual's mask; (B). detecting multiple individuals' masks.

Identifying Face Covering Types

Identifying face covering types relies on training face images with new labels, however, it is difficult to grab well-designed datasets. One existing project has been developed by Wang (2020) in terms of constructing a real-time face mask type detection model to identify what type of mask a person is wearing. The mask type detection model has been trained using TensorFlow object detection API, which can be applied to identify three types of face masks, such as homemade cloth covers, surgical masks, and N95 masks, along with no masks. More than 880 images have been collected and labeled, however, such a dataset is not available. By following the same data collection and labeling process, a new image pool containing 400 labeled face masks types has been created, which is used for training the proposed model in this chapter. It is difficult to cross-validate the training and testing procedure since the training set has limited samples and no alternative datasets are available for training and testing from one to another. Therefore, the proposed face mask types detector has been directly validated with sample images. The testing outputs are shown in Figure 6. All types of face masks have been detected and identified, along with a relatively high accuracy per mask type detection.

Figure 6. Testing outputs for selected face mask types. (A). surgical masks; (B). N95 masks; (C). homemade covers.

Detecting Correctly or Incorrectly Masked Conditions

Similar to the face covering types detector, checking whether individuals wear masks correctly or not requires pre-labeled image sets for training the proposed model. A publicly available dataset, called MaskedFace-Net, has been generated by Cabani et al. (2021), which can be employed for training the correctly/incorrectly masked conditions detection model. The MaskedFace-Net dataset consists of 67,193 correctly masked face images and 69,823 incorrectly masked faces, which is created by using a mask-to-face deformable model. The dataset has been divided into an 80% training set and a 20% validating set. The trained model yields a high accuracy (averagely over 90%) for both training and validating sets. Figure 7 depicts the training accuracy and testing with loss curves of the correctly/incorrectly masked faces detection model proposed.

Figure 7. Training and testing epochs verse loss and accuracy curves for the correctly/ incorrectly masked faces detection model

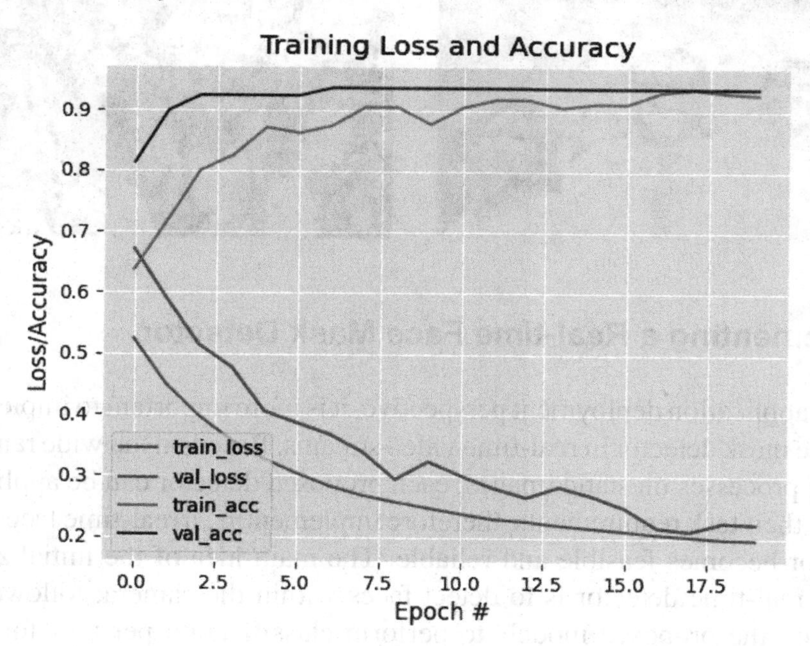

The correctly/incorrectly masked faces detector proposed is then validated by testing its effectiveness given multiple sample images collected from

Google. The results are shown in Figure 8. Several individuals who wear masks incorrectly are not detected during identifying incorrectly masked faces in crowd. The problem happens due to the same reason of failures in detecting face masks in the case of multiple individual detection.

Figure 8. Testing outputs for the selected correctly/incorrectly masked faces. (A). the correctly masked face; (B). the incorrectly masked face; (C). detecting incorrectly masked faces for multiple individuals; (D). checking incorrectly masked faces in crowd.

Implementing a Real-time Face Mask Detector

To the application deployment perspective, it is more important to implement the face mask detector in real-time video streams. Based on the wide range of testing processes on static images, each proposed detector can be applied to satisfy their task requirements, therefore implementing a real-time face mask detector becomes feasible and reliable. The main idea of the initialization of the real-time detector is to detect faces within the camera, followed by applying the proposed models to perform classification per task for each region of interest (ROI). Such a process can be done by using OpenCV with the connection to the webcam video stream. Alternatively, it can be exported to smaller computational hardwares, i.e. Raspberry Pi. Figure 9 shows the example outputs from the deployed real-time face mask detector. The result indicates that the system can detect masked or non-masked persons, mask types, and correctly/incorrectly face masks wearing conditions in real-time. Overall, such a real-time face mask detector is effective for close-up applications, which can be deployed in front of entrances of buildings or other public spaces.

Figure 9. Testing outputs from the real-time face mask detector. (A). correctly wearing; (B). incorrectly wearing; (C). surgical mask; (D). N95; (E). homemade; (F). non-masked face.

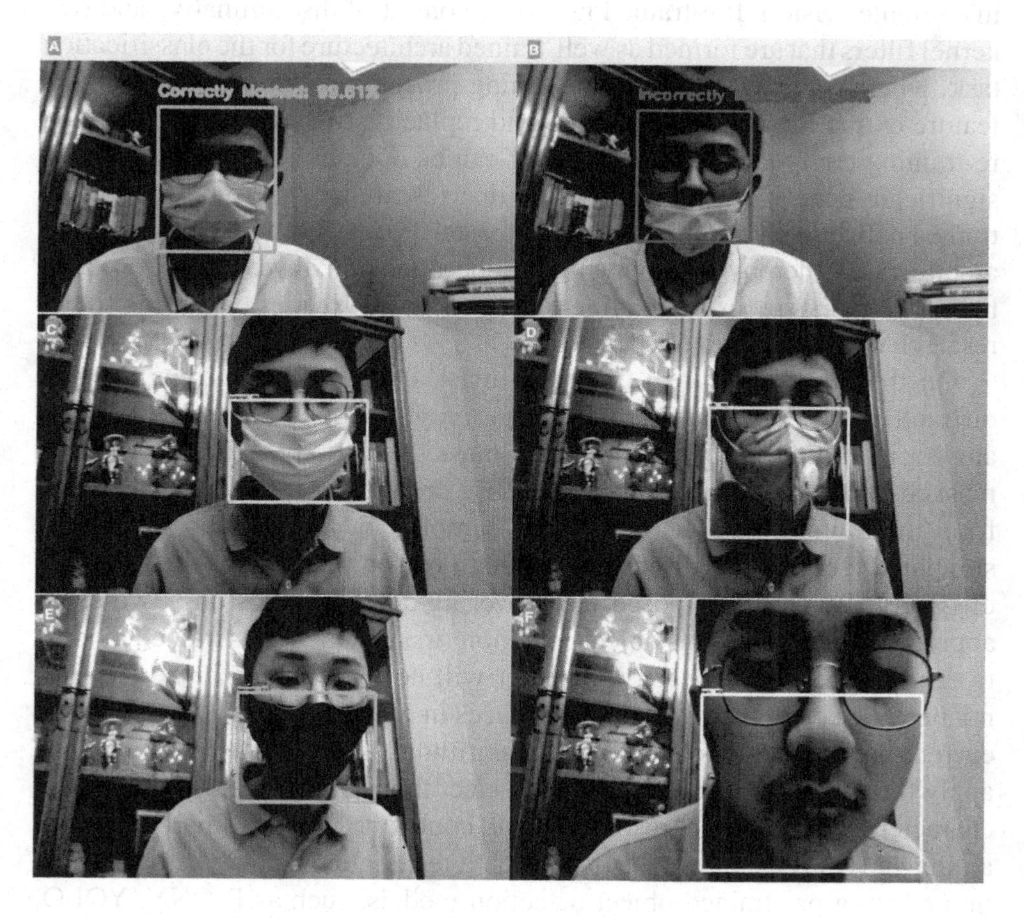

FUTURE RESEARCH DIRECTIONS

One possible research direction is to improve the face mask detection algorithm by using fine-tuning approach. The fine-tuning method plays a very important role in deep transfer learning networks. Followed by the feature extraction, the network can be fine-tuned with several pre-trained networks that either classify or recognize object categories in the image. Moreover, the network using fine-tuning can improve the model accuracy via feature extraction process, which is a modified network that has been widely used in computer vision. Pre-trained networks consist of discriminative and rich kernel filters that are formed as well-trained architecture for the classification task, whereas performing the fine-tuning procedure, instead of the simple feature extraction, allows removing and replacing the architecture, thereby re-training components of the network can be operational. One of the most significant benefit with fine-tuning is that researchers can modify the pre-trained networks by adding more layers based on a variety of existing network architectures, thereby improving the ad-hoc algorithm. Investigating possible hybrid networks by using the fine-tuning method provides one of the future research orientations in this filed.

On the other hand, real-time face mask detection algorithms are still difficult to be deployed for monitoring crowd people, e.g. some persons are not identified in the multiple face masks detection; several incorrectly masked individuals are not identified in the crowd. Except for the limitations from the training dataset (i.e. limited samples for training the model and simulated face covering images), the main reason that caused the technical difficulty is that the proposed face covering detector relies on a two-stage approach, which performs object detection first and then uses trained models to each image. The face mask detector will not be working if the face has not been detected, while detecting all faces in crowd is still challenging with current pre-trained object detection algorithms. One possible solution is to apply the tiny face detection that can detect thousands faces by using novel characterization of scale, resolution, and context to identify small objects in an image (Hu & Ramanan, 2017). Such a technique can also be embedded into existing pre-trained object detection models, such as R-CNN, YOLO, SSD, and MobileNet (Zhu et al., 2017; Yang et al., 2019; Ye et al., 2019; Zhu et al., 2020). Related studies for face mask detection with tiny face are limited due to the difficulty of the training set generation, however, such a topic will be discussed in depth as more related studies would be proposed in the future.

CONCLUSION

As one of the effective non-pharmacological disease control measures, wearing a mask can reduce the risk of getting infected during the COVID-19 pandemic. Cutting-edge technologies have been widely applied to overcome the challenges in terms of monitoring face covering conditions using computer vision and artificial intelligence. In this chapter, typical face mask detection approaches have been investigated with a comprehensive testing procedure using a deep transfer learning algorithm. The proposed model has been applied to deal with several face covering monitor issues, such as detecting whether individuals wear a mask or not, identifying face covering types, and determining correctly/incorrectly masked faces. The testing procedures have been performed in static images and real-time video streams, which yield a high model efficiency matrix per model and each task can be applied in the real-time. The proposed face mask detection module can be further used in multiple real-world applications. Despite the limitation existed in the current face detection algorithm, the possible solution has been discussed in the form of the future research directions. Although the data collection process may be difficult, tiny face detection approach provides an effective solution towards the improvement of the face mask detector.

REFERENCES

Ahmad, I., Moon, I., & Shin, S. J. (2018, January). Color-to-grayscale algorithms effect on edge detection—A comparative study. In *2018 International Conference on Electronics, Information, and Communication (ICEIC)* (pp. 1-4). IEEE. 10.23919/ELINFOCOM.2018.8330719

Bhandary, P. (2020). *Mask Classifier*. https://github.com/prajnasb/observations

Bochkovskiy, A., Wang, C. Y., & Liao, H. Y. M. (2020). *Yolov4: Optimal speed and accuracy of object detection.* arXiv preprint arXiv:2004.10934.

Brey, P. (2004). Ethical aspects of facial recognition systems in public places. *Journal of Information, Communication and Ethics in Society.*

Cabani, A., Hammoudi, K., Benhabiles, H., & Melkemi, M. (2021). MaskedFace-Net–A dataset of correctly/incorrectly masked face images in the context of COVID-19. *Smart Health (Amsterdam, Netherlands)*, *19*, 100144. doi:10.1016/j.smhl.2020.100144 PMID:33521223

Centers for Disease Control and Prevention (CDC). (2020). *Coronavirus (COVID-19)*. https://www.cdc.gov/coronavirus/2019-ncov/index.html

Chowdary, G. J., Punn, N. S., Sonbhadra, S. K., & Agarwal, S. (2020, December). Face mask detection using transfer learning of inceptionv3. In *International Conference on Big Data Analytics* (pp. 81-90). Springer, Cham. 10.1007/978-3-030-66665-1_6

Dalkiran, M. (2020). *LFW Simulated Masked Face Dataset*. https://www.kaggle.com/muhammeddalkran/lfw-simulated-masked-face-dataset

Das, A., Ansari, M. W., & Basak, R. (2020, December). Covid-19 Face Mask Detection Using TensorFlow, Keras and OpenCV. In *2020 IEEE 17th India Council International Conference (INDICON)* (pp. 1-5). IEEE.

Din, N. U., Javed, K., Bae, S., & Yi, J. (2020). A novel GAN-based network for unmasking of masked face. *IEEE Access: Practical Innovations, Open Solutions*, *8*, 44276–44287. doi:10.1109/ACCESS.2020.2977386

Dugdale, C. M., & Walensky, R. P. (2020). Filtration efficiency, effectiveness, and availability of N95 face masks for COVID-19 prevention. *JAMA Internal Medicine*, *180*(12), 1612–1613. doi:10.1001/jamainternmed.2020.4218 PMID:32780097

Eikenberry, S. E., Mancuso, M., Iboi, E., Phan, T., Eikenberry, K., Kuang, Y., Kostelich, E., & Gumel, A. B. (2020). To mask or not to mask: Modeling the potential for face mask use by the general public to curtail the COVID-19 pandemic. *Infectious Disease Modelling*, *5*, 293–308. doi:10.1016/j.idm.2020.04.001 PMID:32355904

Ejaz, M. S., Islam, M. R., Sifatullah, M., & Sarker, A. (2019, May). Implementation of principal component analysis on masked and non-masked face recognition. In *2019 1st international conference on advances in science, engineering and robotics technology (ICASERT)* (pp. 1-5). IEEE. 10.1109/ICASERT.2019.8934543

Faizah, A., Saputro, P. H., Firdaus, A. J., & Dzakiyullah, R. N. R. (2021). Implementation of the Convolutional Neural Network Method to Detect the Use of Masks. *International Journal of Informatics and Information Systems*, *4*(1), 30–37. doi:10.47738/ijiis.v4i1.75

Fasfous, N., Vemparala, M. R., Frickenstein, A., Frickenstein, L., & Stechele, W. (2021). *BinaryCoP: Binary Neural Network-based COVID-19 Face-Mask Wear and Positioning Predictor on Edge Devices*. doi:10.1109/IPDPSW52791.2021.00024

Fouquet, H. (2020). Paris Tests Face-Mask Recognition Software on Metro Riders. *Bloomberg*. https://www.bloomberg.com/news/articles/2020-05-07/paris-tests-face-mask-recognition-software-on-metro-riders

Garrigou, A., Laurent, C., Berthet, A., Colosio, C., Jas, N., Daubas-Letourneux, V., Jackson Filho, J.-M., Jouzel, J.-N., Samuel, O., Baldi, I., Lebailly, P., Galey, L., Goutille, F., & Judon, N. (2020). Critical review of the role of PPE in the prevention of risks related to agricultural pesticide use. *Safety Science*, *123*, 104527. doi:10.1016/j.ssci.2019.104527

Ghosh, S., Das, N., & Nasipuri, M. (2019). Reshaping inputs for convolutional neural network: Some common and uncommon methods. *Pattern Recognition*, *93*, 79–94. doi:10.1016/j.patcog.2019.04.009

Girshick, R. (2015). Fast r-cnn. In *Proceedings of the IEEE international conference on computer vision* (pp. 1440-1448). IEEE.

Girshick, R., Donahue, J., Darrell, T., & Malik, J. (2014). Rich feature hierarchies for accurate object detection and semantic segmentation. In *Proceedings of the IEEE conference on computer vision and pattern recognition* (pp. 580-587). 10.1109/CVPR.2014.81

Güven, J. (2019). *Investigating techniques for improving accuracy and limiting overfitting for YOLO and real-time object detection on iOS*. Academic Press.

Hammoudi, K., Cabani, A., Benhabiles, H., & Melkemi, M. (2020). *Validating the correct wearing of protection mask by taking a selfie: design of a mobile application "CheckYourMask" to limit the spread of COVID-19*. Academic Press.

Howell, A., Havens, L., Swinford, W., & Arroliga, A. (2021). PPE Effectiveness–Yes, the Buck and Virus can Stop Here. *Infection Control and Hospital Epidemiology*, 1–3. doi:10.1017/ice.2021.75 PMID:33602377

Hu, P., & Ramanan, D. (2017). Finding tiny faces. In *Proceedings of the IEEE conference on computer vision and pattern recognition* (pp. 951-959). IEEE.

Jain, V., & Learned-Miller, E. (2010). *Fddb: A benchmark for face detection in unconstrained settings* (Vol. 2, No. 4). UMass Amherst Technical Report.

Jiang, M., Fan, X., & Yan, H. (2020). *Retinamask: A face mask detector.* arXiv preprint arXiv:2005.03950.

Keras Applications. (n.d.). *Keras*. Retrieved May 8, 2021, from https://keras.io/api/applications/

Khan, S., Rahmani, H., Shah, S. A. A., & Bennamoun, M. (2018). A guide to convolutional neural networks for computer vision. *Synthesis Lectures on Computer Vision*, *8*(1), 1–207. doi:10.2200/S00822ED1V01Y201712COV015

Khojasteh, P., Júnior, L. A. P., Carvalho, T., Rezende, E., Aliahmad, B., Papa, J. P., & Kumar, D. K. (2019). Exudate detection in fundus images using deeply-learnable features. *Computers in Biology and Medicine*, *104*, 62–69. doi:10.1016/j.compbiomed.2018.10.031 PMID:30439600

Kong, X., Wang, K., Wang, S., Wang, X., Jiang, X., Guo, Y., ... Ni, Q. (2021). Real-time mask identification for COVID-19: an edge computing-based deep learning framework. *IEEE Internet of Things Journal.*

Liu, S., & Agaian, S. S. (2021, April). COVID-19 face mask detection in a crowd using multi-model based on YOLOv3 and hand-crafted features. In Multimodal Image Exploitation and Learning 2021 (Vol. 11734). International Society for Optics and Photonics.

Liu, W., Anguelov, D., Erhan, D., Szegedy, C., Reed, S., Fu, C. Y., & Berg, A. C. (2016, October). Ssd: Single shot multibox detector. In *European conference on computer vision* (pp. 21-37). Springer.

Loey, M., Manogaran, G., Taha, M. H. N., & Khalifa, N. E. M. (2021a). A hybrid deep transfer learning model with machine learning methods for face mask detection in the era of the COVID-19 pandemic. *Measurement*, *167*, 108288. doi:10.1016/j.measurement.2020.108288 PMID:32834324

Loey, M., Manogaran, G., Taha, M. H. N., & Khalifa, N. E. M. (2021b). Fighting against COVID-19: A novel deep learning model based on YOLO-v2 with ResNet-50 for medical face mask detection. *Sustainable Cities and Society*, *65*, 102600. doi:10.1016/j.scs.2020.102600 PMID:33200063

Lowe, D. G. (1999, September). Object recognition from local scale-invariant features. In *Proceedings of the seventh IEEE international conference on computer vision* (Vol. 2, pp. 1150-1157). IEEE. 10.1109/ICCV.1999.790410

Lowe, D. G. (2004). Distinctive image features from scale-invariant keypoints. *International Journal of Computer Vision, 60*(2), 91–110. doi:10.1023/B:VISI.0000029664.99615.94

Meivel, S., Devi, K. I., Maheswari, S. U., & Menaka, J. V. (2021). Real time data analysis of face mask detection and social distance measurement using Matlab. *Materials Today: Proceedings*. Advance online publication. doi:10.1016/j.matpr.2020.12.1042 PMID:33643853

Militante, S. V., & Dionisio, N. V. (2020, August). Real-Time Facemask Recognition with Alarm System using Deep Learning. In *2020 11th IEEE Control and System Graduate Research Colloquium (ICSGRC)* (pp. 106-110). IEEE. 10.1109/ICSGRC49013.2020.9232610

Nagrath, P., Jain, R., Madan, A., Arora, R., Kataria, P., & Hemanth, J. (2021). SSDMNV2: A real time DNN-based face mask detection system using single shot multibox detector and MobileNetV2. *Sustainable Cities and Society, 66*, 102692. doi:10.1016/j.scs.2020.102692 PMID:33425664

Ngan, M., Grother, P., & Hanaoka, K. (2020). *Ongoing Face Recognition Vendor Test*. FRVT.

Nixon, M., & Aguado, A. (2019). *Feature extraction and image processing for computer vision*. Academic press.

Oumina, A., El Makhfi, N., & Hamdi, M. (2020, December). Control The COVID-19 Pandemic: Face Mask Detection Using Transfer Learning. In *2020 IEEE 2nd International Conference on Electronics, Control, Optimization and Computer Science (ICECOCS)* (pp. 1-5). IEEE.

Pispati, A., Bharadwaj, K. S. S., & Apparao, M. L. V. (2020). *Orchestrating Artificial Intelligence in Exit Strategies from Lockdowns*. Academic Press.

Pollard, M. (2020). Even mask-wearers can be ID'd, China facial recognition firm says. *Reuters*, (March), 9.

Qin, B., & Li, D. (2020). Identifying Facemask-Wearing Condition Using Image Super-Resolution with Classification Network to Prevent COVID-19. *Sensors (Basel), 20*(18), 5236. doi:10.339020185236 PMID:32937867

Redmon, J., Divvala, S., Girshick, R., & Farhadi, A. (2015). *You look only once: unified real-time object detection.* arXiv preprint arXiv:1506.02640.

Redmon, J., & Farhadi, A. (2017). YOLO9000: better, faster, stronger. In *Proceedings of the IEEE conference on computer vision and pattern recognition* (pp. 7263-7271). IEEE.

Redmon, J., & Farhadi, A. (2018). *Yolov3: An incremental improvement.* arXiv preprint arXiv:1804.02767.

Ren, S., He, K., Girshick, R., & Sun, J. (2016). Faster R-CNN: Towards real-time object detection with region proposal networks. *IEEE Transactions on Pattern Analysis and Machine Intelligence*, *39*(6), 1137–1149. doi:10.1109/TPAMI.2016.2577031 PMID:27295650

Sanjaya, S. A., & Rakhmawan, S. A. (2020, October). Face Mask Detection Using MobileNetV2 in The Era of COVID-19 Pandemic. In *2020 International Conference on Data Analytics for Business and Industry: Way Towards a Sustainable Economy (ICDABI)* (pp. 1-5). IEEE. 10.1109/ICDABI51230.2020.9325631

Shen, M., Zu, J., Fairley, C. K., Pagán, J. A., An, L., Du, Z., Guo, Y., Rong, L., Xiao, Y., Zhuang, G., Li, Y., & Zhang, L. (2021). Projected COVID-19 epidemic in the United States in the context of the effectiveness of a potential vaccine and implications for social distancing and face mask use. *Vaccine*, *39*(16), 2295–2302. doi:10.1016/j.vaccine.2021.02.056 PMID:33771391

Varshni, D., Thakral, K., Agarwal, L., Nijhawan, R., & Mittal, A. (2019, February). Pneumonia detection using CNN based feature extraction. In *2019 IEEE International Conference on Electrical, Computer and Communication Technologies (ICECCT)* (pp. 1-7). IEEE. 10.1109/ICECCT.2019.8869364

Wang, Y. (2020). *Which Mask Are You Wearing? Face Mask Type Detection with TensorFlow and Raspberry Pi.* https://towardsdatascience.com/which-mask-are-you-wearing-face-mask-type-detection-with-tensorflow-and-raspberry-pi-1c7004641f1

Wang, Z., Wang, P., Louis, P. C., Wheless, L. E., & Huo, Y. (2021). *WearMask: Fast In-browser Face Mask Detection with Serverless Edge Computing for COVID-19.* arXiv preprint arXiv:2101.00784.

Wen, L., Li, X., & Gao, L. (2019). A transfer convolutional neural network for fault diagnosis based on ResNet-50. *Neural Computing & Applications*, 1–14.

Wu, X., Sahoo, D., & Hoi, S. C. (2020). Recent advances in deep learning for object detection. *Neurocomputing*, *396*, 39–64. doi:10.1016/j.neucom.2020.01.085

Yadav, S. (2020). Deep Learning based Safe Social Distancing and Face Mask Detection in Public Areas for COVID-19 Safety Guidelines Adherence. *International Journal for Research in Applied Science and Engineering Technology*, *8*(7), 1368–1375. doi:10.22214/ijraset.2020.30560

Yang, Z., Xu, W., Wang, Z., He, X., Yang, F., & Yin, Z. (2019, October). Combining yolov3-tiny model with dropblock for tiny-face detection. In *2019 IEEE 19th International Conference on Communication Technology (ICCT)* (pp. 1673-1677). IEEE. 10.1109/ICCT46805.2019.8947158

Ye, F., Ding, M., Gong, E., Zhao, X., & Hang, L. (2019, June). Tiny face detection based on deep learning. In *2019 14th IEEE Conference on Industrial Electronics and Applications (ICIEA)* (pp. 407-412). IEEE. 10.1109/ICIEA.2019.8834282

Zhu, C., Zheng, Y., Luu, K., & Savvides, M. (2017). Cms-rcnn: contextual multi-scale region-based cnn for unconstrained face detection. In *Deep learning for biometrics* (pp. 57–79). Springer. doi:10.1007/978-3-319-61657-5_3

Zhu, J., Li, D., Han, T., Tian, L., & Shan, Y. (2020, August). ProgressFace: Scale-Aware Progressive Learning for Face Detection. In *European Conference on Computer Vision* (pp. 344-360). Springer. 10.1007/978-3-030-58539-6_21

ADDITIONAL READING

Aboah, A. (2021). A vision-based system for traffic anomaly detection using deep learning and decision trees. In *Proceedings of the IEEE/CVF Conference on Computer Vision and Pattern Recognition* (pp. 4207-4212). 10.1109/CVPRW53098.2021.00475

Elgendy, M. (2020). *Deep Learning for Vision Systems*. Simon and Schuster.

Gollapudi, S. (2019). *Learn computer vision using OpenCV: with deep learning CNNs and RNNs*. Apress. doi:10.1007/978-1-4842-4261-2

He, K., Zhang, X., Ren, S., & Sun, J. (2016). Deep residual learning for image recognition. In *Proceedings of the IEEE conference on computer vision and pattern recognition* (pp. 770-778).

Kodali, R. K., & Dhanekula, R. (2021, January). Face Mask Detection Using Deep Learning. In *2021 International Conference on Computer Communication and Informatics (ICCCI)* (pp. 1-5). IEEE.

Loey, M., Manogaran, G., Taha, M. H. N., & Khalifa, N. E. M. (2021). A hybrid deep transfer learning model with machine learning methods for face mask detection in the era of the COVID-19 pandemic. *Measurement, 167*, 108288. doi:10.1016/j.measurement.2020.108288 PMID:32834324

Nagrath, P., Jain, R., Madan, A., Arora, R., Kataria, P., & Hemanth, J. (2021). SSDMNV2: A real time DNN-based face mask detection system using single shot multibox detector and MobileNetV2. *Sustainable Cities and Society, 66*, 102692. doi:10.1016/j.scs.2020.102692 PMID:33425664

O'Mahony, N., Campbell, S., Carvalho, A., Harapanahalli, S., Hernandez, G. V., Krpalkova, L., ... Walsh, J. (2019, April). Deep learning vs. traditional computer vision. In *Science and Information Conference* (pp. 128-144). Springer.

Wang, L., & Sng, D. (2015). *Deep learning algorithms with applications to video analytics for a smart city: A survey.* arXiv preprint arXiv:1512.03131.

Zoph, B., Vasudevan, V., Shlens, J., & Le, Q. V. (2018). Learning transferable architectures for scalable image recognition. In *Proceedings of the IEEE conference on computer vision and pattern recognition* (pp. 8697-8710). 10.1109/CVPR.2018.00907

KEY TERMS AND DEFINITIONS

Computer Vision: An automation technology that makes computers to gain high-level understanding from images and videos throughout acquiring, processing, analyzing, and recognizing digital data by transforming visual images into numerical or symbolic information.

Convolutional Neural Network: A typical deep learning model that is commonly used to image classification, object detection, natural language procession, and predictive analysis. Such a network structure is a regularized version of fully connected networks, which belong to the class of artificial neural network.

Deep Learning: A broad family of machine learning models based on neural networks. Typical deep learning models are deep neural networks, convolutional neural networks, recurrent neural networks, deep belief networks, and deep reinforcement learning.

Deep Transfer Learning: A optimized transfer learning method that can decrease the distribution difference between source and target domains.

Face Mask Detection: A computer-based monitor that detects whether individuals are wearing a mask or not.

Human-Computer Interaction: A study field that focused on designing the interactive tools through applying computer technologies.

Image Recognition: One of the most classical issues in computer vision, image processing, and object detection, which deals with determine whether or not an image contains specific objects, patterns, or features.

Object Detection: A computer vision technique that can recognize objects from image or video.

Chapter 7
Identifying COVID-19 Cases Rapidly and Remotely Using Big Data Analytics

ABSTRACT

Effective screening of COVID-19 enables quick and efficient diagnostic tests and further mitigates the burden on public healthcare systems. Existing smart tools in COVID-19 self-assessment can be applied as the potential solution through analyzing users' responses from either answering several questions about typical symptoms or distinguishing differences of voice patterns between healthy and infected individuals. However, such applications cannot provide a comprehensive understanding of COVID-19 identification from different angles. In this chapter, a smart app framework of the multi-angel self-assessment for COVID-19 is proposed and examined in terms of its feasibility and efficiency using a variety of cutting-edge technologies, including machine learning, unsupervised text clustering, and deep learning. The app consists of three major components that learn users' responses through symptoms, messages, and voices. Experimental results are investigated with data collected from the real world, indicating the app can identify COVID-19 cases efficiently.

INTRODUCTION

As the number of COVID-19 hospitalizations and fatal cases continuously increases, the pandemic challenges healthcare systems globally in many terms,

DOI: 10.4018/978-1-7998-8793-5.ch007

such as shortages in personal protective equipment, significant increases in hospitality demands, the lack of medical capacities, and highly extensive testing among the population. Identifying the COVID-19 cases is the first step through the whole medical decision-making process that will support in forms of determining confirmed cases, generating appropriate disease control policies, allocating healthcare resources, and making clinical decisions and treatment planning. Reverse transcriptase polymerase chain reaction (RT-PCR) tests, as one of the most validated and common-used medical diagnosis for the novel coronavirus, have been applied since the pandemic. However, major challenges remain in COVID-19 tests, for example, shortages in testing capacity, time consuming, and increasing social costs. Traditional diagnostic tests, such as molecular tests and antigen tests, require people to be tested in designated places with a person-to-person contact manner, which may increase the risks of exposure to the virus. Specifically, such a COVID-19 diagnostic test produces many problems in the lockdown scenario, e.g. difficulties of allocating test capacities, challenges of managing test queues, and risks of the increasing human mobility. Despite the at-home test kits available in pharmacies, the tests results are usually available in many days after labs receiving the sample, which may not be applicable for those who need the test results immediately. Individuals do not know under what circumstances they should be tested, typically, when they may not fully understand critical issues related to availabilities, characteristics, and strategies around the COVID-19 test. Such controversies and conflicts contribute to the increased burdens on healthcare systems and delay critical preventive measures on disease control policies.

Encumbered by exponential increase of COVID-19 cases, lockdowns, and travel restrictions, the role of effective screening technologies should be considered in responding to the pandemic crisis. Cutting-edge technologies such as big data and artificial intelligence (AI) have been substantially applied as evidence-based decision support providers in epidemiological modeling and disease control management. Predictive analytics and machine learning techniques have been widely used to assist mining insightful information from various data sources, which can be transformed through knowledge discovering processes for appropriate actions. Such big data predictive analytics frameworks based on machine learning and deep learning have been developed over hybrid data sources with real-world applications for analyzing a variety of diseases, such as dengue fever, SARS, bird flu, H1N1, and MERS

(Bansal et al., 2016; Chae et al., 2018; Souza et al., 2020). The same ideology can be potentially incorporated in designing the effective screening system against COVID-19, especially in the early stage of the outbreak, in assisting medical professionals to identify confirmed cases quickly. Several studies related to predictive approaches with features, such as clinical symptoms, laboratory testing results, and computer tomography (CT) scans, have been proposed in terms of predictive analytics of early symptoms (Tostmann et al., 2020), suspected case identifications with the AI-based diagnosis aid tool (Feng et al., 2020), and automated detectors through deep learning-based CT image analysis (Gozes et al., 2020). Despite the effectiveness of automatic identifications relied on such features, existing models based on clinical information from hospitalized patients are not applicable for the general population because of the limited sample size and the restricted hypothesis. Besides, some clinical data, such as CT images and laboratory test results, rely on medical diagnosis and assay that require individuals be tested onsite with medical professionals, thereby may not be adaptable to the principle of the effective screening for COVID-19 under the lockdown scenario.

In response to the current demand of identifying COVID-19 cases rapidly and remotely, several coronavirus self-checkers have been proposed in the form of online interactive clinical assessment platforms. Coronavirus Self-Checker, deployed by the U.S. Centers for Disease Control and Prevention (CDC), can help people to make decisions on when to seek COVID-19 testing (CDC, 2020). Google has launched the COVID-19 self-assessment tool that enables individuals to determine which types of medical care they should consider for further actions if they suspect themselves or someone else has been infected (Google, 2020). Similar tools have been deployed in the healthcare systems and governments' COVID-19 information hubs, such as Coronavirus Self-Checker and COVID-19 Vaccine FAQ (Johns Hopkins Medicine, 2020), COVID-19 Self-Assessment Tool (Holy Name Medical Center, 2020), and COVID-19 Symptom Checker (The State of New Jersey, 2020). The mechanism of the self-assessment tool is to ask series of questions related to personal information and clinical symptoms, and to provide recommended actions and risk level assessments based on the user's responses. Such self-assessment tools require users to provide a set of general information, such as age, gender, race, life-threatening symptoms (pain in chest, difficulty breathing, new disorientation, dehydration, etc.), COVID-19 cases contact history, tests history, typical COVID-19 symptoms (i.e. fever, cough, difficulty breathing, sore throat, headache, etc.), medical care and living conditions, and underlying diseases (i.e. obesity, diabetes, high blood pressure, cancer, etc.), which are formed as a frequently asked questions (FAQ) system. Although most self-checkers are online and mobile-friendly, their functionalities cannot determine whether an individual has been infected or not. Such a FAQ system is effective for informational purposes only, however, it may not be suitable for automatic diagnosis.

To the user experience perspective, a COVID-19 self-assessment tool should be highly intelligent, operationally remote, and easily interactive. Several existing applications can satisfy such requirements. For instance, the COVID-19 Checkup module embedded into the Symptomate, an online symptoms checker that can determine what kinds of disease a user may have and find out what could be causing it, has been launched to provide users with a rapid and accurate coronavirus diagnosis based on symptoms and some simple questions (Symptomate, 2020). Such an intelligent health assessment tool can determine COVID-19 infected cases using AI and expert knowledge systems, along with results from scientific studies and big data analytics. Symptomate's COVID-19 self-assessment module can be considered as an evolutional version of the FAQ system because of the support of AI and big data. However, its mechanism still relies on symptoms checking process that may not be applicable to all circumstances. One of the significant challenges is to detect asymptomatic infections since people may not suspect they are infected, thereby will not use any self-checking tools. To deal with such a problem, researchers from MIT proposed a deep learning-based model to detect asymptomatic cases through mobile phone recorded voices (Chu, 2020). Such a prototype has been incorporated into a no-cost, convenient, noninvasive prescreening app, which can identify individuals who are likely to be asymptomatic COVID-19 cases (*Record Your Cough*, n.d.). Users just need to cough into the phone after logging in, and get results on whether they have been infected. Motivated by the current challenges and demands for COVID-19 self-assessment tools, this chapter is designed to examine and investigate how to identify COVID-19 suspected and confirmed cases rapidly and effectively through AI and big data. Such a work may help data science professionals and research communities to understand several objectives as follows:

- introducing most recent applications for identifying COVID-19 cases through self-reported information.
- illustrating how artificial intelligence and big data analytics work for developing a smart tool of the multi-angle COVID-19 self-assessment incorporating the mobile app designs in a cloud-based environment.

- investigating how to implement the symptoms-based self-assessment module by analyzing clinical data with machine learning.
- leveraging unsupervised learning to detect COVID-19 suspected and confirmed cases through clustering text data extracted from messages posted by users via the communication and information sharing module.
- examining how deep learning empowers the voice-based self-testing module in terms of distinguishing between healthy and COVID-19 infected individuals.

BACKGROUND

COVID-19 is a highly transmitted disease from human to human, which is estimated that a virus carrier will infect approximately three new individuals (Liu et al., 2020), therefore, identifying infected cases plays the most critical role in terms of targeting individuals who need to be under isolation and quarantine in the early stage of the pandemic. Clinical studies in the early stage of the COVID-19 outbreak indicate that the majority of the confirmed patients have developed at least one of the symptoms, including fever (89%), cough (68%), fatigue (38%), sputum production (34%), and shortness of breath (19%), which are similar symptoms associated with other respiratory diseases (Guan et al., 2020). Therefore, the symptoms-based diagnosis is not sufficient for identifying confirmed cases as it may confuse with other common flu or cold. Medically reliable approaches for diagnosing and screening COVID-19 are nucleic acid tests and CT scans, especially when big data and machine learning can help medical professionals to interpret test and diagnosis results (Bragazzi, et al., 2020). However, such techniques still require a huge amount of resources and in-person contacts, thereby will increase the pressure of the healthcare systems and the risk of the infection when large groups of population are gathering for testing and diagnosis. Despite serval cyber-based self-assessment tools available since the beginning of the COVID-19 outbreak, their functionalities are simple and less intelligent, thereby may not satisfy the current demand.

With the development of big data analytics and AI, tracking self-reported symptoms and predicting potential COVID-19 cases become hotspots in this research domain. Collecting COVID-19 symptoms information is still challenging, whereas it is the foundation of the COVID-19 self-assessment system with AI and big data. Existing studies indicate that such data can be collected from multiple channels, including mobile phone users, publicly accessible databases, smart devices and sensors, and social media platforms. Drew et al. (2020) developed a rapid COVID-19 symptoms tracker application with clinical information collected from users across the UK and the U.S., which offers multiple data analytics on risk factors, case predictions, clinical analysis, and geographical hotspots. Similarly, a novel real-time tracking system of self-reported COVID-19 symptoms has been proposed to predict confirmed cases using data from users of a smartphone app (Menni et al., 2020). Rao & Vazquez (2020) proposed a COVID-19 quick identification framework using AI with analyzing data from a mobile phone-based web survey, which can identify the confirmed cases in the susceptible population under quarantine. A machine learning-based screening of COVID-19 diagnosis method has been presented using clinical information from the Israeli Ministry of Health database, which is publicly available (Zoabi et al., 2021). Quer et al. (2021) developed a wearable COVID-19 detector that can collect smartwatch and activity tracking data with self-reported symptoms and diagnostic results. An info-surveillance system by using machine learning with COVID-19 related information on Twitter has been proposed to classify self-reported symptoms, test availabilities and recovery information (Mackey et al., 2020). Similar studies using data from social media channels cover a range of big data analytics and machine learning algorithms for symptoms-based COVID-19 detections, such as analyzing chronological and geographical features of personal information related to COVID-19 on Twitter (Klein et al., 2020) and exploring clinical symptoms and detecting the coronavirus using Apache Spark with machine learning (Alotaibi et al., 2020).

Despite adequate studies and attempts towards constructing rapid identification tools for detecting COVID-19 cases based on self-reported information, their theoretical hypothesis is based on users or patients who suspect or confirm that they may already have COVID-19, thereby would be under the self-test by themselves based on appeared physical symptoms of the disease. Such the hypothesis may not be held for those who are asymptomatic since they have no discernible symptoms of SARS-CoV-2 (Oran & Topol, 2020). Laguarta et al. (2020) proposed a COVID-19 diagnosis system using recorded cough sounds that are transformed with leveraging acoustic biomarker feature extractors and inputted into a deep learning architecture for identifying a binary pre-screening diagnostic result. Such the model achieves high performance for asymptomatic subjects in terms of its sensitivity (100%) and specificity (83%). Research scientists are considering the extraction of COVID-19 features from respiratory sounds, not only for detecting asymptomatic cases but for pre-screening and diagnosing efficiently (Schuller et al., 2020). Besides, applying deep learning has made a significant contribution to the COVID-19 self-test objective by using voice recognition and speech signal processing. Early screening and diagnosing the novel coronavirus has been implemented by using Automatic Speech Recognition (ASR) with Recurrent Neural Network (RNN) and Long Short-term Memory (LSTM) for classifying cough, breathing, and voice of the patients (Hassan et al., 2020). Similar studies have been proposed for identifying COVID-19 cases and detecting different types of coughs using various machine learning classifiers and deep learning algorithms, such as Artificial Neural Networks (ANN) and Convolutional Neural Networks (CNN) with patients' speech signals (Bales et al., 2020; Alakus & Turkoglu, 2020). However, it is still challenging to distinguish infections between common influenza and SARS-CoV-2, as cough is always a typical symptom of many diseases. Such an issue can be addressed by investigating auditory features of coughs of COVID-19 confirmed cases via sounds and symptoms from crowdsourced data (Brown et al., 2020; Han et al., 2021). Such the ideology has been applied to developing a multi-pronged medical app composed of AI algorithms for deciphering multiple types of coughs of different diseases, including COVID-19 (Imran et al., 2020).

MAIN FOCUS OF THE CHAPTER

Issues, Controversies, Problems

Although above-mentioned apps and tools provide various functionaries for users to implement the COVID-19 self-test, the underlying model for each method only supports the limited functionality, thereby individuals cannot operate multi-angle self-diagnosis through any of the apps. A technologically intensive platform is essential and demanded given the recently urgent situation in regards to implementing the multi-dimensional self assessment tool for identifying COVID-19 cases efficiently. However, no such a tool or application exists since technical difficulties may remain in many aspects, ranging from data collection approaches, model training procedures, functionality integration methods, data flow architecture designs, to app deployments and marketing strategies. This chapter deals with the controversies and problems by introducing a mobile app developed by the Research and Development

(R&D) team at IR Big Data and AI Lab. Such an app can identify COVID-19 suspected and confirmed cases efficiently.

Features of the Chapter

The feature of this chapter aims on introducing three different modules of the proposed system focusing on discussing how to identify COVID-19 suspected and confirmed cases through AI and big data analytics. Three modules together offer a comprehensive understanding of the automatic system for the COVID-19 self-assessment application using a variety of data types, including clinical data, textual information, and human voice signals. The mobile app user interface (UI) has been designed for each module, providing a broad range of functionalities, such as the capability of the symptoms-based self-assessment, the functionality for communication and information sharing, and the mechanism of the voice-based self-testing. The symptoms-based self-assessment module has been designed as a Symptomate-like UI that allows users to evaluate whether they have infected the disease or not by answering serval questions. Questions are generated by analyzing clinical data from hospitalized patients and suspected cases. A smart tool will learn users' responses through a pre-trained model using machine learning. The communication and information sharing module has been designed as a WhatsApp-like UI that allows users to communicate in real-time and offline. The expected functionality of the module allows users to post text and voice messages, share images and documents, and make real-time voice/video calls, whereas initializing the app functionality only concentrated on textual information due to the limited development capacities. The R&D team will then analyze the beta test data that contains signal messages in regards to COVID-19. Users with typical symptoms and/or test positive will be identified automatically through an unsupervised learning algorithm. The voice-based self-testing module records the user's voice through coughing, breathing, and reading a few pre-defined sentences. Such recordings are processed by a deep learning model that has been trained through an existing COVID-19 voice data collection. The module is expected to distinguish the sounds of COVID-19 suspected cases from healthy individuals. Each module will be investigated and discussed throughout the data acquisition, experimental designs, methodologies, experimental results, and potential improvements as the solution and recommendation for implementing a multi-angle assessment system for the COVID-19 self-testing capability.

SOLUTIONS AND RECOMMENDATIONS

Symptoms-based Self-Assessment Module

The symptoms-based self-assessment module relies on deriving the most significant symptoms and factors associated with COVID-19 by analyzing clinical information from patients. Analyzing clinical data and predicting COVID-19 confirmed cases offer an effective solution towards constructing the self-assessment tool via checking symptoms and other critical factors. However, patients' information are sensitive, thereby will not be publicly accessible. Moreover, samples from hospitalized individuals are not effective in screening for COVID-19 due to the limited size of recorded data, hence, analyzing clinical data from hospital or laboratory tests may not be adoptable in the general population. The Israeli Ministry of Health (IMH) provides a dataset which covers individual records of SARS-CoV-2 results through RT-PCR assay of a nasopharyngeal swab. The dataset is publicly available with clinical information recorded on a daily basis and in nationwide scope. Additional information is also accessible, involving clinical symptoms and signs, gender, and age. The age group is divided into two categories based on whether the tested person is under or above 60 years old. Input variables for model training are eight binary features, such as gender (male/female), age 60+ (true/false), cough (true/false), fever (true/false), sore throat (true/false), shortness of breath (true/false), headache (true/false) and known contact with an individual confirmed to have COVID-19 (true/false). A binary response of test results for either positive or negative is given for each record in the dataset. Overall, it contains totally 100,000 records of tested persons, of whom 7452 (7.45%) are test positive. Among the tested positive group, 4152 (55.7%) are male; 6027 (80.9%) are under 60 years old; Numbers (ratios) of appeared symptoms including cough, fever, sore throat shortness of breath, and headache are 3345 (44.9%), 2834 (38.0%), 1177 (15.8%), 550 (7.4%), and 1223 (16.4%), respectively; 3640 (48.8%) are known contact with a person who has confirmed the virus..

To the analytics perspective, such dataset can be applied for both feature importance analysis through statistical inferences and machine learning-based prediction of COVID-19 diagnosis. For instance, Zoabi et al. (2021) performed a machine learning approach to predict COVID-19 cases. The original dataset has been divided into an 80% training set and a 20% validation set. A predictive model has been implemented through utilizing decision tree model with the combination of gradient boosting. The gradient-boosting predictor is pre-trained by the LightGBM Python module and can deal with the missing values for the model training. The model testing is performed on the validation set with area under the receiver operating characteristic curve (auROC) as the model performance standard. Furthermore, feature importance analysis is conducted through calculating SHapley Additive exPlanations (SHAP) values, which benefits complex models. SHAP values indicates the impact of having a specific value for a given feature in comparison to the overall model predictions. Different metrics, such as sensitivity, specific, false-negative rate, false-positive rate, and false discovery rate, have been measured to evaluate the model performance. A bootstrap percentile strategy has been implemented as the re-sampling method to estimate the confidence intervals for the different performance measures. In fact, observations provided by IMH are unbalanced. Records are more exhaustive for those who have been confirmed for COVID-19, while the records for those who tested negative have mislabeling clinical signs and symptoms. Mislabeling is possibly caused by an underestimation of symptoms among those who tested negative. After filtering out symptoms with high bias, the model proposed by authors of the article can reach an auROC of 0.862. This chapter references the same experimental design with different samples obtained from IMH. An auROC of 0.88 has been achieved, which is higher comparing with the result from the existing study, with a slight change in the SHAP, as shown in Figure 1.

Figure 1. ROC curve and SHAP plot for the prospective test set through proposed model training. (A). ROC curve with AUC score. (B). SHAP beeswarm plot where factor "test_indication" and "age_60_and_above" indicate "Contact with confirmed cases" and "Age 60+", respectively.

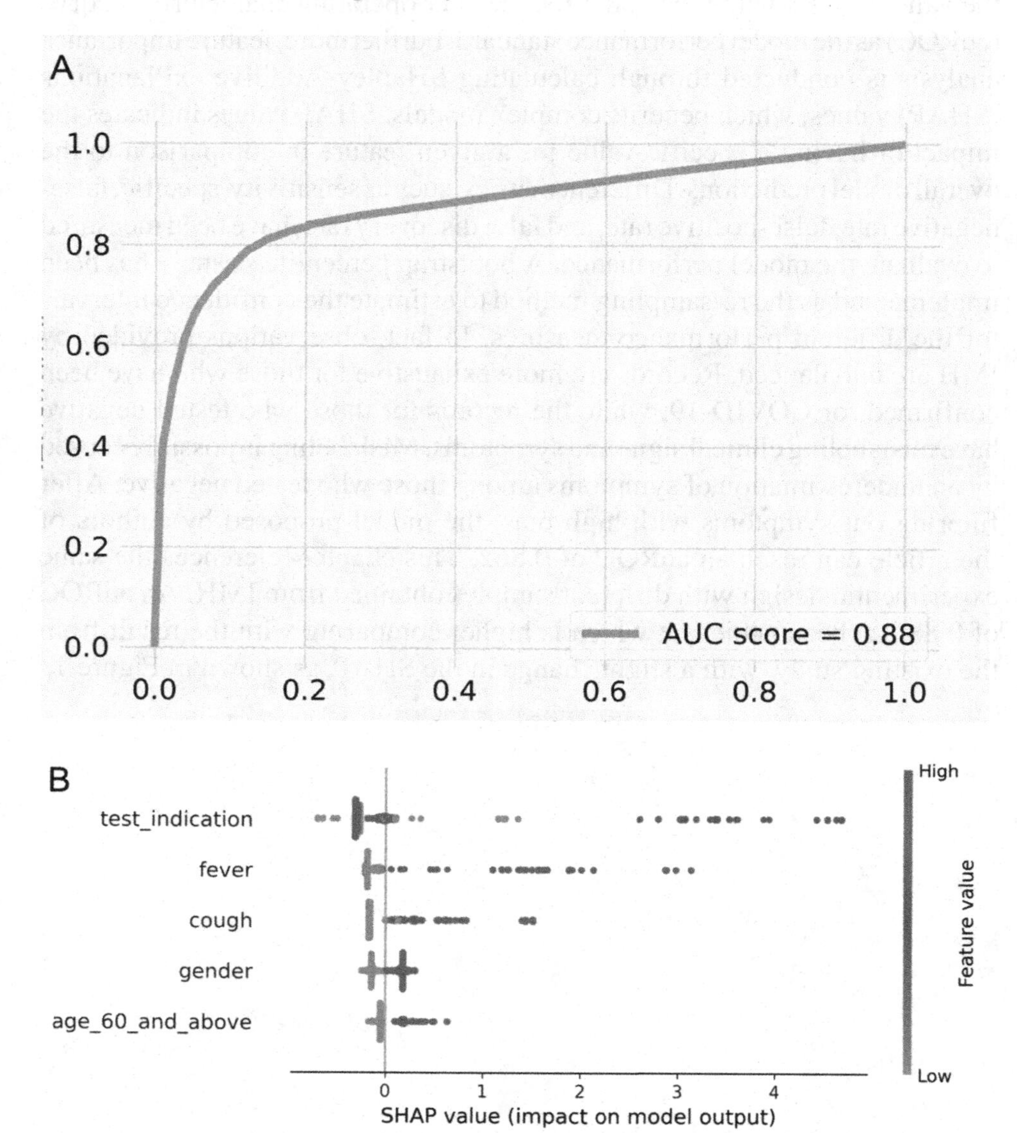

Such a predictive model can be applied for implementing the self-test screening in terms of generating questions for users and identifying COVID-19

cases based on users' responses. Nevertheless, one of the biggest challenges among the experimental design is to fix the unbalance property from the original dataset. Expanding the sample size can solve the bias problem for the data. For example, cough and fever are the major symptoms given by the IMH guidelines. Such attributes are also common symptoms for the group of people who tested negative to coronavirus. The COVID-19 diagnosis prediction system still requires additional samples to improve the whole framework. The public organizations play an important role of the communication of data records to the scientific community. The symptom impact analysis can help to find out the additional symptoms as symptom features for the model in the next stage.

Communication and Information Sharing Module

As per the functionality of the communication and information sharing module, textual information in regards to users' posts can be derived from the cloud-based database. Such a database has been protected as it contains personal information. The R&D team can only access the beta testing back-end for operating the data analytical tasks in cloud, thereby no data will be downloaded or released. Such a dataset, containing totally 34,861 message records, has been collected between May 19, 2020 and April 20, 2021, emphasized on COVID-19-related keywords, including "corona", "coronavirus", "coronavid19", and "covid19". Such keywords have been chosen based upon previous studies according to the association with the COVID-19-related social media conversations (Shoaei & Dastani, 2020; Garcia & Berton, 2021). A subset of 16,308 records has been extracted through filtering the terms that are related to COVID-19 symptoms, testing, and recovery conversations. Such terms include "diagnosed", "fever", "test", "testing kit", "pneumonia", "symptoms", "isolating", "cough", "ER", and "emergency room". Furthermore, hashtags, stop words, and other unnecessary data have been removed from the subset for the better performance in data analytics.

The target is to determine who has been infected by COVID-19 by analyzing textual data from samples in the subset. As an unsupervised machine learning method, Biterm Topic Model (BTM) has been applied to extract themes to fulfill the detection of substance abuse disorder and other public health problems. The messages with the same word-related themes have been grouped into one cluster. The topic of each theme will be gathered into a bag of words and then the discrete probability distribution of themes can be analyzed. The BTM model classifies topic clusters based on word phrases, frequencies, and characteristics, which are relevant to COVID-19-related symptoms, testing, and recovery. Finally, each record has been manually annotated corresponding to the topic clusters. The number of topic clusters is selected by calculating a coherence score to measure the correlation among the messages in the same cluster. If the messages in the cluster are significantly correlated to each other, the coherence score will be higher. The number of the clusters are chosen at 3, 4, 6, 8, and 10, respectively. The optimal value of clusters can be determine based the comparison of coherence score. The coherence score is calculation based on the formula as follows,

$$C\left(t;v^{t}\right)=\sum_{m=2}^{M}\sum_{l=1}^{m-1}\log\left(\frac{D\left(v_{t}^{m},v_{t}^{l}\right)+1}{D\left(v_{l}^{t}\right)}\right)$$

where $C\left(t;v^{t}\right)$ represents the u-mass coherence score, $D(v)$ represents the document frequency for the word type v, and D(v,v0) r$_e$presents the co-document frequency of the word types v and v0. $v\left(t\right)=\left(v_{t}^{1},\ldots,v_{t}^{m}\right)$ i$_s$ defined as a set of topic words with high probability.

Experimentally, the topic clusters have been identified with 5,367 message records from 1,057 active users, which will be further labeled by BTM for detecting the signal messages. The signal messages are divided into four thematic groups, such as self-reporting of symptoms, reporting of symptoms for others, user confirmation COVID-19 positive after testing, and user confirmation COVID-19 negative after testing, as shown in Table 1. Each message has been categorized into two major groups in regards to either symptoms reports and test results. Each message will send a unique ID associated with the users, which helps to track. Manual annotation of 500 messages has been conducted by the R&D team. The optimal number of clusters can be determined as 4. BTM is then applied to identify relevant topic clusters. Finally, 3,712 records posted by 491 different users have been recognized as signal messages associated with the COVID-19-related symptoms, whereas only 121 messages posted by 25 users have been determined as the group of reporting of symptoms for others. For the category of test results, 1,131 messages from 529 users are identified as COVID-19 tested negative, while only 403 messages from 12 users as the confirmed cases have been detected.

Table 1. Numbers of messages related to COVID-19 symptoms, testing, and recovery

Message Category	Theme	Number of Massages	Number of Users
Symptoms reports	Self-reporting of symptoms	3,712	491
	Reporting of symptoms for others	121	25
Test results	User confirmation COVID-19 positive after testing	403	12
	User confirmation COVID-19 negative after testing	1,131	529

Such a BTM-based approach can be deployed for monitoring COVID-19 cases effectively by tracking messages posted through the communication and information sharing module. However, such the method should be improved by increasing the sample size as well as the records for the manual annotation process. As the beta test process is ongoing, more data will be available in the future. Furthermore, users may post the messages that are totally opposite from one to another, therefore, checking the reliability of the users' responses is essential for the next stage of the improvement towards the experimental design.

Voice-based Self Testing Module

Key symptoms, such as dry cough, difficulty in breathing, and loss of speech, have been characterized as the significant signs of COVID-19 by WHO and CDC. Early studies confirmed such conclusion, i.e. a data pool containing 7,178 SARS-CoV-2 positive cases validated the appearance of such symptoms (Menni et al., 2020). Explored by the medical professionals, understanding human voice features such as speech breathing patterns and cough sounds is intricately tied to the change in pathology of the respiratory system (HuBerAND & Stathopoulos, 2015). As a respiratory disease, COVID-19 can be detected by analyzing the patients' voice characteristics. The cough sounds, breath sounds and voice sounds can be used to development voice diagnosis using machine learning techniques. Cough is a reflex mechanism, which can help clear the inhaled and exhaled material in the trachea and main stem bronchi. This reflex is acted by cough receptors. The cough sounds can be generated by air turbulence, vibration between the tissues, and fluid movement in the airways (Polverino et al., 2012). Cough sounds usually experience three phases, including explosive phase, intermediate phase, and voicing phase. Respiratory infections would change the physical structure of the respiratory system. Thus, based on cough sounds, the pathological status can be identified to further make diagnosis. Additionally, another symptom for the most COVID-19 confirmed cases is breathing difficulty. It is possible to cause a spectrogram difference of breath sounds between the healthy and confirmed respondents. The spectrograms of breath sounds of respiratory cycles from a healthy person are usually wide-band spectrograms, which the spectrograms from a COVID-19 positive case would show weakened band spectrograms. Furthermore, the diseases of respiratory system can lead to bio-marker changes in the speech breathing cycles. Vocal fatigue can influence the phonation threshold pressure. The laryngeal dysfunction of speech can be observed through the spectrograms of sustained vowels.

Sharma et al. (2020) launched a voice recognition project, namely Coswara, for the COVID-19 diagnosis. The respiratory sounds were collected from healthy and unhealthy people through crowdsourcing using a web application. The collected sounds were stored in a database and then the database was released for open access. The sound data include nine different types of audios for each respondent, including cough shallow, cough heavy, breathing shallow, breathing deep, vowel /ey/, vowel /i/, vowel /u:/, counting normal, and counting fast. Besides, some related information for each individual was also collected. It covers age, gender, country, state, covid status, test status, vaccine test status, cough, test type, test date, hypertension, ischemic heart disease, asthma, cold, fatigue, others preexist, sore throat, fever, loss of smell, diabetes, data of CT scan, CT score, and diarrhoea. It can be further used for data analysis and classification modeling to fulfill the respiratory disease identification for COVID-19 diagnosis. Based on the evaluation of the model performance, the application will give diagnosis result and even a COVID-19 alert when it recognizes a positive case. The sound data include nine different types of audios for each respondent, including cough shallow, cough heavy, breathing shallow, breathing deep, vowel /ey/, vowel /i/, vowel /u:/, counting normal, and counting fast.

An experiment of voice diagnosis is performed using the data released for Jun 18, 2021 from Coswara dataset, which contained 56 respondents and nine audio records for each respondent. Only 30 respondents have certain records of test status and they are either positive or negative. The audio information of these 30 respondents is used for audio classification through machine learning technique. This experiment only used seven types of audio, related to cough, breathing, and vowel speech. Such a dataset can be applied for initializing a voice-based COVID-19 monitor through recognizing the differences of sounds between healthy and infected individuals. Visualizing voice patterns is the first step towards such attempt. The spectrograms of COVID-19 positive and negative cases are analyzed as presented in Figure 2. It can be observed that some low-frequency sounds are recognized, following a cough of the COVID-19 positive patient.

Figure 2. Spectrogram of COVID-19 recordings from healthy and infected individuals. (A). a typical cough pattern for the positive case. (B). a typical cough pattern for the negative case. (C). a typical breath pattern for the positive case. (D). a typical breath pattern for the negative case. (E). a typical vowel pattern for the positive case. (F). a typical vowel pattern for the negative case.

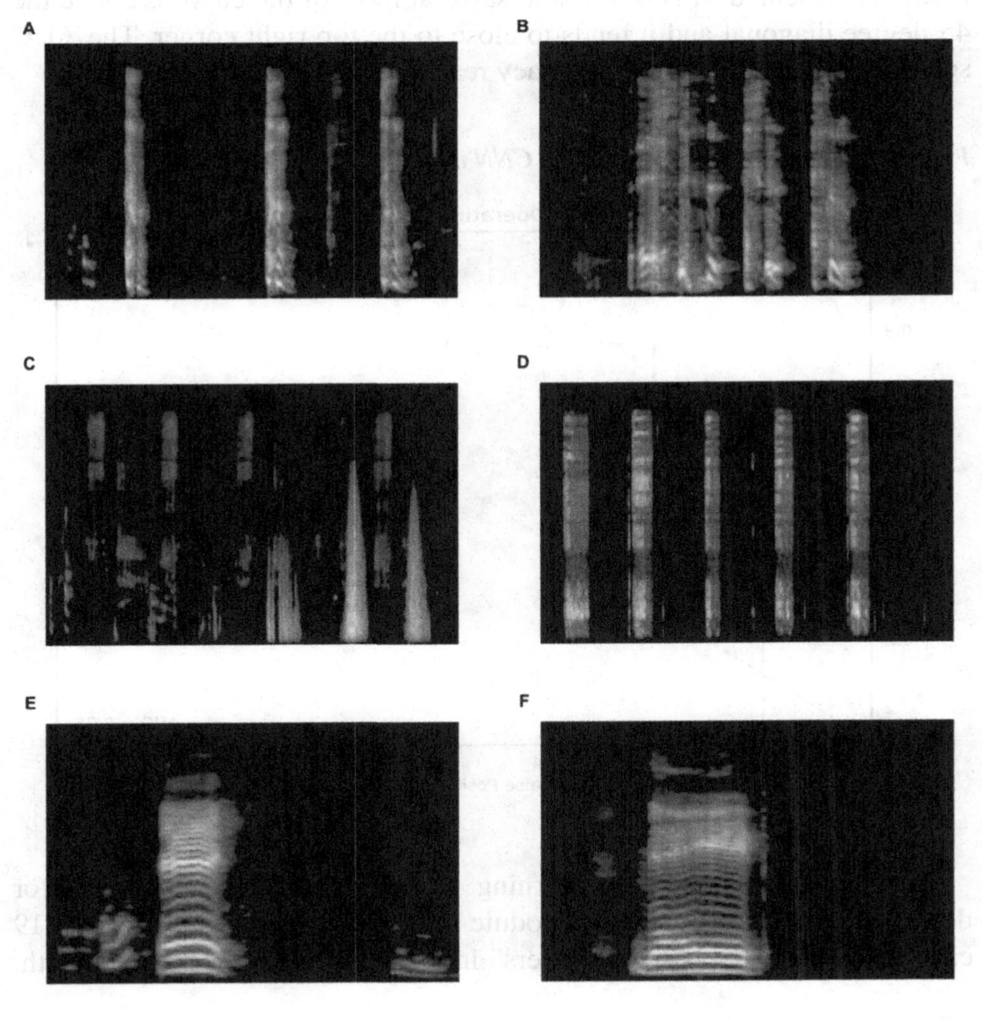

In order to implement the voice diagnosis using machine learning, the spectral variables were extracted as new features from the underlying audio data. The goal of this process is to find out the distinguishable components of an audio signal. The Fourier Transform method was used to convert the time-based signal into the frequency-based features. A Python module, called LibROSA, can be used to conduct the signal processing. The extracted spectral variables include spectral centroid, spectral rolloff, spectral bandwidth, zero-crossing rate, Mel-frequency Cepstral Coefficients (MFCCs), and chroma feature. After feature extraction, totally 210 rows of feature data were generated. The whole feature data was divided into 80% for training set and 20% for testing set. Base on remarkable model performance, CNN model was used to further training the audio classifier. In Figure 3, the ROC curve of CNN model is presented. It is obvious to say that most of the curve is above the 45-degree diagonal and it tends to close to the top right corner. The AUC score achieved 0.75 and the accuracy reached to 76.19%.

Figure 3. ROC curve for training the CNN model

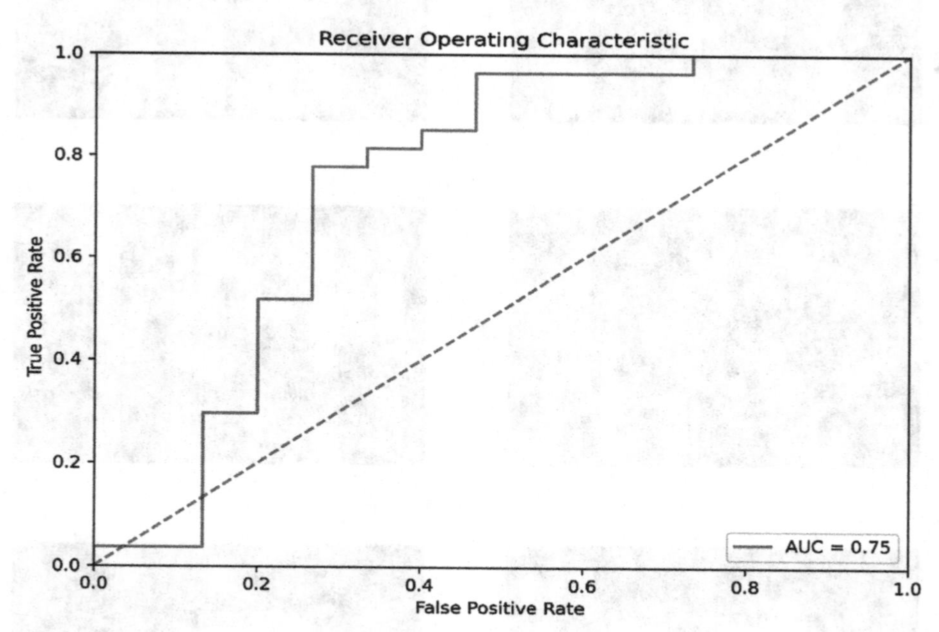

Such a voice-based deep learning architecture can be employed for deploying a self-test screening module in terms of identifying COVID-19 cases automatically based on users' interactions through cough, breath, and conversations. To the application initialization perspective, the model performance can be acceptable. However, a lower accuracy cannot satisfy the requirement of the real-world application. Therefore, improving the model performance seems urgent and demanded. Alternative neural network architectures, such as LSTM, RNN, etc., can be incorporated and investigated. Moreover, the model performance may be increased by fine-tuning the model structure. Besides, collecting more samples may also help the model training and testing process, whereas such a work should be ongoing.

FUTURE RESEARCH DIRECTIONS

Improving the accuracy of the machine learning model applied to the symptoms-based self-assessment module can be the potential research directions in terms of expanding the sample size and solving the bias problem of the dataset. To the predictive analysis perspective, such a diagnostic system relies on adding more high-quality data to increase the model performance. Public authorities play a critical role of the communication in data recording and sharing. Besides, the symptom impact analysis can help to search the unexpected signs and other untypical symptoms as new features that can be used in the predictive models. Such an attempt will not only benefit the improvement of the COVID-19 identification, but help to explore new patterns of the disease as the coronavirus may variate. However, such an idea depends on whether real-time information retrievals can be implemented or not. To the technical perspective, analyzing data derived from the communication and information sharing module can be the supplementation in the form of a new channel to collect update information through the real-time stream from the discussion board of the app. Therefore, implementing a real-time architecture of the data flow by using big data tools (i.e. Spark and Hadoop) will be the next stage of the development. On the other hand, collecting extra variables (i.e. travel history, social status, economic conditions, etc.) may also benefit the model performance, whereas it will require additional data collection strategies, such as web-based questionnaires, mobile phone-based surveys, and phone-call interviews. Such a data collection procedure will be ongoing, and the developers will improve the robustness of the predictive models by training on a larger pool of users.

Despite the effectiveness of the communication and information sharing module, several challenges still remain. First of all, the data collection process is less diversity. The development team only collected data from a single platform with the limited sample size for the annotation process. Such a data collection strategy may cause the experimental results to be biased. Future studies should focus on extending data collection and analytic approaches to multiple channels of the user, i.e. social media activities, diagnostic reports, online surveys, medical records, etc. Furthermore, although the application on a large corpus of words using BTM is effective as the major computational method for clustering users' conversations, such a model relies on the precise annotation to determine whether messages contain a COVID-19 signal or not, which is particularly critical to remove false positive responses from outputs calculated by the model. Further attempts should consider a supervised learning approach based on feature extractions with classifications of conversation characteristics from the user responses.

A user-friendly tool should consider multiple functionalities for users' expected demands as much as possible. For example, rapid test kits have been widely used across pathologies for identifying SARS-CoV-2, which are available for at-home tests as well. However, seemingly easy-to-determine test results are highly subjective due to the interpretations of the visible outputs appeared in the test window that may vary between individuals and products. To deal with such a problem, an AI-based smartphone app, called xRCovid, has been developed using machine learning to identify COVID-19 results from serological rapid diagnostic tests in the form of eliminating human reading ambiguities (Mendels et al., 2021). Using xRCovid can assist people to operate the COVID-19 self-assessment based on interpreting the rapid tests results confidently. Besides, checking body temperature has also become an important factor in identifying whether a person is healthy or not. Most recently, a low cost fever detection system has been developed by the Artificial Intelligence and Visualizations (AiVi) team at North Carolina A&T State University, concentrated on creating an automatic screening framework for checking body temperatures (Ruffin et al., 2021). Similarly, Farady et al. (2020) proposed a mask detection and head temperature classification algorithm with deep learning networks, in which the RetinaNet has been applied to identifying the body temperature by recognizing images from thermal cameras. Both xRCovid and fever detection models can be implemented with image processing and deep learning, which will indicate the further studies towards such a direction.

CONCLUSION

This chapter investigates several state-of-the-art techniques that can be applied to identifying COVID-19 suspected and confirmed cases using self-reporting information from users' responses through the proposed mobile app, which consists of three components, such as the symptoms-based self-assessment module, the communication and information sharing module, and the voice-based self-testing module. Each module has been illustrated and examined through presenting and discussing the feasibility, the data collection process, the experimental design, and potential improvements. Each module has been examined by implementing the data-driven approaches, including supervised learning and unsupervised learning, along with real-world datasets. For the symptoms-based self-assessment module, questions can be created by training a gradient-boosting predictor and by evaluating the factor important matrix through SHAP. Based on clinical data from the IMH, the experimental results indicate that the machine learning-based approach can be applied to predict COVID-19 infections effectively as per users' response by answering several questions generated by analyzing the SHAP plot, with a higher auROC score of 0.88 comparing with the previous study. Moreover, text information derived from the communication and information sharing module contains features and attributes in regards to COVID-19 symptoms and test results, whereas the BTM model can be applied to cluster suspected and confirmed groups based on analyzing the messages collected from the beta testing process. Furthermore, the voice-based self-testing module has been proposed to detect COVID-19 cases through learning the differences of sound patterns between healthy and infected individuals. Based on a voice dataset designed for COVID-19 detection, the experimental result obtains the AUC score and the accuracy at 0.75 and 76.19% respectively, which is acceptable for implementing the module in the sandbox stage. In general, each module detects COVID-19 cases rapidly and effectively, which can be implemented for the multi-angle diagnosis of COVID-19 based on self-reporting responses from users. On the other hand, the app can be improved by collecting extra data, training alternative models, and adding extra modules, hence such works form the opportunity of the future studies.

REFERENCES

Alakus, T. B., & Turkoglu, I. (2020). Comparison of deep learning approaches to predict COVID-19 infection. *Chaos, Solitons, and Fractals*, *140*, 110120. doi:10.1016/j.chaos.2020.110120 PMID:33519109

Alotaibi, S., Mehmood, R., Katib, I., Rana, O., & Albeshri, A. (2020). Sehaa: A big data analytics tool for healthcare symptoms and diseases detection using Twitter, Apache Spark, and Machine Learning. *Applied Sciences (Basel, Switzerland)*, *10*(4), 1398. doi:10.3390/app10041398

Bales, C., Nabeel, M., John, C. N., Masood, U., Qureshi, H. N., Farooq, H., ... Imran, A. (2020, October). Can machine learning be used to recognize and diagnose coughs? In *2020 International Conference on e-Health and Bioengineering (EHB)* (pp. 1-4). IEEE. 10.1109/EHB50910.2020.9280115

Bansal, S., Chowell, G., Simonsen, L., Vespignani, A., & Viboud, C. (2016). Big data for infectious disease surveillance and modeling. *The Journal of Infectious Diseases*, *214*(suppl_4), S375–S379. doi:10.1093/infdis/jiw400 PMID:28830113

Bragazzi, N. L., Dai, H., Damiani, G., Behzadifar, M., Martini, M., & Wu, J. (2020). How big data and artificial intelligence can help better manage the COVID-19 pandemic. *International Journal of Environmental Research and Public Health*, *17*(9), 3176. doi:10.3390/ijerph17093176 PMID:32370204

Brown, C., Chauhan, J., Grammenos, A., Han, J., Hasthanasombat, A., Spathis, D., ... Mascolo, C. (2020, August). Exploring automatic diagnosis of covid-19 from crowdsourced respiratory sound data. In *Proceedings of the 26th ACM SIGKDD International Conference on Knowledge Discovery & Data Mining* (pp. 3474-3484). 10.1145/3394486.3412865

CDC. (2020). *Coronavirus Self-Checker*. https://www.cdc.gov/coronavirus/2019-ncov/symptoms-testing/coronavirus-self-checker.html

Chae, S., Kwon, S., & Lee, D. (2018). Predicting infectious disease using deep learning and big data. *International Journal of Environmental Research and Public Health*, *15*(8), 1596. doi:10.3390/ijerph15081596 PMID:30060525

Chu, J. (2020, October 29). *Artificial intelligence model detects asymptomatic Covid-19 infections through cellphone-recorded coughs: Results might provide a convenient screening tool for people who may not suspect they are infected.* MIT News Office. https://news.mit.edu/2020/covid-19-cough-cellphone-detection-1029

Drew, D. A., Nguyen, L. H., Steves, C. J., Menni, C., Freydin, M., Varsavsky, T., ... Chan, A. T. (2020). Rapid implementation of mobile technology for real-time epidemiology of COVID-19. *Science*, *368*(6497), 1362–1367. doi:10.1126cience.abc0473 PMID:32371477

Farady, I., Lin, C. Y., Rojanasarit, A., Prompol, K., & Akhyar, F. (2020, September). Mask Classification and Head Temperature Detection Combined with Deep Learning Networks. In *2020 2nd International Conference on Broadband Communications, Wireless Sensors and Powering (BCWSP)* (pp. 74-78). IEEE. 10.1109/BCWSP50066.2020.9249454

Feng, C., Huang, Z., Wang, L., Chen, X., Zhai, Y., Zhu, F., & Li, T. (2020). *A novel triage tool of artificial intelligence assisted diagnosis aid system for suspected COVID-19 pneumonia in fever clinics.* MedRxiv.

Garcia, K., & Berton, L. (2021). Topic detection and sentiment analysis in Twitter content related to COVID-19 from Brazil and the USA. *Applied Soft Computing*, *101*, 107057. doi:10.1016/j.asoc.2020.107057 PMID:33519326

Google. (2020). *COVID-19 self-assessment.* https://landing.google.com/screener/covid19

Gozes, O., Frid-Adar, M., Greenspan, H., Browning, P. D., Zhang, H., Ji, W., . . . Siegel, E. (2020). *Rapid ai development cycle for the coronavirus (covid-19) pandemic: Initial results for automated detection & patient monitoring using deep learning ct image analysis.* arXiv preprint arXiv:2003.05037.

Guan, W. J., Ni, Z. Y., Hu, Y., Liang, W. H., Ou, C. Q., He, J. X., Liu, L., Shan, H., Lei, C., Hui, D. S. C., Du, B., Li, L., Zeng, G., Yuen, K.-Y., Chen, R., Tang, C., Wang, T., Chen, P., Xiang, J.,... Zhong, N. S. (2020). Clinical characteristics of coronavirus disease 2019 in China. *The New England Journal of Medicine*, *382*(18), 1708–1720. doi:10.1056/NEJMoa2002032 PMID:32109013

Han, J., Brown, C., Chauhan, J., Grammenos, A., Hasthanasombat, A., Spathis, D., . . . Mascolo, C. (2021, June). Exploring Automatic COVID-19 Diagnosis via voice and symptoms from Crowdsourced Data. In *ICASSP 2021-2021 IEEE International Conference on Acoustics, Speech and Signal Processing (ICASSP)* (pp. 8328-8332). IEEE. 10.1109/ICASSP39728.2021.9414576

Hassan, A., Shahin, I., & Alsabek, M. B. (2020, November). Covid-19 detection system using recurrent neural networks. In *2020 International Conference on Communications, Computing, Cybersecurity, and Informatics (CCCI)* (pp. 1-5). IEEE. 10.1109/CCCI49893.2020.9256562

Holy Name Medical Center. (2020). *Covid-19 Self Assessment Tool*. https://www.holyname.org/covid19/covid-19-self-assesment-tool.aspx

Huberand, J., & Stathopoulos, E. T. (2015). 2 Speech Breathing Across the Life Span and in Disease. The Handbook of Speech Production, 13.

Imran, A., Posokhova, I., Qureshi, H. N., Masood, U., Riaz, M. S., Ali, K., John, C. N., Hussain, M. D. I., & Nabeel, M. (2020). AI4COVID-19: AI enabled preliminary diagnosis for COVID-19 from cough samples via an app. *Informatics in Medicine Unlocked*, *20*, 100378. doi:10.1016/j.imu.2020.100378 PMID:32839734

Johns Hopkins Medicine. (2020). *Coronavirus Self-Checker and COVID-19 Vaccine FAQ*. https://www.hopkinsmedicine.org/coronavirus/covid-19-self-checker.html

Klein, A. Z., Magge, A., O'Connor, K. M., Cai, H., Weissenbacher, D., & Gonzalez-Hernandez, G. (2020). *A Chronological and Geographical Analysis of Personal Reports of COVID-19 on Twitter*. MedRxiv; doi:10.1101/2020.04.19.20069948

Laguarta, J., Hueto, F., & Subirana, B. (2020). COVID-19 Artificial Intelligence Diagnosis using only Cough Recordings. *IEEE Open Journal of Engineering in Medicine and Biology*, *1*, 275–281. doi:10.1109/OJEMB.2020.3026928 PMID:34812418

Liu, Y., Gayle, A. A., Wilder-Smith, A., & Rocklöv, J. (2020). The reproductive number of COVID-19 is higher compared to SARS coronavirus. *Journal of Travel Medicine*, *27*(2), taaa021. doi:10.1093/jtm/taaa021 PMID:32052846

Mackey, T., Purushothaman, V., Li, J., Shah, N., Nali, M., Bardier, C., Liang, B., Cai, M., & Cuomo, R. (2020). Machine learning to detect self-reporting of symptoms, testing access, and recovery associated with COVID-19 on Twitter: Retrospective big data infoveillance study. *JMIR Public Health and Surveillance*, *6*(2), e19509. doi:10.2196/19509 PMID:32490846

Mendels, D. A., Dortet, L., Emeraud, C., Oueslati, S., Girlich, D., Ronat, J. B., Bernabeu, S., Bahi, S., Atkinson, G. J. H., & Naas, T. (2021). Using artificial intelligence to improve COVID-19 rapid diagnostic test result interpretation. *Proceedings of the National Academy of Sciences of the United States of America*, *118*(12), e2019893118. doi:10.1073/pnas.2019893118 PMID:33674422

Menni, C., Valdes, A. M., Freidin, M. B., Sudre, C. H., Nguyen, L. H., Drew, D. A., Ganesh, S., Varsavsky, T., Cardoso, M. J., El-Sayed Moustafa, J. S., Visconti, A., Hysi, P., Bowyer, R. C. E., Mangino, M., Falchi, M., Wolf, J., Ourselin, S., Chan, A. T., Steves, C. J., & Spector, T. D. (2020). Real-time tracking of self-reported symptoms to predict potential COVID-19. *Nature Medicine*, *26*(7), 1037–1040. doi:10.103841591-020-0916-2 PMID:32393804

Oran, D. P., & Topol, E. J. (2020). Prevalence of asymptomatic SARS-CoV-2 infection: A narrative review. *Annals of Internal Medicine*, *173*(5), 362–367. doi:10.7326/M20-3012 PMID:32491919

Polverino, M., Polverino, F., Fasolino, M., Andò, F., Alfieri, A., & De Blasio, F. (2012). Anatomy and neuro-pathophysiology of the cough reflex arc. *Multidisciplinary Respiratory Medicine*, *7*(1), 1–5. doi:10.1186/2049-6958-7-5 PMID:22958367

Quer, G., Radin, J. M., Gadaleta, M., Baca-Motes, K., Ariniello, L., Ramos, E., Kheterpal, V., Topol, E. J., & Steinhubl, S. R. (2021). Wearable sensor data and self-reported symptoms for COVID-19 detection. *Nature Medicine*, *27*(1), 73–77. doi:10.103841591-020-1123-x PMID:33122860

Rao, A. S. S., & Vazquez, J. A. (2020). Identification of COVID-19 can be quicker through artificial intelligence framework using a mobile phone–based survey when cities and towns are under quarantine. *Infection Control and Hospital Epidemiology*, *41*(7), 826–830. doi:10.1017/ice.2020.61 PMID:32122430

Record Your Cough. (n.d.). *MIT Covid-19 Initiative*. Retrieved June 10, 2021, from https://opensigma.mit.edu/

Ruffin, T., Steele, J., Acquaah, Y., Sharma, N., Sarku, E., Tesiero, R., & Gokaraju, B. (2021, March). Non-invasive Low Cost Fever Detection Systems. In *SoutheastCon 2021* (pp. 1–7). IEEE. doi:10.1109/SoutheastCon45413.2021.9401821

Schuller, B. W., Schuller, D. M., Qian, K., Liu, J., Zheng, H., & Li, X. (2020). *Covid-19 and computer audition: An overview on what speech & sound analysis could contribute in the sars-cov-2 corona crisis.* arXiv preprint arXiv:2003.11117.

Sharma, N., Krishnan, P., Kumar, R., Ramoji, S., Chetupalli, S. R., Ghosh, P. K., & Ganapathy, S. (2020). *Coswara—A Database of Breathing, Cough, and Voice Sounds for COVID-19 Diagnosis.* arXiv preprint arXiv:2005.10548. doi:10.21437/Interspeech.2020-2768

Shoaei, M. D., & Dastani, M. (2020). The role of twitter during the COVID-19 crisis: A systematic literature review. *Acta Informatica Pragensia*, *9*(2), 154–169. doi:10.18267/j.aip.138

Souza, J., Leung, C. K., & Cuzzocrea, A. (2020, April). An innovative big data predictive analytics framework over hybrid big data sources with an application for disease analytics. In *International Conference on Advanced Information Networking and Applications* (pp. 669-680). Springer. 10.1007/978-3-030-44041-1_59

Symptomate. (2020). *Symptomate provides you with a fast and accurate health assessment.* https://symptomate.com/

The State of New Jersey. (2020). *New Jersey COVID-19 Information Hub: COVID-19 Symptom Checker.* https://covid19.nj.gov/forms/self

Tostmann, A., Bradley, J., Bousema, T., Yiek, W. K., Holwerda, M., Bleeker-Rovers, C., ten Oever, J., Meijer, C., Rahamat-Langendoen, J., Hopman, J., van der Geest-Blankert, N., & Wertheim, H. (2020). Strong associations and moderate predictive value of early symptoms for SARS-CoV-2 test positivity among healthcare workers, the Netherlands, March 2020. *Eurosurveillance*, *25*(16), 2000508. doi:10.2807/1560-7917.ES.2020.25.16.2000508 PMID:32347200

Zoabi, Y., Deri-Rozov, S., & Shomron, N. (2021). Machine learning-based prediction of COVID-19 diagnosis based on symptoms. *NPJ Digital Medicine*, *4*(1), 1-5.

ADDITIONAL READING

Agbehadji, I. E., Awuzie, B. O., Ngowi, A. B., & Millham, R. C. (2020). Review of big data analytics, artificial intelligence and nature-inspired computing models towards accurate detection of COVID-19 pandemic cases and contact tracing. *International Journal of Environmental Research and Public Health, 17*(15), 5330. doi:10.3390/ijerph17155330 PMID:32722154

Al Ismail, M., Deshmukh, S., & Singh, R. (2021, June). Detection of COVID-19 through the analysis of vocal fold oscillations. In *ICASSP 2021-2021 IEEE International Conference on Acoustics, Speech and Signal Processing (ICASSP)* (pp. 1035-1039). IEEE. 10.1109/ICASSP39728.2021.9414201

Fauziah, Y., Saifullah, S., & Aribowo, A. S. (2020, October). Design Text Mining for Anxiety Detection using Machine Learning based-on Social Media Data during COVID-19 pandemic. In *Proceeding of LPPM UPN "Veteran" Yogyakarta Conference Series 2020–Engineering and Science Series* (Vol. 1, No. 1, pp. 253-261).

Hecker, P., Pokorny, F. B., Bartl-Pokorny, K. D., Reichel, U., Ren, Z., Hantke, S., Eyben, F., Schuller, D. M., Arnrich, B., & Schuller, B. W. (2021). Speaking Corona? Human and Machine Recognition of COVID-19 from Voice. *Proc. Interspeech, 2021*, 1029–1033. doi:10.21437/Interspeech.2021-1771

Li, L. Q., Huang, T., Wang, Y. Q., Wang, Z. P., Liang, Y., Huang, T. B., Zhang, H., Sun, W., & Wang, Y. (2020). COVID-19 patients' clinical characteristics, discharge rate, and fatality rate of meta-analysis. *Journal of Medical Virology, 92*(6), 577–583. doi:10.1002/jmv.25757 PMID:32162702

Manaswi, N. K. (2018). *Deep learning with applications using python: chatbots and face, object, and speech recognition with tensorflow and keras.* Apress. doi:10.1007/978-1-4842-3516-4

Martín-Rojas, R. M., Pérez-Rus, G., Delgado-Pinos, V. E., Domingo-González, A., Regalado-Artamendi, I., Alba-Urdiales, N., Demelo-Rodríguez, P., Monsalvo, S., Rodríguez-Macías, G., Ballesteros, M., Osorio-Prendes, S., Díez-Martín, J. L., & Pascual Izquierdo, C. (2020). COVID-19 coagulopathy: An in-depth analysis of the coagulation system. *European Journal of Haematology, 105*(6), 741–750. doi:10.1111/ejh.13501 PMID:32749010

Patil, T., Pandey, S., & Visrani, K. (2021). A review on basic deep learning technologies and applications. In *Data science and intelligent applications* (pp. 565–573). Springer. doi:10.1007/978-981-15-4474-3_61

Schwab, P., Schütte, A. D., Dietz, B., & Bauer, S. (2020). Clinical predictive models for COVID-19: Systematic study. *Journal of Medical Internet Research*, *22*(10), e21439. doi:10.2196/21439 PMID:32976111

KEY TERMS AND DEFINITIONS

Clinical Analytics: A subject that generates insights for making biomedical decisions by using real-time medical data.

Cloud Computing: An on-demand access that can compute resources of various services through the Internet.

Deep Learning: A broad family of machine learning models based on neural networks. Typical deep learning models are deep neural networks, convolutional neural networks, recurrent neural networks, deep belief networks, and deep reinforcement learning.

Machine Learning: A subject of artificial intelligence that aims at the task of computational algorithms, which allow machines to learning objects automatically through historical data.

Mobile App Development: A process of designing and developing mobile app for mobile devices.

Smart Diagnostics: A technology that can give correct diagnosis quickly through electronic devices, such as mobile and smart watch.

Text Mining: A process of extracting necessary information from unstructured text data for data processing.

Voice Recognition: A technology for machines to understand spoken words and give correct response to the voice.

Chapter 8

Smart Diagnostics of COVID-19 With Data-Driven Approaches

ABSTRACT

The traditional assays and diagnostic methods are time-consuming and expensive. As the COVID-19 pandemic is expected to remain for a while, it is demanded to develop an efficient diagnosis system. This chapter is designed to investigate how to incorporate data-driven approaches to the construction of a smart health framework for COVID-19. Topics cover a broad range of smart diagnosis innovations for supporting current assays and diagnostics, such as data analysis for nucleic acid tests, machine learning-based serological signatures identification, medical image classification using deep learning, and decision support system for automatic diagnosis with clinical information. Each topic has been illustrated and discussed throughout methodologies, data collections, experimental designs and results, limitations, and potential improvements. All applicational potentials have been examined with real-world datasets. The findings conclude that big data and AI work for providing insightful suggestions on multiple diagnostic assays and COVID-19 detection approaches.

DOI: 10.4018/978-1-7998-8793-5.ch008

INTRODUCTION

Since the first SARS-CoV-2 confirmed case was found in Wuhan, China in late December 2019, such a novel coronavirus and its variants have been sweeping around the world for over 18 months as of July 2021. COVID-19 diagnosis plays a critical role in the containment of the disease in terms of the policy-making process through infected case identification, quarantine, and contact tracking. As a respiratory and infectious disease, the regular medical diagnosis of COVID-19 primarily depends on tracking epidemiological history and characteristics (Chang et al., 2020), analyzing clinical symptoms (Li et al., 2020), and confirmed by a variety of medical identification processes, such as nucleic acid amplification tests (Basu et al., 2020), various medical imaging tests including computed tomography (CT) scans (Awulachew et al., 2020), chest x-rays (Durrani et al., 2020), and magnetic resonance imaging (MRI) tests (Langenbach et al., 2020), and a variety of biological tests, i.e. serological assays (Krammer & Simon, 2020), protein tests (Poggiali et al., 2020), enzyme-linked immunoassay (Xiang et al., 2020), and lateral flow antigen detections (Wu et al., 2020a). Early epidemiological studies uncover the clinical characteristics of the novel coronavirus and its infection trails, which have concluded that the COVID-19 diagnostic result cannot be determined solely on the identification of common symptoms, such as fever, cough, difficulty of breath, and fatigue (Guan et al., 2020; Huang et al., 2020). Many of such symptoms can be associated with other respiratory diseases, thereby are not specific and may not be adopted for the precise diagnosis. Besides, the total number of reported infections is underestimated as many mild and asymptomatic cases are not included, whereas such cases are still difficult to detect with a symptoms-based diagnosis (Kobayashi et al., 2020). Therefore, the accurate diagnosis of SARS-CoV-2 is still challenging as the clinical manifestation of such a novel virus is significantly variable from case to case, particularly, with asymptomatic to acute respiratory distress syndrome and multi-organ failures (Girija et al., 2020; Zaim et al., 2020).

With the development of genetics, molecular techniques, and medical image processing, exploring the biological properties of a disease and diagnostic approaches becomes more accurate and efficient. Such ideologies and technologies have been applied to understand the novel coronavirus. Early genetic studies indicate that SARS-CoV-2 is a single-stranded positive RNA genome that encodes an RNA-dependent RNA polymerase and four structural proteins, including the spike surface glycoprotein, a small envelope protein, a matrix protein, and the nucleocapsid protein (Wu et al., 2020b; Walls et al., 2020). A negative result based on real-time polymerase chain reaction (PCR) from the experimental design using nucleic acid test suggests that the origin of the cause of pneumonia is unclear compared with the known pathogen panels (Zhou et al., 2020). Further biometric studies discover that a pathogen with a similar genetic sequence to the beta coronavirus B lineage matches the genome of the bat coronavirus RaTG13, the severe acute respiratory syndrome virus (SARS-CoV), and the Middle East respiratory syndrome virus (MERS-CoV), whereas the similarity of each genome varies from 50% to 96% (Lu et al., 2020). The initial diagnosis of Covid-19 has been led by scanning patients using CT scan. Clinical findings from CT images reveal the differences between healthy lungs and lungs with Covid-19, which are denser, more profuse, and confluent (Ai et al., 2020; Chung et al., 2020). Therefore, nucleic acid tests and initial diagnostics are still essential in the context of Covid-19, providing the knowledge graph for experts to understand the virus efficiently.

A typical testing and diagnostic workflow for SARS-CoV-2 follows a common procedure for the prevention and diagnosis of other infectious diseases. Individuals with appeared symptoms will present at the community test center for sample collection. The collected samples are either tested on-site or transported to the nearby hospitals or lavatories for molecular analysis and sequencing. Positive cases will be addressed by following the disease control policy, i.e. mild cases are self-quarantined at home, while severe cases are hospitalized for further diagnostics and treatments. A variety of specimens such as sputum, bronchial aspirate, bronchoalveolar lavage fluid, oropharyngeal or nasopharyngeal swab, and blood is recommended for the early screening and diagnosis (Zou et al., 2020). Human swab samples contain information in regard to the RNA-depended tests, such as reverse transcriptase real-time PCR (RT-PCR), reverse transcription loop-mediated isothermal amplification (RT-LAMP), clustered regularly interspaced short palindromic repeats (CRISPR), etc., which have been widely used for the mass nucleic acid tests. Blood samples are addressed with diagnostic approaches in disease surveillance, for instance, enzyme-linked immunosorbent assay (ELISA), lateral flow assay (LFA), and chemiluminescence immunoassay (CLIA). Despite effectiveness and operability of such biological tests and assays, each method requires high quality samples, efficient analytics for searching SARS-CoV-2 targets, and sufficient test capacities.

Medical imaging tests have been applied as the clinical diagnosis for COVID-19 in regard to the early detection of the pneumonia. The chest CT images have been widely used as the early detection for screening pneumonia diseases, such as SARS and MERS (Rao et al., 2003; Memish et al., 2014). CT scans have also been considered to be one of the most effective image processing approaches in the diagnosis of COVID-19 for hospitalized individuals due to the lack of test kits and false negative rate based on RT-PCR (Yang & Yan, 2020). Additionally, patients' chest and whole body images from X-rays and MRIs have been applied to analyzing and diagnosing COVID-19 (Durrani et al., 2020; Fields et al., 2020). However, medical image-based technologies have several ambiguities in the COVID-19 diagnosis, i.e. less sensitivity compared with the diagnostic sensitivity in RT-PCR, the likelihood of causing false positive results, risks of overlapping with similar infections such as influenza, SARS, and MERS, and requirements of skilled professionals and high-end equipments (Rai et al., 2021). Despite the technological limitations of chest radiography, the effectiveness for the COVID-19 diagnosis using medical imaging tests is still beneficial to the current diagnostic plan. Nevertheless, analyzing a large number of medical images may increase the working pressure of medical professionals, thereby reducing the efficiency of the diagnostic process. In particular, healthcare workers are under extreme pressures with an exponentially increasing number of hospitalized cases and exposure risks to the virus during the COVID-19 pandemic (Wang et al., 2020).

Motived by the current challenges in smart diagnostic of COVID-19, this chapter targets on illustrating how data-driven approaches help medical professionals to identify positive cases efficiently. Such a framework may further make contributions in hospital management in the form of providing a comprehensive decision support system for COVID-19 diagnosis. The objectives of this chapter are:

- providing a survey of most recent studies in COVID-19 diagnosis with big data, machine learning, and deep learning.
- investigating how cutting-edge technologies, such as clinical data analysis, machine learning classification, medical image processing, and decision support system with big data, work for multiple diagnostic approaches and assays of COVID-19.
- examining application potentials of the framework of COVID-19 smart diagnosis throughout ideology, methodology, and experimental design with real-world datasets.

BACKGROUND

COVID-19 diagnosis plays a crucial role in the case identification, disease control policy-making, and treatment of contagious subjects. One of the most common-used test strategies for identifying COVID-19 is RT-PCR, which has been demonstrated as the gold standard for SARS-CoV-2 testing and diagnosis (Corman et al., 2020). Although RT-PCR is a sophisticated test, its limitations have aroused the attention of researchers and health authorities in regards to requirements of extensive and delicate infrastructures (Loeffelholz & Tang, 2020), enormous demands of testing capacities (Lippi et al., 2020), a higher false positive rate (Surkova et al., 2020), and a lower sensitivity in clinical applications (Ai et al., 2020). Despite inevitable limitations in COVID-19 detection, the RT-PCR test result has been widely used as one of the most important attributes for case identification and risk assessment, along with machine learning (Arpaci et al., 2021; Assaf et al., 2020). In general, such studies are relied on a classification task using various machine learning algorithms with a variety of features, such as demographical data, clinical information, medical assay results, etc. However, rapid diagnostics tools for COVID-19 incorporating RT-PCR are rarely discussed. Nevertheless, analyzing virus DNA sequences with the RT-PCR is still dominate in COVID-19 diagnosis. Besides, other RNA-depended tests, such as RT-LAMP and CRISPR, have been applied in terms of rapid molecular diagnosis and fluorescence detection for COVID-19 with machine learning (de Oliveira et al., 2021; Samacoits et al., 2021).

Alternative testing strategies and methods, such as routine blood tests analysis and case identification based on clinical data, are still invaluable and necessary as they can be applied to validate the RT-PCR tests with well-documented false negative results (West et al., 2020). A variety of machine learning algorithms, such as random forest (RF), logistic regression (LR), decision tree (DT), Naive Bayes (NB), k-nearest neighbors (KNN), support vector machine (SVM), and XGBoost, has been incorporated in constructing COVID-19 rapid diagnostic models. For instance, a RF-based COVID-19 diagnosis model has been proposed for case identification based on routine blood tests (Tschoellitsch et al., 2021). Li et al. (2020) distinguished patients between influenza and SARS-CoV-2 based on results from clinical trials and routine tests using a XGBoost classification model, which achieved a high degree in both sensitivity (92.5%) and specificity (97.9%). A binary classification model using LR has been employed to establish a scoring system for calculating the probability of a COVID-19 positive case with an AUC score of 88.9% (Tordjman et al., 2020). In this study, a pool of clinical datasets was collected and aggregated from different hospitals with suggested biological features of patients who were positively correlated to COVID-19 diagnosis, i.e. lymphocytes, eosinophils, basophils, and neutrophils. Most existing studies in this domain hired a single machine learning model, while others applied multiple algorithms and then picked the optimal one based on the model performance evaluation, or used hybrid models to build new machine learning models, or incorporated with deep learning, e.g. RF was selected as the optimal classifier out of eight different machine learning models (Brinati et al., 2020); a combination of multiple classical machine learning models has been investigated by Goodman-Meza et al. (2020) in the form of increasing COVID-19 inpatient diagnostic capacity using demographic and laboratory attributes; deep learning architectures, such as artificial neural network (ANN), convolutional neural network (CNN), recurrent neural network (RNN), long short-term memory (LSTM), and their combinations, were applied for modeling the COVID-19 detection with features representing clinical data, such as hematocrit, hemoglobin, platelets, and red blood cell count (Alakus & Turkoglu, 2020).

Serological assays have been widely applied as one of the most efficient diagnostic plans in COVID-19 surveillance in terms of detecting the antibodies against SARS-CoV-2 in infected patients. The development of fast and simple test kits based on the identification of antibodies has been discussed in response to the COVID-19 infection due to the challenges of the novel coronavirus infection, such as undetectable viral RNA after the lone incubation period and false negative test results because of improper handling of viral samples (Rai et al., 2021). Detecting antibodies developed in response to viral infection and viral antigen is the fundamental principle of the serological assays, whereas machine learning can be applied as the assistant tool for the COVID-19 surveillance projects based on the survey of recent studies in this field. A multiplex assay has been developed by Rosado et al. (2021) to identify the serological signatures of COVID-19 with RF, which has been chosen to recognize the SARS-CoV-2 infection with the combination of antibody responses to multiple antigens. Similarly, the same methods and models were selected with the multiplex serological assay data collected from French hospitals to classify patients with COVID-19 infections, of which antibody responses to trimeric spike protein with a sensitivity of 91.6% and a specificity of 99.1% (Rosado et al., 2020). Mowery et al. (2020) presented an improved serology analysis concentrated on achieving superior classification performance by integrating multiple LFA tests using machine learning. Such machine learning-based serological assays based on antibody responses to multiple antigens can be used as the complement of the smart diagnosis of COVID-19 in terms of providing efficient and robust serological classification.

Medical image-based diagnostic methods have been explored as the supplemental tests and diagnosis plans for COVID-19. The chest CT scan is one of the first live image processing techniques in identifying pneumonia diseases. Besides, x-rays has also been widely applied for pneumonia-related illnesses. Chest radiography has been deployed for the COVID-19 diagnostics in hospitals, whereas medical professionals may need to pay much attentions on screening its output images. Although medical image auto-detections have not been recommended for the primary diagnosis of COVID-19, image recognition methods could assist in monitoring the early detection of COVID-19, and potentially shorten the time of diagnosis. Existing studies illustrate such an attempted approach can be implemented by leveraging machine learning and deep learning. Mahdy et al. (2020) proposed an automatic COVID-19 lung image classification system using x-ray images with multi-level thresholding and SVM, which obtained the sensitivity of 95.8%, the specificity of 99.7%, and the accuracy of 97.5%. Similarly, a machine learning-based frequency domain algorithm, named as FFT-Gabor, has been applied in image classification of CT scans as the complementary digitalized test of COVID-19 cases (Al-Karawi et al., 2020). Kana et al. (2020) developed a web-based COVID-19 diagnostic tool using machine learning on chest radiographs, which allows users to achieve the test result in seconds with drag-and-drop throughout its web graphical user interface. Some studies focus on extracting COVID-19 chest image features and image classifications using deep learning models, i.e. the CNN-based feature extraction plus SVM as the classifier for detecting COVID-19 cases using x-ray images (Ismael & Şengür, 2021) and an attention-based COVID-19 chest image classification using VGG-16 (Sitaula & Hossain, 2021). Moreover, transfer learning has also been applied for the image classification task in the context of COVID-19 detections, e.g. an automatic diagnostic tool based on transfer learning with CNN for detecting COVID-19 cases from x-ray images (Apostolopoulos & Mpesiana, 2020), the transfer learning-based COVID-19 detector with VGG-19 using images from both x-rays and CT scans (Horry et al., 2020), and the self-supervised super sample decomposition for transfer learning model (4S-DT) for x-ray image classification task of COVID-19 detections (Abbas et al., 2021). Such models and techniques have achieved high degrees of model performance in terms of accuracy, sensitivity, specificity, and AUC, with less time on predictions once the model has been trained, which can be applied in constructing the smart diagnosis system for COVID-19.

MAIN FOCUS OF THE CHAPTER

Issues, Controversies, Problems

Despite adequate studies proposed for diagnosing COVID-19, no one focuses on smart health systematically through a comprehensive understanding of how data analytics and machine learning/deep learning can assist in the construction of the automatic diagnostic system for COVID-19. A smart COVID-19 diagnostic system should involve analytical works throughout clinical results from different assays. However, the controversy of the current studies lays on the insufficient and exclusive investigation in this domain. Most studies proposed the data-driven approaches based on restricted factors, i.e. applied only clinical data. Others developed machine learning and/or deep learning models for COVID-19 classification only depending on single attribute, e.g. the medical image. Therefore, such methods may not satisfy the requirement of the comprehensive smart diagnostic system for COVID-19.

Features of the Chapter

This chapter is proposed to illustrate the smart diagnostics of COVID-19 from multiple aspects of detection approaches, such as nucleic acid test, serological signatures identification, medical image-based diagnosis, and automatic diagnosis with clinical information. This chapter aims on how such approaches can be implemented efficiently by incorporating data analytics, machine learning, and deep learning. Each method has been investigated in terms of ideology and methodology based on related research. All application potentials have been examined with real-word datasets, along with discussions on experimental results, limitations, and improvements.

SOLUTIONS AND RECOMMENDATIONS

Smart Nucleic Acid Amplification Test

As the "gold standard" for clinical diagnostic detection for COVID-19, the nucleic acid amplification test (NAAT) is one of the most effective assays and preferred test to identify early viral infections. Common types of NAAT-based test approaches include RT-PCR, RT-LAMP, and CRISPR-based diagnosis. With the applications of data-driven methods in smart diagnosis, NAAT can be reshaped with the digitalized transformation in terms of data analytics for identifying COVID-19 rapidly. Despite unavailable data acquisition for each approach, several existing studies illustrate how data-driven approaches can be applied for making different types of NAAT efficiently. Hence, investigating such studies in theory may also benefit the COVID-19 smart diagnostic framework.

Gomes et al. (2020) developed a pseudo-convolutional machine learning approach for representing DNA sequences obtained by the RT-PCR test. Such a technique can identify virus sequences from a large scale of database with high degrees of sensitivity and specificity, thereby can be applied to optimize the current molecular diagnosis of COVID-19 in terms of matching SARS-CoV-2 DNA sequences faster and effective. Similar studies indicate machine learning algorithms can be applied in providing evidence of associations between SARS-CoV-2 and bat betacoronaviruses (Randhawa et al., 2020a) and implementing the rapid classification of pathogens using intrinsic genomic signatures of COVID-19 (Randhawa et al., 2020b). Although machine learning can improve the analytical tasks of DNA sequences identification, such approaches are still based on the RT-PCR framework with the inevitable limitations. Therefore, alternative methods are highly recommended and essential in terms of the establishment of smart diagnostic tools for COVID-19.

Mohanty et al. (2021) discussed the spectral tools for the rapid SARS-CoV-2 detection, including colorimetric RT-LAMP and matrix-assisted laser desorption/ionization (MALDI) mass spectrometry (MS), using machine learning, such as NB, KNN, DT, SVM, and RF. Such algorithms can be used for potential spectral techniques to overcome the challenges of RT-PCR in terms of constructing a rapid testing framework for COVID-19 detections. Similar rapid detection systems have also been proposed using a broad range of MALDI-based and MS-based approaches with machine learning algorithms and statistical analysis, e.g. a MALDI-TOF-based serum peptidome profiling using logistic regression (Yan et al., 2021). a hybrid detection tool of MALDI-TOF-MS with multivariate analysis and machine learning (Rocca et al., 2020), and a machine learning-based spectral feature selection for a MS-based system of rapid detection of COVID-19 (Cardozo et al., 2020). To the mass COVID-19 testing and diagnostic perspectives, such methods can achieve better accuracy, less time of implementation, lower testing and diagnostic capacity with detecting infected cases directly from nasal swab specimens without the requirement of the RNA isolation stage. Therefore, RT-LAMP and MALDI-MS assays combined with machine learning can guide the potential applications of rapid testing and diagnostic systems for COVID-19.

On the other hand, the CRISPR-based NAAT has been applied as the alternative testing and diagnostic strategy for ongoing surveillance of COVID-19. The experimental designs with machine learning can help to provide fast, effective, and specific detection tools. Metsky et al. (2020) proposed a CRISPR-based experimental design for COVID-19 detection using machine learning. Such a detection system can be used for screening SARS-CoV-2 with a CRISPR-Cas 13 detection framework and the high-performing COVID-19 assay. Similar studies indicate that CRISPR-based assay designs can be deployed for rapid and sensitive diagnostics of COVID-19 in terms of visual detections and sequence matching methods (Ding et al., 2020; Lucia et al., 2020). However, one of the most significant limitation of CRISPR systems is low guide RNA (gRNA) activity, therefore high-accurate prediction of such activity is extremely critical. Ameen et al. (2021) applied deep learning algorithms to predict the gRNA activity. A variety of machine learning and deep learning models have been investigated in this study, of which the optimal model increases the prediction performance up to 40%. Such a model can help in selecting the best gRNA for COVID-19 identification within the CRISPR-based system.

Serological Signatures Identification With Machine Learning

The antibody response is induced by infection with severe acute respiratory syndrome, which can change over time. Multiplex serological assays are proposed to measure IgG and IgM antibody responses to a range of nucleoprotein antigens and COVID-19 spikes. Antibody samples are usually in the form of serum collected from hospitalized patients and health workers, along with RT-PCR test results. Multiplex assays for identifying IgG and IgM antibody reactions against candidate COVID-19 antigens should be optimized with considering nucleoprotein constructs, spike proteins, seasonal flu types, and other viruses (i.e. H1N1, adenovirus type 40, rubella virus, SARS, MERS, etc.). Such an experimental dataset has been publicly accessible through an existing study and the GitHub repository (Rosado et al, 2021). Serum samples have been collected from adult patients and hospital workers between January 24 to April 7, 2020, along with negative control samples selected from pre-pandemic panels. All samples were addressed under a viral inactivation protocol that measures the effect on the measurement of antibody levels. IgG and IgM antibody measurements against different COVID-19 antigens were detected by an optimized multiplex assay that operates on over 95 well plates using multiple proteins measurements derived from fluorescent-based detection systems with median fluorescent intensity (MFI).

Diagnostic sensitivity can be defined as the ratio of confirmed infections by RT-PCR with a higher antibody level given a certain seropositivity cutoff for antibody responses to a single antigen, whereas the proportion of negative controls with lower antibody levels is determined as the diagnostic specificity. A receiver operating characteristic (ROC) analysis can be applied to evaluate the trade-off between sensitivity and specificity. Classification tasks will be employed for identifying patients with the COVID-19 infection combined antibody responses to multiple antigens, which can be done by using a variety of machine learning algorithms, including NB, KNN, SVM, DT, RF, etc. with splitting training and testing sets and N-fold cross-validation. The Wilson method can be used to calculate the binomial confidence intervals (CIs). A two-sided t test can be applied to measure the difference of antibody responses, while the correlations between the antibody responses can be quantified by calculating the Spearman's correlation. McNemar's test can be employed to investigate the differences in classification performance. Antibody kinetics provide deeper and predictive understanding of the characterization of the virus within a mixed-effects analytic framework, which can be fitted in statistical and mathematical models, i.e. a Bayesian model using Markov Chain Monte Carlo (MCMC) process with priors data estimated from long-term experimental deigns of antibody kinetics following infection with other viruses. The multiplex diagnostic tests can achieve better classification performance than single antigen tests as the number of days after symptom increases. The multiple antigen classification can be performed with the multiplex data using the random forest algorithm which achieves the optimal model performance. Statistical adjustment was used to solve the bias in seroprevalence due to inadequate sensitivity and specificity. The best values of sensitivity and specificity can reduce the negative effect of the expected error in seroprevalence.

Experimental results illustrate that antibody responses are relatively lower in negative control samples than in infected samples conformed by RT-PCR (t test p value < 0.0001), for all COVID-19 biomarkers, including antigens, IgG and IgM for both. The ROC analysis based on RF has been applied to investigate the trade-off between sensitivity and specificity. The results indicate that the optimal biomarker is anti-trimeric spike IgG (S^{tri}IgG) with the sensitivity of 92% and the specificity of 99%. Such a biomarker also provides the best classification performance than any others (McNemar's test p value ~ 0). Figure 1 shows an example out of 156 pairwise comparisons of antibody responses. Each point denotes a measured antibody response from a sample from two different hospitals (green and dark purple) and healthcare workers in Strasbourg (orange). Negative control samples are included from France (light blue), Thailand (dark blue), and Peru blood donors (light purple). The similar clustering distribution of antibody responses demonstrated in these plots is that negative control samples gathered at the bottom and the serum samples gathered at the top and in the center.

Figure 1. Pairwise comparison of anti-S^{tri}IgG and anti-RBD_{v2}IgG

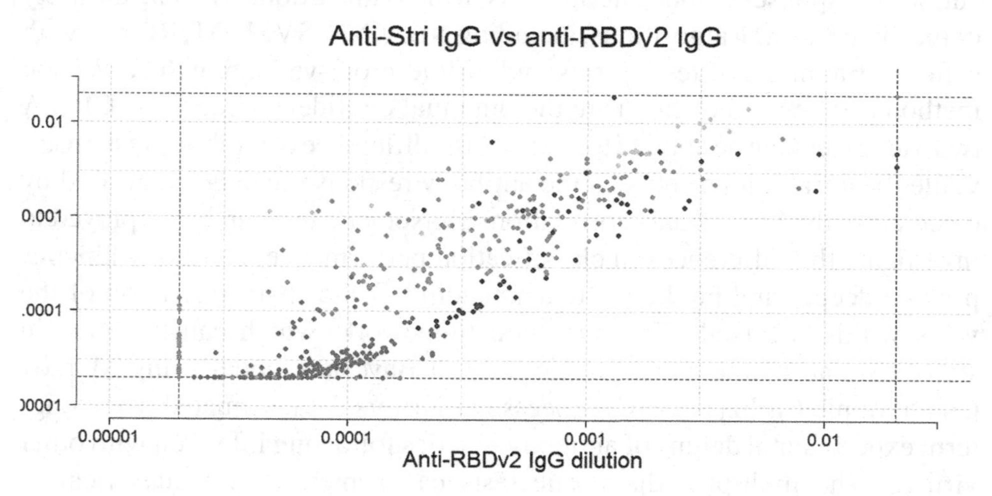

The RF-based classifier performance of multiplex combinations of antibody responses has been investigated with ROC curves in Figure 2A.

Despite diminishing returns in sensitivity, the model performance has been significantly improved by adding more data from extra biomarker families, such as SARS-CoV-2 nucleoprotein (NP), SARS-CoV-2 spike glycoprotein receptor-binding domain (RBD), and SARS-CoV-2 spike glycoprotein S2 domain. The model has been incorporated with such biomarkers, including RBD_{v2}IgG, NP_{v1}IgG, S2IgG, RBD_{v1}IgM, and NP_{v1}IgM. Figure 2B indicates that the sensitivity is less than 95% with a single biomarker, whereas the model provides a nearly 99% in sensitivity by combining all six biomarkers.

Figure 2. RF-based classifier performance with model improvement. (A). ROC curves for multi-biomarker classification; (B). model improvement by adding additional biomarkers.

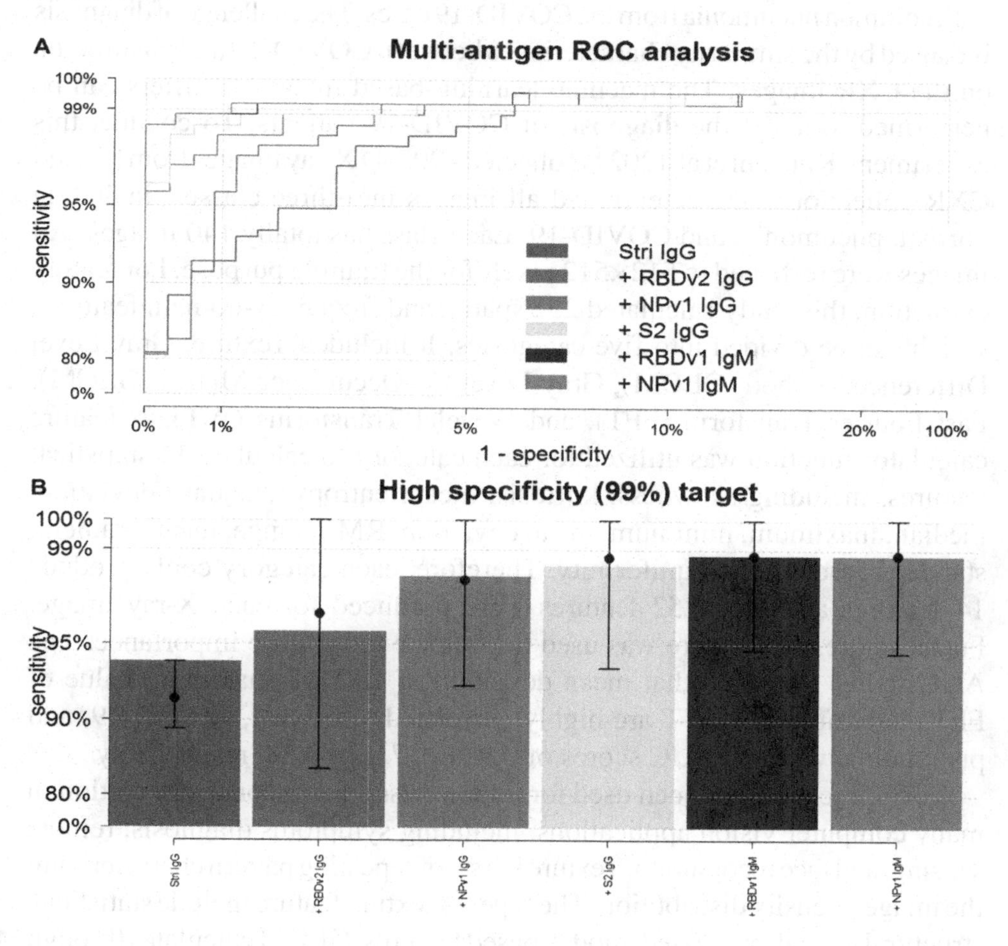

Although such a machine learning-based approach can be applied in smart diagnosis for COVID-19, some limitations are still existed in terms of the absence of samples and the problem of assay designs. Firstly, data collected from patients who have symptoms onset with RT-PCR-confirmed asymptomatic is the absence of samples, thereby affecting the model performance. Secondly, irrelevant antigens are included in the assay, rather than emphasizing on more closely related betacoronaviruses. Therefore, multiple types of such antigens should be the inclusion that may improve the performance of the classification.

Chest Image Classification With Deep Learning

Chest-X ray (CXR) radiography is a useful method to distinguish the patients with common pneumonia from the COVID-19 cases. The challenge of diagnosis is caused by the similarity characteristics between COVID-19 and pneumonia on the CXR images. The machine learning-based image classifiers can be performed to assist the diagnosis of COVID-19 patients. To conduct this experiment, Khuzani et al. (2021) collected 420 2-D X-ray images from labeled CXR collections, and categorized all images into three classes, including normal, pneumonia, and COVID-19. Each class has totally 140 images. All images were reshaped to 512×512 pixels for the training purpose. For feature extraction, this study calculated 252 spatial and frequency-domain features, which can be divided into five categories. It included Texture, Gray Level Difference Method (GLDM), Gray-Level Co-Occurrence Matrix (GLCM), Fast Fourier Transform (FFT), and Wavelet Transforms (WT). A feature calculator function was utilized for each category to calculate 14 statistical features, including skewness, kurtosis, mean, entropy, standard deviation, median, maximum, minimum, mean deviation, RMS, range, mean gradient, standard gradient, and uniformity. Therefore, each category could produce 14 features and total 252 features were produced for each X-ray image. Furthermore, AUC score was used to evaluate the feature importance. The AUC scores indicated that mean deviation of GLDM, maximum value of FFT, and kurtosis of WT are highly correlated to normal, COVID-19, and pneumonia, with the AUC scores of 0.91, 0.87, and 0.88, respectively.

Texture feature has been used for image classification or segmentation in many computer vision applications, including symptoms diagnosis, remote sensing, and face recognition. Texture is a set of repeating pattern characterizing the image intensity distribution. The types of texture feature include statistical, structural, signal-processed, model-based features. GLDM calculates through

Gray Level Difference Method Probability Density Functions to extract statistical texture features of the image. GLCM is another statistical method of texture feature extraction. GLCM can measure the spatial relationship of pixels through calculating the frequency for the pairs of pixel getting specific values and for the certain spatial relationship appearing in the image. FFT can transform a signal from special domain to a feature in frequency domain through making the Discrete Fourier Transform (DFT) matrix divided by a product of sparse factors. FFT algorithm can reach better accuracy in faster speed than DFT can do. WT performs the mathematical function to transform continuous-time signal to various scale components. WT is more advantageous than Fourier Transform (FT), since WT can deal with the signals both in time and frequency domains. The Continuous Wavelet Transform (CWT) is usually applied to analyze time-frequency-based problems. To identify three target labels, an image classifier was built with a multi-layer neural network, which has two hidden layers of 128 and 16 neurons and one output classifier layer. The training process went through 33 epochs. The classification model achieved 100% sensitivity and 96% precision for the COVID-19 classification.

This chapter references the same ideology with different data sources. CXR and CT images can be retrieved from the public open-source, including the COVID-19 image data collection proposed by the University of Montreal (Cohen et al., 2020) and CXR images available from Kaggle (Mooney, 2019). The first dataset is available at GitHub, containing more than 500 CXR and CT images of individuals who are suspected or confirmed cases of COVID-19 or other infections by viral and bacterial pneumonias, such as MERS, SARS, and ARDS. Such a dataset was collected from hospitals and other public information pools with the ad-hoc orientation for the AI-driven COVID-19 diagnostic using medical image processing. The second dataset can be downloaded from Kaggle which provides a structured data container for each image category (i.e. pneumonia or normal), having 5,863 CXR images in total. CXR images were collected from retrospective cohorts of children from a hospital in Guangzhou, China, with the evaluation set checked by a third-party expert group. Despite the limitations of the Kaggle dataset (i.e. noisy inputs, unbalanced samples, and incorrect labels), such a dataset is a complementary data pool for training a COVID-19 detector through CXR and CT images. Characteristic patterns of both COVID-19 positive and negative CXR can be illustrated by visualizing the samples, as shown in Figure 3. The typical pattern of COVID-19 positive cases based on CXR illustrates the presence of patchy and confluent, ground-glass opacity, and lower lung zone distribution on the image, suggesting that such characteristic features of

COVID-19 in medical image visualization can be used in connection between clinical judgement and AI-driven image classification to guid the diagnosis.

Figure 3. COVID-19 CXR image samples. (A). an example CXR image for the positive case; (B). an example CXR image for the negative case.

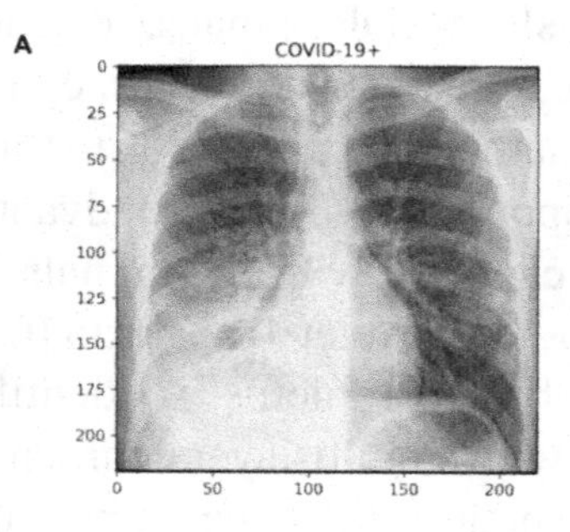

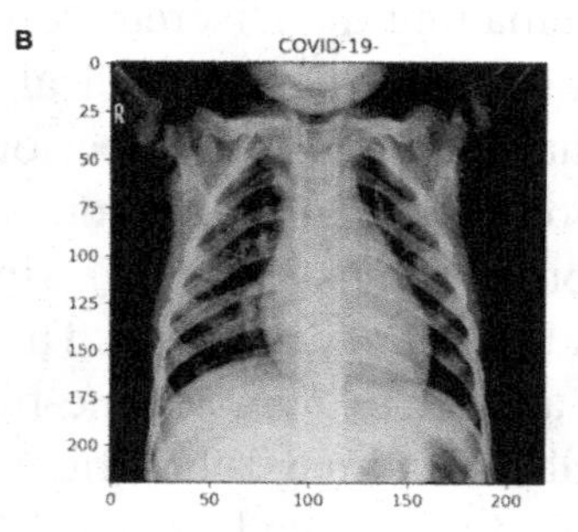

As per the regular image classification process, COVID-19 CXR images in the combined dataset have been augmented due to the limited sample size, followed by splitting the new dataset into a 80% training set and a 20% testing set. The optimal model has been determined by referring to the result from the above-mentioned study, followed by implementing the fine-tuned network to the image classifier using TensorFlow and Keras in Python. The proposed model can obtain 91.3% in accuracy with a sensitivity of 100% and a specificity of 80%. Although CXR training data applied in this chapter is limited, the proposed model is not overfitting based on the accuracy and training loss plot, as shown in Figure 4A. The experimental results suggest that the proposed model can be applied to implement the automatic COVID-19 diagnostics through CXR images, as illustrated in Figure 4B. Therefore, such a detector can classify COVID-19 infections from regular individuals effectively.

Figure 4. Visualizing model evaluation measurements and classification outputs. (A). accuracy and training loss on COVID-19 CXR image classification. (B). positive (left) and negative (right) samples determined by the proposed model.

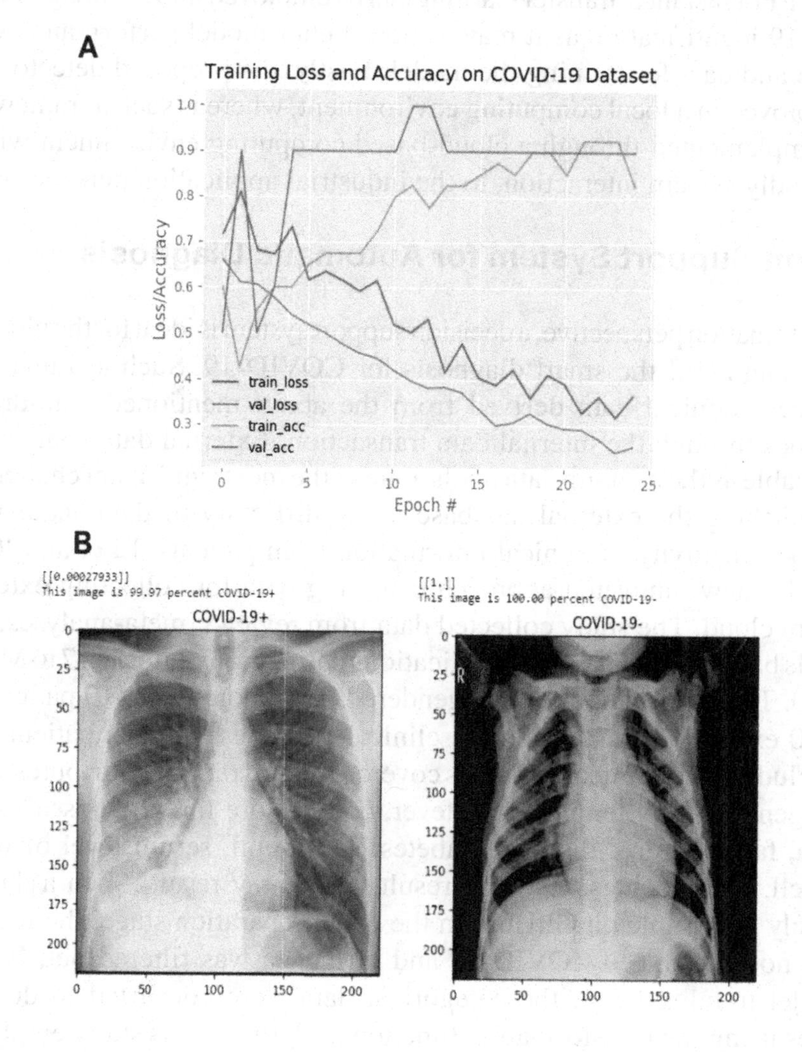

Despite the effectiveness of the proposed model in COVID-19 detection, several problems still remain throughout data collection, model selection, and system deployment. Firstly, the limited samples in the dataset lead to lower model performance in specificity, indicating that the model can only classify 80% of individuals who do not have COVID-19. Collecting more data will benefit the model improvement, however, such a job requires additional datasets that are publicly assessable and have been confirmed with medical professionals. Secondly, this chapter relies on a single deep learning-based model, however, alternative approaches should also be investigated and examined. For instance, transfer learning can be employed into the image-based COVID-19 identification as it may achieve better model performance with less time and data for training the model. Lastly, the proposed detector has been deployed in a local computing environment, whereas such a framework can be implemented through a cloud-based computing environment with a user-friendly system interaction, to the industrial application perspectives.

Decision Support System for Automatic Diagnosis

To the automation perspective, a decision support system is ideal for the ultimate decision engine of the smart diagnosis for COVID-19. Such a framework can process clinical data derived from the above-mentioned data-driven approaches through the internal data transaction. External data sources are also valuable as the back-up data pools. One of the most significant challenges for establishing the external database is the difficulty of data acquisition due to the sensitivity of clinical information from patients. Li et al. (2020) proposed a new innovated approach using big data for collecting external data from cloud. The study collected data from reviews, meta-analyses, and editorials based on nearly 1,500 publications dating from January 17 to March 23, 2020. The clinical dataset was gendered from more than 410 patients in over 150 existing studies. Multiple clinical variables for each patient data were included. The clinical variables covered a board range of attributes, such as age, gender, body temperature, fever, cough, sore throat, nausea vomit, diarrhea, fatigue, renal disease, diabetes, neutrophil, serum level of white blood cell, lymphocytes, CT scan result, and X-ray result. Such a dataset is publicly assessable on GitHub. In the data preparation stage, the records that are not relevant to COVID-19 and influenza was filtered out. Before the model training stage, the categorical data were converted to dummy variables using the transformation function in Python. This study employed XGBoost algorithm to achieve the patient classification through training on the clinical variables. The whole dataset was divided into a 80% training set and a 20% test set. A 5-fold cross-validation operation was performed with 70 boosting iterations. Bayesian optimization was used to find out the optimal hyperparameters, including max depth, gamma, learning rate, and the number of estimators, for XGBoost. The classification results were evaluated through calculating the AUC score and analyzing the ROC curve. The experimental results indicated that XGBoost achieved a sensitivity, a specificity, an AUC, and a F-1 score of 92.5%, 97.9%, 0.977, and 0.929, respectively. Since XGBoost were suspected to overfitting, additional three machine learning models, such as ridge regression, random forest, and lasso regression, were also examined with an AUC of 96.6%, 95.3%, and 96.3%, respectively. However, none of them achieved the model performance as high as XGBoost.

This chapter references the same experimental design with the updated dataset available from the ongoing data collection process. Instead of training multiple models, XGBoost has been chosen as the example to illustrate how such an approach works for implementing a decision support system for automatic diagnosis of COVID-19. Figure 5A shows the ROC curve of the XGBoost classification result, indicating the model effectiveness for the classification task, along with a higher AUC score achieved at 0.982 due to the increased size of samples. In feature importance analysis presented in Figure 5B, the top 10 important features in the prediction model are age, lymphocytes, neutrophil, cough, fever, fatigue, CT scan result, normal serum level of white blood cell, low serum level of white blood cell, and no cough. Again, such a result is sightly different comparing to the previous study. A decision tree structure with 6 levels of the XGBoost model is presented in Figure 5C, which is a representation of the full model.

Figure 5. Experimental results based on XGBoost classification of COVID-19 and influenza individuals. (A). ROC curve. (B). feature importance ranking for classification (C). 6-level of decision tree structure.

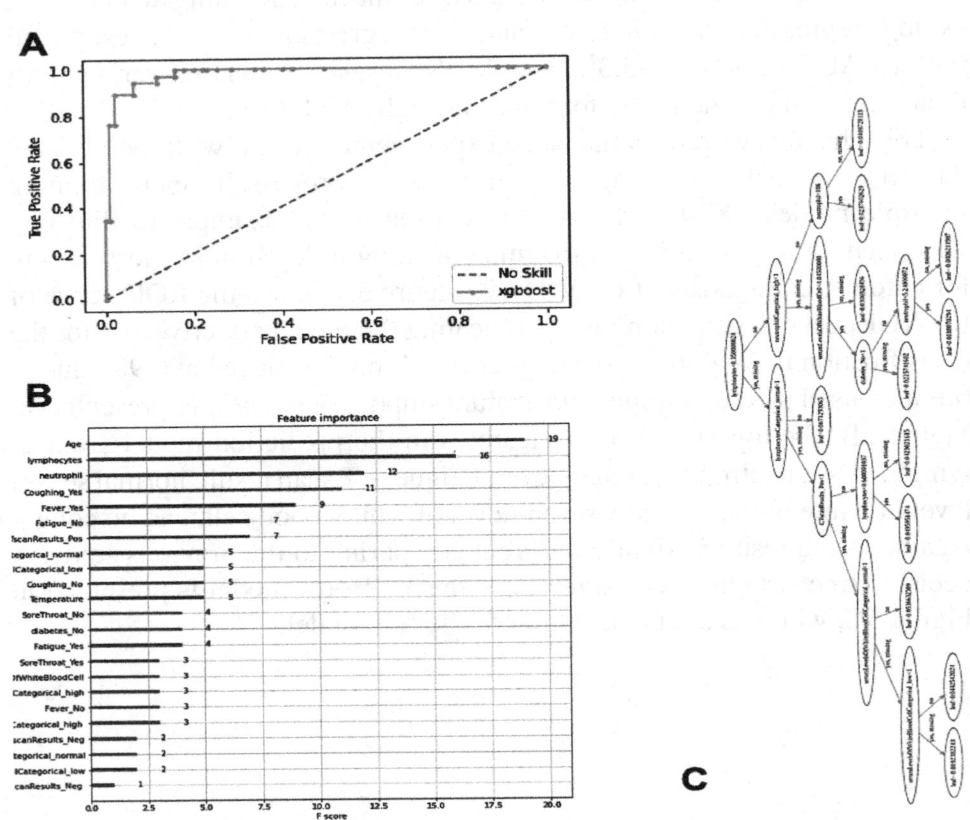

Although XGBoost can achieve a high degree in model performance on patient classification, some limitations still exist. Most of the open-access clinical reports do not include individual-level patient data. Therefore, the attainable data for model training is seriously limited. The number of influenza patients is greater than COVID-19 patients, thereby causing the model trained on the unbalanced input. Such a problem may influence the performance of XGBoost model training due to the unbalanced distribution in the training set. Additionally, such an experimental design only incorporates H1N1 influenza signs and symptoms due to the difficulty of the data acquisition. The strains active in the current influenza season should be further considered. Furthermore, the dataset focuses on certain cases of COVID-19 patients, thereby it may not be able to represent the situation of the general population.

The sample risk should be avoided in the further stage. Nevertheless, the machine learning-based classification can be applied to implementing a decision support system for diagnosing COVID-19 automatically in terms of the case prediction, the feature importance analysis, and the decision tree construction. Therefore, such an approach can be involved into the smart health system designed for COVID-19 diagnosis, particularly when medical test results are measured from other smart diagnostic frameworks, such as data-driven NAATs, smart serological signatures identification, and deep learning-based medical image classifications.

FUTURE RESEARCH DIRECTIONS

The future research directions will focus on how to incorporate existing digitalized healthcare objects into the smart diagnosis system for COVID-19. Existing studies have discussed potential applications of two additional assays, such as ELISA and Point-of-Care assay, with big data and AI. However, such proposed tools have not yet been applied for COVID-19. The gap in between will provide more research opportunities towards this direction. Firstly, ELISA is a biochemistry assay that can detect the existence of antibodies, antigens, proteins in biological samples. Machine learning and image analysis techniques can be used to fulfill the combination of high sensitivity and precision with a short assay. Deep learning has been used to count the numbers of capturing beads and the enzyme active beads for a certain cytokine observation. CNN-based image processing is applied to deal with multiplexed digital immunoassay. One challenge of applying the digital immunoassay approach is to give a fast and reasonable data analysis without human supervision. In addition, another challenge is to discern multi-color-filled images and empty microwells and to detect signals accurately even when the image defects appear and image focus shifts. Gou et al. (2019) proposed a machine learning method to solve these challenges, which can avoid human supervision for error correction. Moreover, to apply multi-color images for high-throughput analysis, a dual-pathway parallel-computing algorithm was developed by Song et al. (2021). The basic concept of this novel algorithm is CNN visualization for image processing. Such a CNN-based network architecture contains two parts. One is the downsampling process for category classification, which has two convolution layers and a max-pooling layer. Another is the upsampling process for pixel segmentation, which has a transposed convolution layer with ReLU, a softmax layer and a pixel classification layer. To evaluate the effectiveness of the CNN method, a comparison with global thresholding and segmentation (GTS) is performed. The images for each source were used to train the neural network. Secondly, Point-of-Care Testing (POCT) has applied data analytics, artificial intelligence, and machine learning techniques to increase the diagnosis efficiency. Hardware components for POCT include electrochemical and photonic systems and the mobile application. Mobile phone usually can play the role of reading or connecting. Priye et al. (2017) proposed the Zika virus detection, which is a mobile detection tool to collect crude human sample matrices, containing blood, saliva, and urine. Additionally, this mobile tool can also effectively detect dengue and chikungunya viruses. The smartphone camera is used as a light source to image the reaction tubes on isothermal hot plate with sensors that can measure and record the surface temperature. Besides, a paper-based blood type detector has been developed by Guan et al. (2014). The installed smartphone app can be used to analyze the result of sensor and display the detection result of the blood type on the smartphone screen as a text. Moreover, the phone-based POCT can be further improved by the combination of sensing platforms. The sensing platforms includes Optosensors, Electrochemical sensors, and mechanical sensors. Optosensors usually use smartphone camera as an optical and mechanical component. Electrochemical sensors can measure biomarker concentration through reduction-oxidation reactions. Mechanical sensing platforms can measure particle concentration by the adsorption of biomolecules and the changes of resonance frequency. Moreover, O'Sullivan et al. (2019) proposed a survey on developments in transduction, connectivity and deep learning for POCT. Such a study provides a comprehensive vision for guiding the construction of the smart health system in the form of the digitalized POCT with AI.

CONCLUSION

The urgency of COVID-19 has compiled the state-of-the-art technologies in smart diagnosis with the increasing number of innovations in this field. The support of data-driven approaches in COVID-19 diagnostics is irreplaceable. The attempts in existing data analysis, machine learning, deep learning, medical image processing, along with the decision support system in which they are applied to construct the smart health framework, have been investigated and examined in detail. The findings illustrate how big data and AI can be configured to tackle the current situations in COVID-19 diagnosis, providing deep and experimental insight on multiple diagnostic assays and COVID-19 detection approaches incorporating nucleic acid test, serological assay, X-ray and CT image-based diagnosis, and automatic diagnosis with clinical data. A broader picture of how other technologies, such as smart ELISA and digitalized POCT, has been also discussed as the supplementary of the smart diagnostic system for COVID-19.

REFERENCES

Abbas, A., Abdelsamea, M. M., & Gaber, M. M. (2021). 4S-DT: Self-Supervised Super Sample Decomposition for Transfer Learning With Application to COVID-19 Detection. *IEEE Transactions on Neural Networks and Learning Systems*.

Ai, T., Yang, Z., Hou, H., Zhan, C., Chen, C., Lv, W., Tao, Q., Sun, Z., & Xia, L. (2020). Correlation of chest CT and RT-PCR testing for coronavirus disease 2019 (COVID-19) in China: A report of 1014 cases. *Radiology*, *296*(2), E32–E40. doi:10.1148/radiol.2020200642 PMID:32101510

Al-Karawi, D., Al-Zaidi, S., Polus, N., & Jassim, S. (2020). *Machine learning analysis of chest CT scan images as a complementary digital test of coronavirus (COVID-19) patients*. MedRxiv; doi:10.1101/2020.04.13.20063479

Alakus, T. B., & Turkoglu, I. (2020). Comparison of deep learning approaches to predict COVID-19 infection. *Chaos, Solitons, and Fractals*, *140*, 110120. doi:10.1016/j.chaos.2020.110120 PMID:33519109

Ameen, Z. S. I., Ozsoz, M., Mubarak, A. S., Al Turjman, F., & Serte, S. (2021). C-SVR Crispr: Prediction of CRISPR/Cas12 guideRNA activity using deep learning models. *Alexandria Engineering Journal*, *60*(4), 3501–3508. doi:10.1016/j.aej.2021.02.007

Apostolopoulos, I. D., & Mpesiana, T. A. (2020). Covid-19: Automatic detection from x-ray images utilizing transfer learning with convolutional neural networks. *Physical and Engineering Sciences in Medicine*, *43*(2), 635–640. doi:10.100713246-020-00865-4 PMID:32524445

Arpaci, I., Huang, S., Al-Emran, M., Al-Kabi, M. N., & Peng, M. (2021). Predicting the COVID-19 infection with fourteen clinical features using machine learning classification algorithms. *Multimedia Tools and Applications*, *80*(8), 11943–11957. doi:10.100711042-020-10340-7 PMID:33437173

Assaf, D., Gutman, Y. A., Neuman, Y., Segal, G., Amit, S., Gefen-Halevi, S., Shilo, N., Epstein, A., Mor-Cohen, R., Biber, A., Rahav, G., Levy, I., & Tirosh, A. (2020). Utilization of machine-learning models to accurately predict the risk for critical COVID-19. *Internal and Emergency Medicine, 15*(8), 1435–1443. doi:10.100711739-020-02475-0 PMID:32812204

Awulachew, E., Diriba, K., Anja, A., Getu, E., & Belayneh, F. (2020). Computed tomography (CT) imaging features of patients with COVID-19: Systematic review and meta-analysis. *Radiology Research and Practice, 2020*, 2020. doi:10.1155/2020/1023506 PMID:32733706

Basu, A., Zinger, T., Inglima, K., Woo, K. M., Atie, O., Yurasits, L., See, B., & Aguero-Rosenfeld, M. E. (2020). Performance of Abbott ID Now COVID-19 rapid nucleic acid amplification test using nasopharyngeal swabs transported in viral transport media and dry nasal swabs in a New York City academic institution. *Journal of Clinical Microbiology, 58*(8), e01136–e20. doi:10.1128/JCM.01136-20 PMID:32471894

Brinati, D., Campagner, A., Ferrari, D., Locatelli, M., Banfi, G., & Cabitza, F. (2020). Detection of COVID-19 infection from routine blood exams with machine learning: A feasibility study. *Journal of Medical Systems, 44*(8), 1–12. doi:10.100710916-020-01597-4 PMID:32607737

Cardozo, K. H. M., Lebkuchen, A., Okai, G. G., Schuch, R. A., Viana, L. G., Olive, A. N., Lazari, C. S., Fraga, A. M., Granato, C. F. H., Pintão, M. C. T., & Carvalho, V. M. (2020). Establishing a mass spectrometry-based system for rapid detection of SARS-CoV-2 in large clinical sample cohorts. *Nature Communications, 11*(1), 1–13. doi:10.103841467-020-19925-0 PMID:33273458

Chang, C. M., Tan, T. W., Ho, T. C., Chen, C. C., Su, T. H., & Lin, C. Y. (2020). COVID-19: Taiwan's epidemiological characteristics and public and hospital responses. *PeerJ, 8*, e9360. doi:10.7717/peerj.9360 PMID:32551205

Chung, M., Bernheim, A., Mei, X., Zhang, N., Huang, M., Zeng, X., Cui, J., Xu, W., Yang, Y., Fayad, Z. A., Jacobi, A., Li, K., Li, S., & Shan, H. (2020). CT imaging features of 2019 novel coronavirus (2019-nCoV). *Radiology, 295*(1), 202–207. doi:10.1148/radiol.2020200230 PMID:32017661

Cohen, J. P., Morrison, P., Dao, L., Roth, K., Duong, T. Q., & Ghassemi, M. (2020). *Covid-19 image data collection: Prospective predictions are the future.* arXiv preprint arXiv:2006.11988.

Corman, V. M., Landt, O., Kaiser, M., Molenkamp, R., Meijer, A., Chu, D. K., Bleicker, T., Brünink, S., Schneider, J., Schmidt, M. L., Mulders, D. G. J. C., Haagmans, B. L., van der Veer, B., van den Brink, S., Wijsman, L., Goderski, G., Romette, J.-L., Ellis, J., Zambon, M., ... Drosten, C. (2020). Detection of 2019 novel coronavirus (2019-nCoV) by real-time RT-PCR. *Eurosurveillance*, *25*(3), 2000045. doi:10.2807/1560-7917.ES.2020.25.3.2000045 PMID:31992387

de Oliveira, K. G., Estrela, P. F. N., de Melo Mendes, G., Dos Santos, C. A., de Paula Silveira-Lacerda, E., & Duarte, G. R. M. (2021). Rapid molecular diagnostics of COVID-19 by RT-LAMP in a centrifugal polystyrene-toner based microdevice with end-point visual detection. *Analyst (London)*, *146*(4), 1178–1187. doi:10.1039/D0AN02066D PMID:33439160

Ding, X., Yin, K., Li, Z., & Liu, C. (2020). All-in-One dual CRISPR-cas12a (AIOD-CRISPR) assay: A case for rapid, ultrasensitive and visual detection of novel coronavirus SARS-CoV-2 and HIV virus. *bioRxiv*. doi:10.1101/2020.03.19.998724

Durrani, M., Inam ul Haq, U. K., & Yousaf, A. (2020). Chest X-rays findings in COVID 19 patients at a University Teaching Hospital-A descriptive study. *Pakistan Journal of Medical Sciences, 36*(S4), S22.

Fields, B. K., Demirjian, N. L., Dadgar, H., & Gholamrezanezhad, A. (2020, November). Imaging of COVID-19: CT, MRI, and PET. In Seminars in Nuclear Medicine. WB Saunders.

Girija, A. S., Shankar, E. M., & Larsson, M. (2020). Could SARS-CoV-2-induced hyperinflammation magnify the severity of coronavirus disease (CoViD-19) leading to acute respiratory distress syndrome? *Frontiers in Immunology*, *11*, 1206. doi:10.3389/fimmu.2020.01206 PMID:32574269

Gomes, J. C., Silva, L. H. D. S., Ferreira, J., Júnior, A. A., dos Santos Rocha, A. L., Castro, L., ... dos Santos, W. P. (2020). Optimizing the molecular diagnosis of Covid-19 by combining RT-PCR and a pseudo-convolutional machine learning approach to characterize virus DNA sequences. *bioRxiv*. doi:10.1101/2020.06.02.129775

Goodman-Meza, D., Rudas, A., Chiang, J. N., Adamson, P. C., Ebinger, J., Sun, N., Botting, P., Fulcher, J. A., Saab, F. G., Brook, R., Eskin, E., An, U., Kordi, M., Jew, B., Balliu, B., Chen, Z., Hill, B. L., Rahmani, E., Halperin, E., & Manuel, V. (2020). A machine learning algorithm to increase COVID-19 inpatient diagnostic capacity. *PLoS One*, *15*(9), e0239474. doi:10.1371/journal.pone.0239474 PMID:32960917

Gou, T., Hu, J., Zhou, S., Wu, W., Fang, W., Sun, J., Hu, Z., Shen, H., & Mu, Y. (2019). A new method using machine learning for automated image analysis applied to chip-based digital assays. *Analyst (London)*, *144*(10), 3274–3281. doi:10.1039/C9AN00149B PMID:30990486

Guan, L., Tian, J., Cao, R., Li, M., Cai, Z., & Shen, W. (2014). Barcode-like paper sensor for smartphone diagnostics: An application of blood typing. *Analytical Chemistry*, *86*(22), 11362–11367. doi:10.1021/ac503300y PMID:25301220

Guan, W. J., Ni, Z. Y., Hu, Y., Liang, W. H., Ou, C. Q., He, J. X., & Zhong, N. S. (2020). *Clinical characteristics of 2019 novel coronavirus infection in China*. MedRxiv; doi:10.1101/2020.02.06.20020974

Horry, M. J., Chakraborty, S., Paul, M., Ulhaq, A., Pradhan, B., Saha, M., & Shukla, N. (2020). COVID-19 detection through transfer learning using multimodal imaging data. *IEEE Access: Practical Innovations, Open Solutions*, *8*, 149808–149824. doi:10.1109/ACCESS.2020.3016780 PMID:34931154

Huang, C., Wang, Y., Li, X., Ren, L., Zhao, J., Hu, Y., Zhang, L., Fan, G., Xu, J., Gu, X., Cheng, Z., Yu, T., Xia, J., Wei, Y., Wu, W., Xie, X., Yin, W., Li, H., Liu, M., ... Cao, B. (2020). Clinical features of patients infected with 2019 novel coronavirus in Wuhan, China. *Lancet*, *395*(10223), 497–506. doi:10.1016/S0140-6736(20)30183-5 PMID:31986264

Ismael, A. M., & Şengür, A. (2021). Deep learning approaches for COVID-19 detection based on chest X-ray images. *Expert Systems with Applications*, *164*, 114054. doi:10.1016/j.eswa.2020.114054 PMID:33013005

Kana, E. B. G., Kana, M. G. Z., Kana, A. F. D., & Kenfack, R. H. A. (2020). A web-based diagnostic tool for COVID-19 using machine learning on chest radiographs (CXR). medRxiv. doi:10.1101/2020.04.21.20063263

Khuzani, A. Z., Heidari, M., & Shariati, S. A. (2021). COVID-Classifier: An automated machine learning model to assist in the diagnosis of COVID-19 infection in chest x-ray images. *Scientific Reports*, *11*(1), 1–6. PMID:33414495

Kobayashi, T., Jung, S. M., Linton, N. M., Kinoshita, R., Hayashi, K., Miyama, T., ... Nishiura, H. (2020). *Communicating the risk of death from novel coronavirus disease (COVID-19).* Academic Press.

Krammer, F., & Simon, V. (2020). Serology assays to manage COVID-19. *Science, 368*(6495), 1060–1061. doi:10.1126cience.abc1227 PMID:32414781

Langenbach, M. C., Hokamp, N. G., Persigehl, T., & Bratke, G. (2020). MRI appearance of COVID-19 infection. *Diagnostic and Interventional Radiology, 26*(4), 377–378. doi:10.5152/dir.2020.20152 PMID:32352920

Li, L. Q., Huang, T., Wang, Y. Q., Wang, Z. P., Liang, Y., Huang, T. B., Zhang, H., Sun, W., & Wang, Y. (2020). COVID-19 patients' clinical characteristics, discharge rate, and fatality rate of meta-analysis. *Journal of Medical Virology, 92*(6), 577–583. doi:10.1002/jmv.25757 PMID:32162702

Li, W. T., Ma, J., Shende, N., Castaneda, G., Chakladar, J., Tsai, J. C., Apostol, L., Honda, C. O., Xu, J., Wong, L. M., Zhang, T., Lee, A., Gnanasekar, A., Honda, T. K., Kuo, S. Z., Yu, M. A., Chang, E. Y., Rajasekaran, M. R., & Ongkeko, W. M. (2020). Using machine learning of clinical data to diagnose COVID-19: A systematic review and meta-analysis. *BMC Medical Informatics and Decision Making, 20*(1), 1–13. doi:10.118612911-020-01266-z PMID:32993652

Lippi, G., Simundic, A. M., & Plebani, M. (2020). Potential preanalytical and analytical vulnerabilities in the laboratory diagnosis of coronavirus disease 2019 (COVID-19). *Clinical Chemistry and Laboratory Medicine, 58*(7), 1070–1076. doi:10.1515/cclm-2020-0285 PMID:32172228

Loeffelholz, M. J., & Tang, Y. W. (2020). Laboratory diagnosis of emerging human coronavirus infections–the state of the art. *Emerging Microbes & Infections, 9*(1), 747–756. doi:10.1080/22221751.2020.1745095 PMID:32196430

Lu, R., Zhao, X., Li, J., Niu, P., Yang, B., Wu, H., Wang, W., Song, H., Huang, B., Zhu, N., Bi, Y., Ma, X., Zhan, F., Wang, L., Hu, T., Zhou, H., Hu, Z., Zhou, W., Zhao, L., ... Tan, W. (2020). Genomic characterisation and epidemiology of 2019 novel coronavirus: Implications for virus origins and receptor binding. *Lancet, 395*(10224), 565–574. doi:10.1016/S0140-6736(20)30251-8 PMID:32007145

Lucia, C., Federico, P. B., & Alejandra, G. C. (2020). An ultrasensitive, rapid, and portable coronavirus SARS-CoV-2 sequence detection method based on CRISPR-Cas12. *bioRxiv*. doi:10.1101/2020.02.29.971127

Mahdy, L. N., Ezzat, K. A., Elmousalami, H. H., Ella, H. A., & Hassanien, A. E. (2020). *Automatic x-ray covid-19 lung image classification system based on multi-level thresholding and support vector machine*. MedRxiv; doi:10.1101/2020.03.30.20047787

Memish, Z. A., Al-Tawfiq, J. A., Assiri, A., AlRabiah, F. A., Al Hajjar, S., Albarrak, A., Flemban, H., Alhakeem, R. F., Makhdoom, H. Q., Alsubaie, S., & Al-Rabeeah, A. A. (2014). Middle East respiratory syndrome coronavirus disease in children. *The Pediatric Infectious Disease Journal*, *33*(9), 904–906. doi:10.1097/INF.0000000000000325 PMID:24763193

Metsky, H. C., Freije, C. A., Kosoko-Thoroddsen, T. S. F., Sabeti, P. C., & Myhrvold, C. (2020). CRISPR-based surveillance for COVID-19 using genomically-comprehensive machine learning design. *bioRxiv*. doi:10.1101/2020.02.26.967026

Mohanty, A., Fatrekar, A. P., Krishnan, S., & Vernekar, A. A. (2021). A Concise Discussion on the Potential Spectral Tools for the Rapid COVID-19 Diagnosis. *Results in chemistry*, 100138.

Mooney, P. (2019). *Chest X-Ray Images*. https://www.kaggle.com/paultimothymooney/chest-xray-pneumonia

Mowery, C. T., Marson, A., Song, Y. S., & Ye, C. J. (2020). Improved covid-19 serology test performance by integrating multiple lateral flow assays using machine learning. medRxiv. doi:10.1101/2020.07.15.20154773

O'Sullivan, S., Ali, Z., Jiang, X., Abdolvand, R., Ünlü, M. S., Plácido da Silva, H., ... Holzinger, A. (2019). Developments in transduction, connectivity and AI/machine learning for point-of-care testing. *Sensors (Basel)*, *19*(8), 1917. doi:10.339019081917 PMID:31018573

Poggiali, E., Zaino, D., Immovilli, P., Rovero, L., Losi, G., Dacrema, A., Nuccetelli, M., Vadacca, G. B., Guidetti, D., Vercelli, A., Magnacavallo, A., Bernardini, S., & Terracciano, C. (2020). Lactate dehydrogenase and C-reactive protein as predictors of respiratory failure in CoVID-19 patients. *Clinica Chimica Acta*, *509*, 135–138. doi:10.1016/j.cca.2020.06.012 PMID:32531257

Priye, A., Bird, S. W., Light, Y. K., Ball, C. S., Negrete, O. A., & Meagher, R. J. (2017). A smartphone-based diagnostic platform for rapid detection of Zika, chikungunya, and dengue viruses. *Scientific Reports*, *7*(1), 1–11. doi:10.1038rep44778 PMID:28317856

Rai, P., Kumar, B. K., Deekshit, V. K., Karunasagar, I., & Karunasagar, I. (2021). Detection technologies and recent developments in the diagnosis of COVID-19 infection. *Applied Microbiology and Biotechnology*, *105*(2), 1–15. doi:10.100700253-020-11061-5 PMID:33394144

Randhawa, G. S., Soltysiak, M. P., El Roz, H., de Souza, C. P., Hill, K. A., & Kari, L. (2020a). Machine learning analysis of genomic signatures provides evidence of associations between Wuhan 2019-nCoV and bat betacoronaviruses. BioRxiv.

Randhawa, G. S., Soltysiak, M. P., El Roz, H., de Souza, C. P., Hill, K. A., & Kari, L. (2020b). Machine learning using intrinsic genomic signatures for rapid classification of novel pathogens: COVID-19 case study. *PLoS One*, *15*(4), e0232391. doi:10.1371/journal.pone.0232391 PMID:32330208

Rao, T. A., Paul, N., Chung, T., Mazzulli, T., Walmsley, S., Boylan, C. E., Provost, Y., Herman, S. J., Weisbrod, G. L., & Roberts, H. C. (2003). Value of CT in assessing probable severe acute respiratory syndrome. *AJR. American Journal of Roentgenology*, *181*(2), 317–319. doi:10.2214/ajr.181.2.1810317 PMID:12876004

Rocca, M. F., Zintgraff, J. C., Dattero, M. E., Santos, L. S., Ledesma, M., Vay, C., Prieto, M., Benedetti, E., Avaro, M., Russo, M., Nachtigall, F. M., & Baumeister, E. (2020). A combined approach of MALDI-TOF mass spectrometry and multivariate analysis as a potential tool for the detection of SARS-CoV-2 virus in nasopharyngeal swabs. *Journal of Virological Methods*, *286*, 113991. doi:10.1016/j.jviromet.2020.113991 PMID:33045283

Rosado, J., Pelleau, S., Cockram, C., Merkling, S. H., Nekkab, N., Demeret, C., Meola, A., Kerneis, S., Terrier, B., Fafi-Kremer, S., de Seze, J., Bruel, T., Dejardin, F., Petres, S., Longley, R., Fontanet, A., Backovic, M., Mueller, I., & White, M. T. (2021). Multiplex assays for the identification of serological signatures of SARS-CoV-2 infection: An antibody-based diagnostic and machine learning study. *The Lancet Microbe*, *2*(2), e60–e69. doi:10.1016/S2666-5247(20)30197-X PMID:33521709

Rosado, J., Pelleau, S., Cockram, C., Merkling, S. H., Nekkab, N., Demeret, C., & White, M. T. (2020). *Serological signatures of SARS-CoV-2 infection: implications for antibody-based diagnostics*. MedRxiv; doi:10.1101/2020.05.07.20093963

Samacoits, A., Nimsamer, P., Mayuramart, O., Chantaravisoot, N., Sitthi-Amorn, P., Nakhakes, C., Luangkamchorn, L., Tongcham, P., Zahm, U., Suphanpayak, S., Padungwattanachoke, N., Leelarthaphin, N., Huayhongthong, H., Pisitkun, T., Payungporn, S., & Hannanta-Anan, P. (2021). Machine Learning-Driven and Smartphone-Based Fluorescence Detection for CRISPR Diagnostic of SARS-CoV-2. *ACS Omega*, *6*(4), 2727–2733. doi:10.1021/acsomega.0c04929 PMID:33553890

Sitaula, C., & Hossain, M. B. (2021). Attention-based VGG-16 model for COVID-19 chest X-ray image classification. *Applied Intelligence*, *51*(5), 2850–2863. doi:10.100710489-020-02055-x PMID:34764568

Song, Y., Zhao, J., Cai, T., Stephens, A., Su, S. H., Sandford, E., Flora, C., Singer, B. H., Ghosh, M., Choi, S. W., Tewari, M., & Kurabayashi, K. (2021). Machine learning-based cytokine microarray digital immunoassay analysis. *Biosensors & Bioelectronics*, *180*, 113088. doi:10.1016/j.bios.2021.113088 PMID:33647790

Surkova, E., Nikolayevskyy, V., & Drobniewski, F. (2020). False-positive COVID-19 results: Hidden problems and costs. *The Lancet. Respiratory Medicine*, *8*(12), 1167–1168. doi:10.1016/S2213-2600(20)30453-7 PMID:33007240

Tordjman, M., Mekki, A., Mali, R. D., Saab, I., Chassagnon, G., Guillo, E., Burns, R., Eshagh, D., Beaune, S., Madelin, G., Bessis, S., Feydy, A., Mihoubi, F., Doumenc, B., Mouthon, L., Carlier, R.-Y., Drapé, J.-L., & Revel, M. P. (2020). Pre-test probability for SARS-Cov-2-related infection score: The PARIS score. *PLoS One*, *15*(12), e0243342. doi:10.1371/journal.pone.0243342 PMID:33332360

Tschoellitsch, T., Dünser, M., Böck, C., Schwarzbauer, K., & Meier, J. (2021). Machine learning prediction of sars-cov-2 polymerase chain reaction results with routine blood tests. *Laboratory Medicine*, *52*(2), 146–149. doi:10.1093/labmed/lmaa111 PMID:33340312

Walls, A. C., Park, Y. J., Tortorici, M. A., Wall, A., McGuire, A. T., & Veesler, D. (2020). Structure, function, and antigenicity of the SARS-CoV-2 spike glycoprotein. *Cell*, *181*(2), 281–292. doi:10.1016/j.cell.2020.02.058 PMID:32155444

Wang, H., Liu, Y., Hu, K., Zhang, M., Du, M., Huang, H., & Yue, X. (2020). Healthcare workers' stress when caring for COVID-19 patients: An altruistic perspective. *Nursing Ethics*, *27*(7), 1490–1500. doi:10.1177/0969733020934146 PMID:32662326

West, C. P., Montori, V. M., & Sampathkumar, P. (2020, June). COVID-19 testing: The threat of false-negative results. *Mayo Clinic Proceedings*, *95*(6), 1127–1129. doi:10.1016/j.mayocp.2020.04.004 PMID:32376102

Wu, A., Peng, Y., Huang, B., Ding, X., Wang, X., Niu, P., Meng, J., Zhu, Z., Zhang, Z., Wang, J., Sheng, J., Quan, L., Xia, Z., Tan, W., Cheng, G., & Jiang, T. (2020b). Genome composition and divergence of the novel coronavirus (2019-nCoV) originating in China. *Cell Host & Microbe*, *27*(3), 325–328. doi:10.1016/j.chom.2020.02.001 PMID:32035028

Wu, J. L., Tseng, W. P., Lin, C. H., Lee, T. F., Chung, M. Y., Huang, C. H., Chen, S.-Y., Hsueh, P.-R., & Chen, S. C. (2020a). Four point-of-care lateral flow immunoassays for diagnosis of COVID-19 and for assessing dynamics of antibody responses to SARS-CoV-2. *The Journal of Infection*, *81*(3), 435–442. doi:10.1016/j.jinf.2020.06.023 PMID:32553841

Xiang, J., Yan, M., Li, H., Liu, T., Lin, C., Huang, S., & Shen, C. (2020). *Evaluation of enzyme-linked immunoassay and colloidal gold-immunochromatographic assay kit for detection of novel coronavirus (SARS-Cov-2) causing an outbreak of pneumonia (COVID-19)*. MedRxiv; doi:10.1101/2020.02.27.20028787

Yan, L., Yi, J., Huang, C., Zhang, J., Fu, S., Li, Z., Lyu, Q., Xu, Y., Wang, K., Yang, H., Ma, Q., Cui, X., Qiao, L., Sun, W., & Liao, P. (2021). Rapid detection of COVID-19 using MALDI-TOF-based serum peptidome profiling. *Analytical Chemistry*, *93*(11), 4782–4787. doi:10.1021/acs.analchem.0c04590 PMID:33656857

Yang, W., & Yan, F. (2020). Patients with RT-PCR-confirmed COVID-19 and normal chest CT. *Radiology*, *295*(2), E3–E3. doi:10.1148/radiol.2020200702 PMID:32142398

Zaim, S., Chong, J. H., Sankaranarayanan, V., & Harky, A. (2020). COVID-19 and multi-organ response. *Current Problems in Cardiology*, *45*(8), 100618. doi:10.1016/j.cpcardiol.2020.100618 PMID:32439197

Zhou, P., Yang, X. L., Wang, X. G., Hu, B., Zhang, L., Zhang, W., ... Shi, Z. L. (2020). A pneumonia outbreak associated with a new coronavirus of probable bat origin. *Nature, 579*(7798), 270-273.

Zou, L., Ruan, F., Huang, M., Liang, L., Huang, H., Hong, Z., Yu, J., Kang, M., Song, Y., Xia, J., Guo, Q., Song, T., He, J., Yen, H.-L., Peiris, M., & Wu, J. (2020). SARS-CoV-2 viral load in upper respiratory specimens of infected patients. *The New England Journal of Medicine*, *382*(12), 1177–1179. doi:10.1056/NEJMc2001737 PMID:32074444

ADDITIONAL READING

Al-Turjman, F. (Ed.). (2021). *Artificial Intelligence and Machine Learning for COVID-19* (Vol. 924). Springer Nature. doi:10.1007/978-3-030-60188-1

Alonso-Betanzos, A., & Bolón-Canedo, V. (2018). Big-data analysis, cluster analysis, and machine-learning approaches. *Sex-Specific Analysis of Cardiovascular Function*, 607-626.

Bankman, I. (Ed.). (2008). *Handbook of medical image processing and analysis*. Elsevier.

Cleophas, T. J., Zwinderman, A. H., & Cleophas-Allers, H. I. (2013). *Machine learning in medicine* (Vol. 9). Springer.

Haq, A. U., Li, J. P., Khan, J., Memon, M. H., Nazir, S., Ahmad, S., Khan, G. A., & Ali, A. (2020). Intelligent machine learning approach for effective recognition of diabetes in E-healthcare using clinical data. *Sensors (Basel)*, *20*(9), 2649. doi:10.339020092649 PMID:32384737

Kautish, S. (2021). *Computational Intelligence Techniques for Combating COVID-19*. Springer Nature. doi:10.1007/978-3-030-68936-0

Kwekha-Rashid, A. S., Abduljabbar, H. N., & Alhayani, B. (2021). Coronavirus disease (COVID-19) cases analysis using machine-learning applications. *Applied Nanoscience*, 1–13. doi:10.100713204-021-01868-7 PMID:34036034

Razzak, M. I., Naz, S., & Zaib, A. (2018). Deep learning for medical image processing: Overview, challenges and the future. *Classification in BioApps*, 323-350.

Wang, W., Liang, D., Chen, Q., Iwamoto, Y., Han, X. H., Zhang, Q., & Chen, Y. W. (2020). Medical image classification using deep learning. In *Deep Learning in Healthcare* (pp. 33–51). Springer. doi:10.1007/978-3-030-32606-7_3

Zhou, S. K., Greenspan, H., & Shen, D. (Eds.). (2017). *Deep learning for medical image analysis*. Academic Press.

KEY TERMS AND DEFINITIONS

Clinical Data Analysis: An analysis method that is used to apply on clinical data, including electronic health records, patient registries, disease records, clinical trial records, etc.

Decision Support System: A computer-based framework that can process and analyze the large scale of data for extracting useful knowledges and information, which can be applied to solve problems in decision-making.

Deep Learning: A broad family of machine learning models based on neural networks. Typical deep learning models are deep neural networks, convolutional neural networks, recurrent neural networks, deep belief networks, and deep reinforcement learning.

Machine Learning: A subject of artificial intelligence that aims at the task of computational algorithms, which allow machines to learning objects automatically through historical data.

Medical Image Processing: A processing method that is applied for medical images of human body tissues or organs, such as CT scan and MRI scan, in order to perform diagnosis and clinical analysis.

RT-PCR: A technique that can compound transcription of RNA into DNA and apply polymerase chain reaction to increase DNA targets.

Serological Signatures Identification: An antibody-based diagnostic that uses serological signature to identify infected cases.

Smart Diagnostics: A technology that can give correct diagnosis quickly through electronic devices, such as mobile and smart watch.

Chapter 9
The Role of Big Data Analytics in Drug Discovery and Vaccine Development Against COVID-19

ABSTRACT

Scientific studies related to information on possible treatments and vaccines have been growing with the development of the COVID-19 pandemic. The research databases are publicly available, which provides a solid resource in supporting the global research community. However, challenges remain in terms of searching the insightful information quickly for the purpose of finding the right treatments and vaccines in the current situation. Artificial intelligence technologies can help to build tools in order to search, rank, extract, and aggregate useful results from enormous databases. This chapter presents a systematic review for investigating current research in drug discovery and vaccine development for COVID-19 throughout protein structural basis analysis and visualization, machine learning- and deep learning-based models, and a big data-driven approach. The survey study indicates that applied big data and AI can generate new insights in support of the ongoing fight against COVID-19 in terms of developing new drugs and vaccines efficiently.

DOI: 10.4018/978-1-7998-8793-5.ch009

INTRODUCTION

During the last few decades, biological researchers have been challenged by the development of drug and vaccine in terms of advanced systems for the delivery of therapeutic agents in regards to maximum efficiency and minimum risks. The cost efficiency and medical safety in discovering new medicines and vaccines has been incorporated in the biological industry throughout the whole process of medical product designs. Traditional methods, such as virtual screening (VS) and molecular docking, have been applied to handle operational challenges and technical difficulties, however, such techniques impose another aspect of issues, such as inefficiency and inaccuracy, thereby a surge in the implementation of novel approaches. With the development of artificial intelligence and big data, state-of-the-art technologies, such as machine learning, deep learning, 3D visualization, and natural language processing (NLP), have been widely applied to overcome problems and hurdles encountered in classical computational ideologies for developing and discovering new drugs and vaccines through complex stages, including target selection, candidate validation, therapeutic screening and optimization, clinical trials, and manufacturing tests and practices. Nevertheless, massive controversies and problems in identifying the effectiveness of the medication against a specific disease have been imposed throughout all stages of the drug and vaccine development and discovery. The most significant issue is how to manage the effectiveness of the products, the cost of development, and the efficiency of the industrial process (Zhang et al., 2017), whereas using cutting-edge technologies can provide all essential solutions in the form of the implementation of AI and big data to deal with the problems of scrutinizing the data complexity, reducing the time consumption, minimizing the cost in the pharmaceutical industry and healthcare providers. Such an efficient and scientific manner can be shared with the current situation in the COVID-19 pandemic.

Recently, AI models and big data analytics have been applied in drug and vaccine development, which has been extensively discussed by biological industrial professionals and research communities (Harrer et al., 2019). Based on ML and DL principles, a large scale of avenue of computational methods has been proposed for chemical compounds identification and validation, target identification, peptide synthesis, drug screening and monitoring, drug toxicity evaluation, physiochemical property testing, vaccine efficiency evaluation and repositioning (Jabbari & Rezaei, 2019). VS of compounds from chemical and biological libraries become efficient with the advantages of AI and big data, which can be applied in computing hundreds billion compounds in minutes. With the development of microarray, RNA sequencing (RNA-seq) and high-throughput sequencing (HTS) techniques have been widely applied in drug discovery and vaccine development, which can generate a large volume of biomedical data every day. Such data repositories can be used in finding suitable biological products in terms of the identification of appropriate targets, such as genes and proteins. Big data approaches have made the computational processes much easier by using advanced machine learning and deep learning algorithms with data visualization, which can be applied in feature extraction, patterns recognition, and structure visualization through analyzing and processing the large scale datasets in the biomedical industry (Palanisamy & Thirunavukarasu, 2019). Understanding disease mechanisms is the first step of target identification, of which the gene expression has been widely applied to find genes responsible for the disease, whereas a large amount of data for various disorders has been made by RNA-seq and HTS. Such a process can be implemented by analyzing gene signatures using gene expression data with machine learning for exploring new biomarkers and potential targets (van IJzendoorn et al., 2019). Such gene expression data can be obtained from the big repositories, such as The Cancer Genome Atlas (TCGA), NCBI Gene Expression Omnibus (GEO), Arrayexpress, genome-wide association studies (GWAS), and The National Cancer Institute Genomic Data Commons (NCIGDC), which can be used in analyzing gene expression signatures, determining the interrelation of genomic variants with particular complex disorders, and promising therapeutic targets from the next-gen sequencing technology (Gupta et al., 2021).

During the COVID-19 pandemic, it is urgent to develop novel drugs and appropriate vaccines to reduce the spread of the virus. Although biomedical professionals and research scientists are working to trail error tactics for COVID-19 drug and vaccine discovery, HTS failed to meet such a urgent demand due to the inefficiency of the lab-based operations, whereas VS has been implemented as one of the most popular approaches in potent compounds discovery (Kandeel & Al-Nazawi, 2020). Herein, two subgroups of VS drug design and discovery, structure-based and ligand-based, are still the dominated methods that computationally target the biomolecules of SARS-CoV-2 through DNA, protein, RNA, and lipid (Broom et al., 2020). Therefore, VS provides a fast, low-cost, and efficient strategy for identifying antiviral candidates in regards to COVID-19 by accessing to experimentally and computationally determined protein structures (Zhang et al., 2020a; Keshavarzi Arshadi et al., 2020). On the other hand, developing an appropriate vaccine using the conventional discovery method is still costly, and may take a few years to find the right candidate against the specified pathogen, while such a strategy may not be suitable for the situation of COVID-19 due to the urgency and the variants. The idea of Reverse Vaccinology (RV), a genome-based vaccine development method provides a highly efficient status in vaccine design as it no longer requires identifying vaccine targets (Bullock et al., 2020; Soria-Guerra et al., 2015). Besides, the antigens of the putative target protein can be identified without the limitation to be isolated from bacterial cultures (Bowman et al., 2011). Such advantages may lead biomedical scientists to implement RV prediction projects for developing COVID-19 vaccines.

Over the last decade, machine learning and deep learning algorithms have revolutionized therapy discovery and vaccine design in general (Lavecchia, 2019), along with the evolutional RV virtual frameworks, which have been categorized as rule-based screening methods using AI and big data (Naz et al., 2019). Moreover, machine learning and deep learning models have been incorporated in leaning protein patterns and automating feature extraction from medical data (Zhavoronkov et al., 2019). Such advantages of using AI and big data may also benefit the drug design and vaccine development for COVID-19. Motivated by the current urgency, this chapter provides a comprehensive survey of AI and big data-based methods for COVID-19 drug discovery and vaccine development. The objectives of this chapter are:

- illustrating the protein structural basis of anti-COVID-19 drug and vaccine development using 3D visualization and protein analysis;
- investigating AI-based methods for COVID-19 drug discovery and vaccine design using machine learning and deep learning;
- examining big data-driven approaches in constructing informative systems of COVID-19 treatments and therapy designs based on bioinformatics, knowledge graph, and NLP.

BACKGROUND

AI algorithms along with big data analytics have revolutionized many domains in biological science and medical engineering in recent years. Particularly, drug discovery and vaccine development are two fields that have been impacted significantly by incorporating machine learning and deep learning.

Chen et al. (2018) investigated the most popular deep learning models that can be applied in drug discovery, in which the power of machine learning and deep learning, combined with processing a large volume of biological information, has contributed to the applicational perspectives of drug designs in terms of compound property and activity prediction, de novo design, reaction prediction and retro-synthetic analysis, and ligand-protein interaction prediction. In the perspective of predictive analysis, one of the most popular and reliable tools for drug discovery and design is Graph Convolutional Neural Networks (GCNN), which enables to address graphical features and pattern recognitions through analyzing the relative bioinformatic information within the features (Kearnes et al., 2016). Such an ideology has also been applied in a number of pharmacological research domains and drug development methods, including protein interface estimation (Zitnik et al., 2018), reactivity prediction (Coley et al., 2019), drug-target interactions (Torng & Altman, 2019), and drug property analysis (Bazgir et al., 2019). Over the last few years, the usage of NLP has contributed significantly in drug discovery and design using sequence-based approaches, such as proteomics, genomics, and transcriptomics, in terms of the context-based transformers that have been applied in sequences feature extractions with cutting-edge text mining tools (Devlin et al., 2018). Such tools have also been used in a variety of research topics, such as predicting drug-target interactions, modeling protein sequences, and analyzing retro-synthetic reactions, in the form of extracting features from sequences on the context, location, and merger of the biological tokens (Belinkov & Glass, 2019; Shin et al., 2019; Choromanski et al., 2020). Besides, machine learning and deep learning models, such as recurrent neural network (RNN), long short-term memory (LSTM), and encoder-decoder networks, have been widely implemented in pre-trained models on molecules or protein sequences in terms of secondary structure prediction (Wang et al., 2017), quantitative structure-activity relationship modeling (Chakravarti & Alla, 2019), and protein function prediction (Cao et al., 2018). Moreover, the application of deep learning has contributed to the development of de novo design, which has evolved from the classical method of ligand-based models and molecule creations. Such approaches have been incorporated in the applied deep learning algorithms, including Generative Adversarial Networks (GANs), reinforcement learning, and variational auto-encoders, in terms of data-driven molecule generations, atomic sequences creations, ligand-based designs, and the generation of unique molecules with higher diversity (Guimaraes et al., 2017; De Cao & Kipf, 2018; Simonovsky & Komodakis, 2018).

Over the past few years, AI models and big data approaches have also been applied in the vaccine development domain in the form of the improvement of therapy design, safety management, and reliability evaluation. A support vector machine (SVM) classifier has been proposed in identifying bacterial protective antigens (BPAs) from non-BPAs with a high stander of the model performance, which represents the implementation of machine learning in RV methods with a solid result for antigen prediction (Heinson et al., 2017). Rahman et al. (2019) presented a novel machine learning-based protective antigen identification using random forest (RF) model, along with the Recursive Feature Elimination (RFE) method and minimum redundancy maximum relevance (mRMR) criterion for optimal feature extraction. Such a proposed model, namely Antigenic, has been implemented as one of the most effective open-source RV programs for BPA discovery. Similar tools, such as Vaxign, VaxiJen, VacSol, Jenner-predict, etc., have also been examined and investigated based on the comparison of criteria and programs applied for the selected potential vaccine candidates (PVCs), computational runtime, and model performances (Dalsass et al., 2019). More recently, Ong et al. (2020) presented a supervised learning-based BPA prediction tool, called Vaxign-ML, which has been frequently used due to its superior performance in BPA predictions with the open-source accessibility hosted in a public web server. Essentially, such machine learning models for BPA prediction are testaments to the implementation of computational and mathematical approaches in RV, consisting of data preparation and augmentation, feature extraction and selection, model training and cross-validation, together with model performance evaluation to predict vaccine candidates against the viral pathogens of a specific infectious disease. Additionally, AI-based methods have been used as the feature extractors for identifying biological, structural, and physiochemical features based on RV, bioinformatics, and immunological approaches, such as IEDB and BlastP. Such an ideology has been incorporated in the identification of vaccine targets discovery and vaccine designs against SARS-CoV-2 using deep learning (Abbasi et al., 2020). Besides, recent studies in this field demonstrate that graph-based features have contributed most in representing the antibody. A rapid method in discovering peptides and antibody sequences has been proposed to predict potential neutralizing antibodies for COVID-19 using graph featurization with a set of machine learning models, including RF, SVM, XGBoost, etc., which can be applied in screening thousands of hypothetical antibody sequences in seconds (Magar et al., 2021). Herein, deep learning autoencoders have also shown the capability to improve the feature extraction process in SARS-CoV-2 gene profiling for mutations and transitions (Miyake et al., 2021).

The most recent studies of COVID-19 drug discovery and vaccine development indicate that applications of AI and big data have improved the traditional techniques through protein-based and RNA-based approaches. VS of both repurposed drug candidates and new vaccine targets based on protein identification has been incorporated in the implementations of machine learning and deep learning. Beck et al. (2020) predicted commercially used antiviral drugs for COVID-19 by using a pre-trained deep learning-based drug-target interaction model, which can be applied in identifying available drugs that may act on viral proteins of the coronavirus. A network bioinformatics analysis has been proposed to provide insight into drug repurposing for SARS-CoV-2, of which 30 potential drugs have been examined through identifying multiple genes, human disease pathways, and modules related to COVID-19 (Li et al., 2021). Kadioglu et al. (2021) combined VS, molecular docking, and machine learning to predict novel drug candidates against SARS-CoV-2 by interacting with the COVID-19 target proteins, such as the spike protein, the nucleocapsid protein, and 2'-o-ribose methyltransferase. Such an ideology has been applied in vaccine development by proposing the prediction of prognostic biomarkers associated with COVID-19 using machine learning (Sardar et al., 2021). Besides, understanding conserved RNA structural elements has been demonstrated a critically functional role in drug design and vaccine discovery for COVID-19. Rangan et al. (2020) identified more than 100 RNA genome conservations and secondary structures in SARS-CoV-2 related viruses, along with the predictive analysis for regions of the viral genome. An in silico map of the COVID-19 RNA structure has been proposed by deducing the RNA structural landscape of the COVID-19 transcriptome using a novel bioinformatics method, ScanFold, which can be applied to screening a large scale of potential drug candidates for RNA-binding small molecules (Andrews et al., 2020). The identification of new drug and vaccine targets has also been benefited from analyzing the changes in RNA information with AI and big data, e.g. analysis of therapeutic targets for COVID-19 and drug discovery by data-driven approaches (Wu et al., 2020). Deep sequence modeling has been widely applied in COVID-19 mRNA vaccine development, e.g. a predictive analysis for COVID-19 mRNA sequences responsible for degradation based on bidirectional GRU and LSTM models (Qaid et al., 2021).

MAIN FOCUS OF THE CHAPTER

Issues, Controversies, Problems

The technological improvement from using AI and big data plays a crucial role in COVID-19 drug design and vaccine development for several reasons, such as the automatic feature extraction capability, the generative and predictive ability, and the big data availability. Therefore, the application of AI and big data in therapy discovery for COVID-19 is essential to meet a timely and effective response to the coronavirus. Besides, understanding COVID-19 molecular mechanism and target selection also plays an important role in drug discovery and vaccine development, thereby combining computational methods and biological science will make significant insights and contributions to this field. The applications of machine learning, deep learning, and big data have contributed to many subfields of drug design and vaccine development over the past decades, whereas how such techniques work for the current situation of immediate COVID-19 therapy discovery is still ambiguous. One of the biggest controversies is represented by the unbalanced relationship between the urgency of the situation and the innovated technologies. Despite a number of existing survey studies in this domain, a systematic review for COVID-19 drug design and vaccine development using AI and big data is still essential due to the massive research studies in this field.

Features of the Chapter

Despite the widely applied methods using computationally biomedical designs in drug discovery and vaccine development, many challenges are still remaining with traditional data-driven approaches in terms of cost efficiency, model complexity, and reliability. With the development of AI, such bottlenecks have been broken through by using advanced machine learning and deep learning algorithms in the form of the cutting-edge technologies in drug development. This chapter will conduct a systematic survey study to investigate how big data and AI can help in drug discovering and vaccine development for COVID-19 throughout three major groups of methods, such as protein structural basis, AI-based models, and big data-driven approaches, along with the data and code accessibility for each study.

SOLUTIONS AND RECOMMENDATIONS

Protein Structural Basis Methods

Prior to COVID-19, analyzing protein structural basis has been widely employed in drug discovery and vaccine design. Billings et al. (2019) applied the ProSPr architecture to train on 64´64 crops using data from the CATH S35. The whole dataset was divided into training, validation, and test sets, respectively. A deep convolutional neural network (CNN) model was trained to predict some molecular structure restraints, including residue distances, residue contacts, dihedral angles, etc. To perform the model training process, the residue distances were converted into contacts. The contact accuracy was measured by precision. Based on all the restraints, protein distance predictions could be achieved and a 3D structure of target protein can be built. Similarly, AlQuraishi (2019) designed a differentiable model for protein structure learning. He used ProteinNet dataset, which contains the raw protein sequence and the PSSM and information content profiles of the protein. The protein sequence is decoded by a one-hot vector. The PSSM values of protein is converted through sigmoid transformation to 0 or 1. LSTM model was used to train on the sequence vector input. Recurrent geometric network is used to predict the 3D structure of protein through adjusting bond angles and angle rotation of chemical bonds connecting various amino acids. However, some limitations are still existed in the study on energy model learning, the use of single-scale atomic, etc.

Such ideologies have been applied in protein structural basis of anti-SARS-CoV-2 drug discovery and vaccine design. Sedova et al. (2020) developed a web-based Coronavirus3D server, which can be used as a tool to analyze the 3D architecture of SARS-CoV-2 virus proteins. Coronavirus3D was developed with the Protael package and 2dmol.js library for 3D visualizations. Besides, users can observe the SARS-CoV-2 genome with an interactive perspective. Based on the boundary information on the corresponding proteins, the structures of SARS-CoV-2 could be presented and the mutation frequency of amino-acid can be visualized through a histogram on the server page. The lower level panels have an option to color the chain based on the mutation frequency. ul Qamar et al. (2020) applied the GISAIF database, which contains the data of whole-genome sequences of SARS-CoV-2. From the whole-genome sequences, gene sequences of $3CL^{pro}$ can be extracted and translated into protein sequences through ExPASy server. Through utilizing Modeller v9.11, comparative homology modeling can perform the structural analysis on SARS-CoV-2 3CL^{pro}. In this study, molecular operating environment (MOE) was implemented for molecular docking, ligand-protein interaction and drug likeness analysis. The docking results, the binding behaviors and the stability of compounds were examined and analyzed by molecular dynamics simulations. It can be concluded from the sequence alignment results that 3CL^{pro} was conserved in SARS-CoV-2. Through utilizing the MOE2019 software, Contini (2020) implemented receptor models. To correct PDB inconsistencies and manage the protonation state, the structure preparation module of MOE was carried out. The receptor could be minimized through the backbone restraints and ligand constrains. In this study, two different databases were involved and processed by MOE module for establishing the 3D structure and minimizing the molecule geometry. The combined database contained 3118 various molecules, which were all saved in SDF format. For virtual screening, Amber was used to select and score for the molecular dynamics of complexes. In the process of screenings, the ligand database was involved with the proper setting parameters, such as center and speed for 6LU7. Moreover, the binding energy can also be calculated based on the protocol. Bai et al. (2021) applied a software for 3D drug design, called MolAICal, to perform 3D drug design, based on artificial intelligence and classical algorithm methods. MolAICal module implemented genetic algorithm, which was trained by FDA-approved drug fragments and Vinardo score from PDBbind database, and deep learning generative model, which was trained by drug-like molecules from ZINC database. Besides, MolAICal can generate different ligands with higher 3D structural similarity with crystal ligand of SARS-CoV-2M^{pro}. The results has shown that MolAICal is a powerful tool for novel drug design and creation. Moreover, AlphaFold is an AI-based tool developed by Google DeepMind, which is pre-trained on PDB structural data (Senior et al., 2020). AlphaFold can forecast the three-dimensional structure of protein in the amino acid sequence through two major steps. Firstly, AlphaFold can convert an amino acid sequence of a protein into distance matrix and torsion angle matrix by CNN algorithm. Secondly, the three-dimentsional structure of a protein can be generated based on these two references, through performing a gradient descent optimization method. In general, AlphFold can achieve high-accuracy structures for most modeling domains. Castro et al. (2021) evaluated MHC-II constraints related to the neutralizing antibody response to a mutationally-constrained B cell epitope by visualizing and analyzing spike proteins. Such a study can be referenced in vaccine development for COVID-19. Methods and main achievements obtained from such studies are summarized in Table 1, along with the data and/or code accessibility for each study.

Table 1. Summary of studies in protein structural basis of anti-COVID-19 drug discovery

Reference	Methods	Main Achievements	Data/Code Availability
Sedova et al. (2020)	3D protein structure visualization, genome analysis, web server design, information integration	website integration on the COVID-19 mutations with information about 3D structures of its proteins, allowing users to visually analyze the mutations in the 3D context.	freely available at https://coronavirus3d.org.
ul Qamar et al. (2020)	Sequence analysis, protein structural analysis, molecular docking, molecular dynamics simulations	Construct phylogenetic tree using mamixmum likelihood method, 3D structure prediction of the SARS-CoV-2 3CLpro enzyme	Data will be available on request.
Contini (2020)	Receptor preparation, database preparation, virtual screening	Virtual screening on coronavirus proteins, protease inhibitors used in HIV infections may benefit for COVID-19 therapy	Not available
Bai et al. (2021)	3D drug design, generative model, protein structural preparation, virtual screening	Generate novel ligands with good binding scores and proper XLOGP values	The study results of SARS-CoV-2M^{pro} are shared freely as a reference at https://github.com/MolAICal/COVID-19/tree/master/mpro . Project is available at https://molaical.github.io/ .
Senior et al. (2020)	3D protein structural visualization, neural network, gradient descent algorithm	Prediction of protein structures, create a method to characterize the protein shape	Training, validation and test data splits (CATH domain codes) are available from https://github.com/deepmind/deepmind-research/tree/master/alphafold_casp13 Code is available at https://github.com/deepmind/deepmind-research/tree/master/alphafold_casp13
Castro et al. (2021)	Spike protein analysis, structure analysis, supertype analysis, motif analysis, BLAST analysis	Prediction of MHC-II affinity for 15mer peptides connecting to the RBM B cell epitope, evaluate MHC-II constraints towards antibody	Data and code are available at https://github.com/cartercompbio/SARS_CoV_2_T-B_co-op.

Besides, informative databases have been proposed for research purposes of COVID-19 drug discovery. Alsulami et al. (2021) generated 3D models for SARS-CoV-2 using MODELLER version. To measure the impacts of mutations, multiple tools, such as mCSM-Stability, mCSM-PPI, DeepDDG, PROVEAN, MAESTRO, and I-Mutant, are applied to categorize the mutation and quantify the mutation for the predicted protein. Their study also developed a web interface, which is written in HTML5 and CSS, to query the created SARS-CoV-2 3D database. Chen et al. (2021) developed DockCoV2, a SARS-CoV-2 database, to predict the binding affinity of FDA and Taiwan NHI. DockCoV2 can make drug discovery easier through applying a computational representation of molecular docking. PubChem compound identifiers were chosen as the key identifier for DockCoV2. Besides, the docking utility used in this study is AutoDock Vina, which can implement virtual screening pipeline. Mei et al. (2021) conducted a research to collect all the web resources on drug discovery for COVID-19 treatment. The resources cover many aspects, including antiviral drugs, vaccines, and monoclonal antibodies. Based on the current researches for therapies, the study further discuss the challenges for solving the related medical problems of coronavirus.

AI-based Approaches

Over the last decade, AI models have been frequently applied in drug design and vaccine development. Current studies suggest that machine learning algorithms can be applied in COVID-19 drug development. Bennett et al. (2020) utilized machine learning technique and empirical methods to perform drug discovery. Their study designed atomistic MD simulations and calculate on 15,000 small molecule free energies for water-to-cyclohexane transfer. Based on the prepared large dataset, a spatial graph neural network model was trained to predict the free energies of transfer and reached the reliable accuracy. 3D convolutional neural network and shallow learning had less accuracy. Their results showed the data processed by MD simulation can optimize the predictions and benefit the transferability of the model for different molecules. Xu et al. (2020) trained machine learning classifiers on 2030 natural compounds. Six machine learning methods were involved, which include random forest, support vector machine, k-nearest neighbors, naive bayes, decision tree, and logistic regression. Through the evaluation of ROC curves, their study concluded that logistic regression has the best performance. Moreover, logistic regression with 2048-bit fingerprint reached a

AUC score of 0.976, which is better than other machine learning classifiers. For molecular docking, the lowest binding score indicated that the molecular has the optimal binding affinity. Genheden et al. (2020) developed a software, called AiZynthFinder, based on a Monte Carlo tree and neural network algorithms and can monitor the impact on retrosynthesic predictions. The input of this tool could be a molecule that has been converted to purchasable precursors and AiZynthFinder will output a set of precursors that is unbrokenable by the algorithm. This software was developed completely by Python packages, including TensorFlow, RDKit, and NetworkX.

On the other hand, deep learning models have been widely employed in drug and vaccine development for COVID-19. Gastegger et al. (2020) applied SchNet for Orbitals (SchNOrb), a deep convolutional neural network model, to optimize representation of quasi-atomic minimal basis. The model can predict molecular orbital energies with high accuracy and enable chemical bonding analysis through employment of derived attributes. Furthermore, based on the experimental results, the researches discussed further about the future direction on quantum chemical workflow. Beck et al. (2020) proposed a deep learning model, called Molecule Transformer-Drug Target Interaction (MT-DTI) to recognize accessible drug that can be used to target the proteins of SARS-CoV-2. The pre-trained deep learning model in MT-DTI can perform accurate prediction on binding affinities, relying on chemical sequences and amino acid sequences of a corresponding protein. From the performance comparison between MT-DTI and the traditional 3D structure-based docking algorithm methods, it showed that MT-DTI can achieve better prediction results. Thus, their study further used MT-DTI for repurposing the FDA-approved drugs for SARS-CoV-2 proteins. Pham et al. (2021) proposed a neural network-based approach, called DeepEC, to predict the gene expression profile through model the different chemical structures. They used several various biological datasets to train the model, containing L1000, STRING, DrugBank and COVID-19 transcriptome data for patients. This study performed bayesian-based peak deconvolution on L1000 dataset, extracted human protein-protein interaction network from STRING dataset, and found out the Anotomical Therapeutic Chemical labels and drug targets from DrugBank dataset. The major based line models used in this study covered linear models, vanilla neural network, KNN approach, and tensor-train weight optimization. Zhang et al. (2020b) designed a deep learning-based drug screening approach, which can help novel drug to combat the 2019-nCoV. This study used homology modeling to establish a 3D model of protein for 3C-like protease. Compared with traditional methods, deep learning-based approach based on DFCNN made implementing virtual screening in large scale, with high accuracy, possible. The data applied in this study was the data of RNA sequences, retrieved from Global Initiative on Sharing AII Influenza Data (GISAID) database. Furthermore, the amino acid sequence is converted from RNA sequence further, using a web translator tool, namely Expasy. Ton et al. (2020) proposed a rapid identification model that can detect potential inhibitors of COVID-19 protease by using deep neural networks. A novel deep learning-based platform has also been developed to predict docking scores of Glide, thereby enables developing anti-COVID-19 vaccines in a short time. Methods and main achievements obtained from such studies are summarized in Table 2, along with the data and/or code accessibility for each study.

Table 2. Summary of studies in COVID-19 drug development using machine learning and deep learning

Reference	Methods	Main Achievements	Data/Code Availability
Bennett et al. (2020)	Atomistic molecular dynamics simulations, machine learning, spatial graph neural network model, 3D convolutional neural network, shallow learning	Predict the free energies of transfer	Not available.
Xu et al. (2020)	Random forest, Support Vector Machine, K-Nearest Neighbors, Naive Bayes, Decision Tree, Logistic Regression, molecular docking	Anti-COVID-19 inhibitors, build electrostatic interaction between Rutin and crystal structure of COVID-19 3CLpro	Data is available at https://www.frontiersin.org/articles/10.3389/fmolb.2020.556481/full#supplementary-material
Genheden et al. (2020)	Monte Carlo tree search, artificial neural network	Create an open-source software for retrosynthetic planning	The software is available at https://www.github.com/MolecularAI/aizynthfinder. Data is available at https://doi.org/10.6084/m9.figshare.12334577.v1 .
Gastegger et al. (2020)	Deep neural network, electronic wave function	Predict molecular orbital energies, chemical bonding analysis	Not available.
Beck et al. (2020)	Amino acid sequences, molecule transformer-drug target interaction, AutoDock Vina	Drug-target interaction analysis, prediction on the binding affinity between potential drug and the SARS-CoV-2 proteins	Data is available at https://www.sciencedirect.com/science/article/pii/S2001037020300490#:~:text=Download%20all%20supplementary%20files%20included%20with%20this%20article.

Continued on following page

Table 2. Continued

Reference	Methods	Main Achievements	Data/Code Availability
Pham et al. (2021)	DeepCE architecture, linear models, vanilla neural network, KNN, tensor-train weight optimization	Predict gene expression profiles	The Bayesian-based peak deconvolution LINCS L1000 dataset is available at https://github.com/njpipeorgan/L1000-bayesian. The training, development, and testing gene expression sets used in our study, gene expression profiles generated from DeepCE for all drugs in DrugBank are available at https://github.com/pth1993/DeepCE. DeepCE source code and its usage instructions are available in Github (https://github.com/pth1993/DeepCE) and Zenodo (https://doi.org/10.5281/zenodo.3978774).
Zhang et al. (2020b)	DFCNN, homology modeling, ligand virtual screening, Expasy	Ligand interaction analysis, molecular docking, deep learning-based drug screening	The amino acid sequence is translated from the RNA sequence by Translate web tool, available at https://web.expasy.org/translate/. The ligand database is taken from https://www.chemdiv.com/.
Ton et al. (2020)	Deep neural network, virtual screening	Rapid estimation of docking score, interaction analysis in the SARS-CoV-2 pocket	List of the top 1,000 identified compounds, as well as docking results in SDF format are publicly available at https://drive.google.com/drive/folders/1xgA8ScPRqIunxEAXFrUEkavS7y3tLIMN?usp=sharing.

Big Data-Driven Methods

With the development of big data technologies, discovering appropriate drugs and vaccines for COVID-19 can be addressed with a variety of big data approaches, such as knowledge graph and NLP, whereas integrating bioinformatic information and data is the foundation. Feng et al. (2021) developed an integrated bioinformatics platform, which is a viral-associated disease-specific chem-genomics knowledge-based (Virus-CKB) system. Such a system can be applied in computational systems of the pharmacology-target mapping that predict the FDA-approved drugs for COVID-19. Many research outcomes, such as 65 antiviral drugs, 107 viral-related targets, and 189 accessible 3D crystal or cryo-EM structures, and 2698 chemical agents, have been reported using Virus-CKB, which is deployed in web applications for predicting protein targets and visualizing major outputs, such as HTDocking, TargetHunter, BBB predictor, NGL Viewer, etc. Rajput et al. (2021) proposed a web server, namely DrugRepV, by collecting data of repurposed drugs designed for antiviral activity from existing studies. Such the database can be applied in exploring COVID-19 drug targets and clinical trials as it contains nearly 8,500 entires with biological, chemical, clinical, and structural information of multiple viruses responsible to the COVID-19 pandemic.

Implementing knowledge-based system becomes popular in drug discovering and design for COVID-19. A novel knowledge-based integration system, called COVID-KOP, has been proposed based on the existing Reasoning Over Biomedical Objects linked in Knowledge Oriented Pathways (ROBOKOP). Such a system has been applied in retrieving recent biomedical literatures on COVID-19 through the biomedical knowledge graph, which can be employed in drug discovering and design against COVID-19 by constructing respective confirmatory pathways of drug action (Korn et al., 2021). Prior to COVID-KOP, Korn et al. (2020) also developed a web application, namely COKE, which is a knowledge extractor that can be applied in extracting, curating, and annotating essential drug-target relationships from existing studies on COVID-19. COKE can process the COVID Open Research Dataset (CORD-19) to identify drug-protein pairs, unique proteins, and approved drugs for SARS-CoV-2 for exploring a large scale of knowledge-based insights of COVID-19 drug discovering and development. Besides, NLP plays an important role in COVID-19 drug and vaccine development. Bose et al. (2021) proposed a NLP-based scientific reference system, which can applied in identifying research articles that reflect the short-term threats of COVID-19, such as transmissibility, health risks, and treatment plans, as well as efforts towards long-term immunization systems and challenges of drug and vaccine development and design. Zhang et al. (2021) proposed a literature-based discovery (LBD) method to identify drug candidates for COVID-19. Such a LBD approach can be used in identifying an informative and accurate subset of semantic triples using filtering rules and BERT, which have been applied to construct a knowledge graph to predict drug repurposing candidates for COVID-19. Methods and main achievements obtained from such studies are summarized in Table 3, along with the data and/or code accessibility for each study.

Table 3. Summary of studies in COVID-19 drug development using bioinformatics, knowledge graph, and NLP

Reference	Methods	Main Achievements	Data/Code Availability
Feng et al. (2021)	Blood-brain barrier prediction, HTDocking, potential target prediction, Spider plot-molecule visualization	Virus-CKB server for COVID-19 bioinformation, interaction network plotting,	The Virus-CKB server is accessible at https://www.cbligand.org/g/virus-ckb. An established molecular database prototype DAKB-GPCRs is available at https://www.cbligand.org/dakb-gpcrs/.
Rajput et al. (2021)	big data, database enrichment, network-based analysis, web server development, PHP, Javascript, Pearl, HTML, MySQL	Build repurposed drug database, repurposed drug analysis, interaction network visualization	Data is available at https://bioinfo.imtech.res.in/manojk/drugrepv/.
Korn et al. (2021)	Database management, web interface development	Build a knowledgebase web server for biomedical information concerning COVID-19	COVID-KOP is freely accessible at https://covidkop.renci.org/. Code is available at https://github.com/NCATS-Gamma/robokop.
Korn et al. (2020)	big data, data curation and integration, quantitative structure-activity relationship modeling, Python programming, Flask framework, jQuery library, HTML	Develop the COKE web portal, compare drug-target associations, drug analysis in clinical trials of COVID-19	COKE is freely available at https://coke.mml.unc.edu/. The code is available at https://github.com/DnlRKorn/CoKE.
Bose et al. (2021)	NLP, term-frequency-inverse document frequency, word cloud	COVID-19 abstract word cloud, frequency visualization of 25 most mentioned countries, document analysis	Not available.
Zhang et al. (2021)	Knowledge graph completion algorithms, SemRep, PubMedBERT model, text mining	Develop a drug repurposing approach, construct a biomedical knowledge graph	Code and data are available at https://github.com/kilicogluh/lbd-covid.

FUTURE RESEARCH DIRECTIONS

Despite great successes in using AI and big data on drug design and vaccine development, many challenges are still remaining in terms of high-quality data acquisition and complicated data preprocessing, which can be addressed in future studies. Open-source data is not easy in the biomedical industry, while the research community can collect and share data for the research purpose, thereby data collection and preprocessing will provide a lot of opportunities to future researches. Some pharmaceutical companies have taken the action to start sharing the clinical data with others, whereas such an initiative has obstructed by some technical difficulties, e.g. a uniform data format, which can be solved by proposing an algorithm-based attempt that handles sparse data. However, such an ideology should be investigated and examined experimentally with the real case studies in the future studies. Moreover, AI and big data will also help in managing the deployment and testing procedures of biological products, which can be also investigated in the future studies. Safety and reliability are two most critical aspects in biomedical therapy discovery and development. Public immunization platforms, such as the Vaccine Adverse Event Reporting System (VAERS) and the Vaccine Safety Databank (VSD), are two major registries for recording, ranking, and tracking safety of biomedical products. The tradeoff between the evaluation of safety and efficiency has been improved by implementing data-driven approaches, i.e. computational simulation, AI modeling, and big data analytics (George & Georrge, 2019). The application of NLP has also contributed to the identification of adverse events, e.g. identifying Tdap-related vaccines using data from VSD (Zheng et al., 2019). Similarly, the final stage in drug design and discovery requires the selected drug candidates to be safe for human consumption by observing the drug-side effectiveness combined with the non-toxic confirmation. Such an issue can be addressed by using AI-based implementations for detecting the potential prolongation of QT intervals and cardio-toxicity of the candidate drug designed for COVID-19 (Li et al., 2020).

CONCLUSION

As the COVID-19 pandemic has been rapidly transformed into a global challenge, a major issue for the biomedical and pharmaceutical industry is the problem of costs and efficiency to discover a new drug and to develop an effective vaccine. In this regard, applications of AI and big data provide great opportunities on cost reduction, efficiency improvement, and time saving in the drug design and vaccine development for COVID-19. The survey study proposed in this chapter demonstrates that advances in AI approaches along with improving computational architecture and accessibility of big data make significant contribution to drug discovery and vaccine development for COVID-19 by leveraging multiple machine learning and deep learning methods capable of filtering and screening relabel therapies efficiently. Benefitting from supervised learning models and big data analytics, virtual filtering and de novo design techniques have been improved with higher accuracy and less producing time. Therefore, such a systematic review may provide more insightful information and multiple visions in drug discovery and vaccine development for COVID-19.

REFERENCES

Abbasi, B. A., Saraf, D., Sharma, T., Sinha, R., Singh, S., Gupta, P., ... Gupta, A. (2020). *Identification of vaccine targets & design of vaccine against SARS-CoV-2 coronavirus using computational and deep learning-based approaches*. Academic Press.

AlQuraishi, M. (2019). End-to-end differentiable learning of protein structure. *Cell Systems*, *8*(4), 292–301. doi:10.1016/j.cels.2019.03.006 PMID:31005579

Alsulami, A. F., Thomas, S. E., Jamasb, A. R., Beaudoin, C. A., Moghul, I., Bannerman, B., Copoiu, L., Vedithi, S. C., Torres, P., & Blundell, T. L. (2021). SARS-CoV-2 3D database: Understanding the coronavirus proteome and evaluating possible drug targets. *Briefings in Bioinformatics*, *22*(2), 769–780. doi:10.1093/bib/bbaa404 PMID:33416848

Andrews, R. J., Peterson, J. M., Haniff, H. S., Chen, J., Williams, C., Grefe, M., ... Moss, W. N. (2020). An in silico map of the SARS-CoV-2 RNA Structurome. *bioRxiv*. doi:10.1101/2020.04.17.045161

Bai, Q., Tan, S., Xu, T., Liu, H., Huang, J., & Yao, X. (2021). MolAICal: a soft tool for 3D drug design of protein targets by artificial intelligence and classical algorithm. *Briefings in Bioinformatics, 22*(3), bbaa161.

Bazgir, O., Zhang, R., Dhruba, S. R., Rahman, R., Ghosh, S., & Pal, R. (2019). *REFINED (REpresentation of features as images with NEighborhood Dependencies): A novel feature representation for convolutional neural networks*. arXiv preprint arXiv:1912.05687.

Beck, B. R., Shin, B., Choi, Y., Park, S., & Kang, K. (2020). Predicting commercially available antiviral drugs that may act on the novel coronavirus (SARS-CoV-2) through a drug-target interaction deep learning model. *Computational and Structural Biotechnology Journal*, *18*, 784–790. doi:10.1016/j.csbj.2020.03.025 PMID:32280433

Beck, B. R., Shin, B., Choi, Y., Park, S., & Kang, K. (2020). Predicting commercially available antiviral drugs that may act on the novel coronavirus (SARS-CoV-2) through a drug-target interaction deep learning model. *Computational and Structural Biotechnology Journal, 18*, 784–790. doi:10.1016/j.csbj.2020.03.025 PMID:32280433

Belinkov, Y., & Glass, J. (2019). Analysis methods in neural language processing: A survey. *Transactions of the Association for Computational Linguistics, 7*, 49–72. doi:10.1162/tacl_a_00254

Bennett, W. D., He, S., Bilodeau, C. L., Jones, D., Sun, D., Kim, H., Allen, J. E., Lightstone, F. C., & Ingólfsson, H. I. (2020). Predicting small molecule transfer free energies by combining molecular dynamics simulations and deep learning. *Journal of Chemical Information and Modeling, 60*(11), 5375–5381. doi:10.1021/acs.jcim.0c00318 PMID:32794768

Billings, W. M., Hedelius, B., Millecam, T., Wingate, D., & Della Corte, D. (2019). ProSPr: Democratized implementation of alphafold protein distance prediction network. *bioRxiv, 830273*. doi:10.1101/830273

Bose, P., Roy, S., & Ghosh, P. (2021). A Comparative NLP-Based Study on the Current Trends and Future Directions in COVID-19 Research. *IEEE Access: Practical Innovations, Open Solutions, 9*, 78341–78355. doi:10.1109/ACCESS.2021.3082108 PMID:34786315

Bowman, B. N., McAdam, P. R., Vivona, S., Zhang, J. X., Luong, T., Belew, R. K., Sahota, H., Guiney, D., Valafar, F., Fierer, J., & Woelk, C. H. (2011). Improving reverse vaccinology with a machine learning approach. *Vaccine, 29*(45), 8156–8164. doi:10.1016/j.vaccine.2011.07.142 PMID:21864619

Broom, A., Rakotoharisoa, R. V., Thompson, M. C., Zarifi, N., Nguyen, E., Mukhametzhanov, N., ... Chica, R. A. (2020). Evolution of an enzyme conformational ensemble guides design of an efficient biocatalyst. *bioRxiv*. doi:10.1101/2020.03.19.999235

Bullock, J., Luccioni, A., Pham, K. H., Lam, C. S. N., & Luengo-Oroz, M. (2020). Mapping the landscape of artificial intelligence applications against COVID-19. *Journal of Artificial Intelligence Research, 69*, 807–845. doi:10.1613/jair.1.12162

Cao, R., Freitas, C., Chan, L., Sun, M., Jiang, H., & Chen, Z. (2017). ProLanGO: Protein function prediction using neural machine translation based on a recurrent neural network. *Molecules (Basel, Switzerland)*, *22*(10), 1732. doi:10.3390/molecules22101732 PMID:29039790

Castro, A., Ozturk, K., Zanetti, M., & Carter, H. (2021). In silico analysis suggests less effective MHC-II presentation of SARS-CoV-2 RBM peptides: Implication for neutralizing antibody responses. *PLoS One*, *16*(2), e0246731. doi:10.1371/journal.pone.0246731 PMID:33571241

Chakravarti, S. K., & Alla, S. R. M. (2019). Descriptor free QSAR modeling using deep learning with long short-term memory neural networks. *Frontiers in Artificial Intelligence, 2*, 17.

Chen, H., Engkvist, O., Wang, Y., Olivecrona, M., & Blaschke, T. (2018). The rise of deep learning in drug discovery. *Drug Discovery Today*, *23*(6), 1241–1250. doi:10.1016/j.drudis.2018.01.039 PMID:29366762

Chen, T. F., Chang, Y. C., Hsiao, Y., Lee, K. H., Hsiao, Y. C., Lin, Y. H., Tu, Y.-C. E., Huang, H.-C., Chen, C.-Y., & Juan, H. F. (2021). DockCoV2: A drug database against SARS-CoV-2. *Nucleic Acids Research*, *49*(D1), D1152–D1159. doi:10.1093/nar/gkaa861 PMID:33035337

Choromanski, K., Likhosherstov, V., Dohan, D., Song, X., Gane, A., Sarlos, T., . . . Weller, A. (2020). *Masked language modeling for proteins via linearly scalable long-context transformers.* arXiv preprint arXiv:2006.03555.

Coley, C. W., Jin, W., Rogers, L., Jamison, T. F., Jaakkola, T. S., Green, W. H., Barzilay, R., & Jensen, K. F. (2019). A graph-convolutional neural network model for the prediction of chemical reactivity. *Chemical Science (Cambridge)*, *10*(2), 370–377. doi:10.1039/C8SC04228D PMID:30746086

Contini, A. (2020). *Virtual screening of an FDA approved drugs database on two COVID-19 coronavirus proteins.* Academic Press.

Dalsass, M., Brozzi, A., Medini, D., & Rappuoli, R. (2019). Comparison of open-source reverse vaccinology programs for bacterial vaccine antigen discovery. *Frontiers in Immunology*, *10*, 113. doi:10.3389/fimmu.2019.00113 PMID:30837982

De Cao, N., & Kipf, T. (2018). *MolGAN: An implicit generative model for small molecular graphs.* arXiv preprint arXiv:1805.11973.

Devlin, J., Chang, M. W., Lee, K., & Toutanova, K. (2018). *Bert: Pre-training of deep bidirectional transformers for language understanding*. arXiv preprint arXiv:1810.04805.

Feng, Z., Chen, M., Liang, T., Shen, M., Chen, H., & Xie, X. Q. (2021). Virus-CKB: An integrated bioinformatics platform and analysis resource for COVID-19 research. *Briefings in Bioinformatics*, *22*(2), 882–895. doi:10.1093/bib/bbaa155 PMID:32715315

Gastegger, M., McSloy, A., Luya, M., Schütt, K. T., & Maurer, R. J. (2020). A deep neural network for molecular wave functions in quasi-atomic minimal basis representation. *The Journal of Chemical Physics*, *153*(4), 044123. doi:10.1063/5.0012911 PMID:32752663

Genheden, S., Thakkar, A., Chadimová, V., Reymond, J. L., Engkvist, O., & Bjerrum, E. (2020). AiZynthFinder: A fast, robust and flexible open-source software for retrosynthetic planning. *Journal of Cheminformatics*, *12*(1), 1–9. doi:10.118613321-020-00472-1 PMID:33292482

George, A., & Georrge, J. J. (2019). Viroinformatics: Databases and tools. *Recent Trends in Science and Technology*, *2019*, 117–126.

Guimaraes, G. L., Sanchez-Lengeling, B., Outeiral, C., Farias, P. L. C., & Aspuru-Guzik, A. (2017). *Objective-reinforced generative adversarial networks (ORGAN) for sequence generation models*. arXiv preprint arXiv:1705.10843.

Gupta, R., Srivastava, D., Sahu, M., Tiwari, S., Ambasta, R. K., & Kumar, P. (2021). Artificial intelligence to deep learning: Machine intelligence approach for drug discovery. *Molecular Diversity*, *25*(3), 1–46. doi:10.100711030-021-10217-3 PMID:33844136

Harrer, S., Shah, P., Antony, B., & Hu, J. (2019). Artificial intelligence for clinical trial design. *Trends in Pharmacological Sciences*, *40*(8), 577–591. doi:10.1016/j.tips.2019.05.005 PMID:31326235

Heinson, A. I., Gunawardana, Y., Moesker, B., Hume, C. C. D., Vataga, E., Hall, Y., Stylianou, E., McShane, H., Williams, A., Niranjan, M., & Woelk, C. H. (2017). Enhancing the biological relevance of machine learning classifiers for reverse vaccinology. *International Journal of Molecular Sciences*, *18*(2), 312. doi:10.3390/ijms18020312 PMID:28157153

Jabbari, P., & Rezaei, N. (2019). Artificial intelligence and immunotherapy. *Expert Review of Clinical Immunology*, *15*(7), 689–691. doi:10.1080/1744666X.2019.1623670 PMID:31157571

Kadioglu, O., Saeed, M., Greten, H. J., & Efferth, T. (2021). Identification of novel compounds against three targets of SARS CoV-2 coronavirus by combined virtual screening and supervised machine learning. *Computers in Biology and Medicine*, *133*, 104359. doi:10.1016/j.compbiomed.2021.104359 PMID:33845270

Kandeel, M., & Al-Nazawi, M. (2020). Virtual screening and repurposing of FDA approved drugs against COVID-19 main protease. *Life Sciences*, *251*, 117627. doi:10.1016/j.lfs.2020.117627 PMID:32251634

Kearnes, S., McCloskey, K., Berndl, M., Pande, V., & Riley, P. (2016). Molecular graph convolutions: Moving beyond fingerprints. *Journal of Computer-Aided Molecular Design*, *30*(8), 595–608. doi:10.100710822-016-9938-8 PMID:27558503

Keshavarzi Arshadi, A., Salem, M., Collins, J., Yuan, J. S., & Chakrabarti, D. (2020). DeepMalaria: Artificial intelligence driven discovery of potent antiplasmodials. *Frontiers in Pharmacology*, *10*, 1526. doi:10.3389/fphar.2019.01526 PMID:32009951

Korn, D., Bobrowski, T., Li, M., Kebede, Y., Wang, P., Owen, P., Vaidya, G., Muratov, E., Chirkova, R., Bizon, C., & Tropsha, A. (2021). COVID-KOP: Integrating emerging COVID-19 data with the ROBOKOP database. *Bioinformatics (Oxford, England)*, *37*(4), 586–587. doi:10.1093/bioinformatics/btaa718 PMID:33175089

Korn, D., Pervitsky, V., Bobrowski, T., Alves, V., Schmitt, C., Bizon, C., & Tropsha, A. (2020). *COVID-19 Knowledge Extractor (COKE): a tool and a web portal to extract drug-target protein associations from the CORD-19 corpus of scientific publications on COVID-19*. ChemRxiv; doi:10.26434/chemrxiv.13289222.v1

Lavecchia, A. (2019). Deep learning in drug discovery: Opportunities, challenges and future prospects. *Drug Discovery Today*, *24*(10), 2017–2032. doi:10.1016/j.drudis.2019.07.006 PMID:31377227

Li, J., Shao, J., Wang, C., & Li, W. (2020). The epidemiology and therapeutic options for the COVID-19. *Precision Clinical Medicine*, *3*(2), 71–84. doi:10.1093/pcmedi/pbaa017

Li, X., Yu, J., Zhang, Z., Ren, J., Peluffo, A. E., Zhang, W., Zhao, Y., Wu, J., Yan, K., Cohen, D., & Wang, W. (2021). Network bioinformatics analysis provides insight into drug repurposing for COVID-19. *Medicine in Drug Discovery*, *10*, 100090. doi:10.1016/j.medidd.2021.100090 PMID:33817623

Magar, R., Yadav, P., & Farimani, A. B. (2021). Potential neutralizing antibodies discovered for novel corona virus using machine learning. *Scientific Reports*, *11*(1), 1–11. doi:10.103841598-021-84637-4 PMID:33664393

Mei, L. C., Jin, Y., Wang, Z., Hao, G. F., & Yang, G. F. (2021). Web resources facilitate drug discovery in treatment of COVID-19. *Drug Discovery Today*.

Miyake, J., Sato, T., Baba, S., Nakamura, H., Niioka, H., & Nakazawa, Y. (2021). Cluster Analysis of SARS-CoV-2 Gene using Deep Learning Autoencoder: Gene Profiling for Mutations and Transitions. *bioRxiv*. doi:10.1101/2021.03.16.435601

Naz, K., Naz, A., Ashraf, S. T., Rizwan, M., Ahmad, J., Baumbach, J., & Ali, A. (2019). PanRV: Pangenome-reverse vaccinology approach for identifications of potential vaccine candidates in microbial pangenome. *BMC Bioinformatics*, *20*(1), 1–10. doi:10.118612859-019-2713-9 PMID:30871454

Ong, E., Wang, H., Wong, M. U., Seetharaman, M., Valdez, N., & He, Y. (2020). Vaxign-ML: Supervised machine learning reverse vaccinology model for improved prediction of bacterial protective antigens. *Bioinformatics (Oxford, England)*, *36*(10), 3185–3191. doi:10.1093/bioinformatics/btaa119 PMID:32096826

Palanisamy, V., & Thirunavukarasu, R. (2019). Implications of big data analytics in developing healthcare frameworks–A review. *Journal of King Saud University-Computer and Information Sciences*, *31*(4), 415–425. doi:10.1016/j.jksuci.2017.12.007

Pham, T. H., Qiu, Y., Zeng, J., Xie, L., & Zhang, P. (2021). A deep learning framework for high-throughput mechanism-driven phenotype compound screening and its application to COVID-19 drug repurposing. *Nature Machine Intelligence*, *3*(3), 247–257. doi:10.103842256-020-00285-9 PMID:33796820

Qaid, T. S., Mazaar, H., Alqahtani, M. S., Raweh, A. A., & Alakwaa, W. (2021). Deep sequence modelling for predicting COVID-19 mRNA vaccine degradation. *PeerJ. Computer Science*, *7*, e597. doi:10.7717/peerj-cs.597 PMID:34239977

Rahman, M. S., Rahman, M. K., Saha, S., Kaykobad, M., & Rahman, M. S. (2019). Antigenic: An improved prediction model of protective antigens. *Artificial Intelligence in Medicine*, *94*, 28–41. doi:10.1016/j.artmed.2018.12.010 PMID:30871681

Rajput, A., Kumar, A., Megha, K., Thakur, A., & Kumar, M. (2021). DrugRepV: A compendium of repurposed drugs and chemicals targeting epidemic and pandemic viruses. *Briefings in Bioinformatics*, *22*(2), 1076–1084. doi:10.1093/bib/bbaa421 PMID:33480398

Rangan, R., Zheludev, I. N., Hagey, R. J., Pham, E. A., Wayment-Steele, H. K., Glenn, J. S., & Das, R. (2020). RNA genome conservation and secondary structure in SARS-CoV-2 and SARS-related viruses: A first look. *RNA (New York, N.Y.)*, *26*(8), 937–959. doi:10.1261/rna.076141.120 PMID:32398273

Sardar, R., Sharma, A., & Gupta, D. (2021). Machine learning assisted prediction of prognostic biomarkers associated with COVID-19, using clinical and proteomics data. *Frontiers in Genetics*, *12*, 12. doi:10.3389/fgene.2021.636441 PMID:34093642

Sedova, M., Jaroszewski, L., Alisoltani, A., & Godzik, A. (2020). Coronavirus3D: 3D structural visualization of COVID-19 genomic divergence. *Bioinformatics (Oxford, England)*, *36*(15), 4360–4362. doi:10.1093/bioinformatics/btaa550 PMID:32470119

Senior, A. W., Evans, R., Jumper, J., Kirkpatrick, J., Sifre, L., Green, T., Qin, C., Žídek, A., Nelson, A. W. R., Bridgland, A., Penedones, H., Petersen, S., Simonyan, K., Crossan, S., Kohli, P., Jones, D. T., Silver, D., Kavukcuoglu, K., & Hassabis, D. (2020). Improved protein structure prediction using potentials from deep learning. *Nature*, *577*(7792), 706–710. doi:10.103841586-019-1923-7 PMID:31942072

Shin, B., Park, S., Kang, K., & Ho, J. C. (2019, October). Self-attention based molecule representation for predicting drug-target interaction. In *Machine Learning for Healthcare Conference* (pp. 230-248). PMLR.

Simonovsky, M., & Komodakis, N. (2018, October). Graphvae: Towards generation of small graphs using variational autoencoders. In *International conference on artificial neural networks* (pp. 412-422). Springer. 10.1007/978-3-030-01418-6_41

Soria-Guerra, R. E., Nieto-Gomez, R., Govea-Alonso, D. O., & Rosales-Mendoza, S. (2015). An overview of bioinformatics tools for epitope prediction: Implications on vaccine development. *Journal of Biomedical Informatics*, *53*, 405–414. doi:10.1016/j.jbi.2014.11.003 PMID:25464113

Ton, A. T., Gentile, F., Hsing, M., Ban, F., & Cherkasov, A. (2020). Rapid identification of potential inhibitors of SARS-CoV-2 main protease by deep docking of 1.3 billion compounds. *Molecular Informatics*, *39*(8), 2000028. doi:10.1002/minf.202000028 PMID:32162456

Torng, W., & Altman, R. B. (2019). Graph convolutional neural networks for predicting drug-target interactions. *Journal of Chemical Information and Modeling*, *59*(10), 4131–4149. doi:10.1021/acs.jcim.9b00628 PMID:31580672

ul Qamar, M. T., Alqahtani, S. M., Alamri, M. A., & Chen, L. L. (2020). Structural basis of SARS-CoV-2 3CLpro and anti-COVID-19 drug discovery from medicinal plants. *Journal of Pharmaceutical Analysis, 10*(4), 313-319.

van IJzendoorn, D. G., Szuhai, K., Briaire-de Bruijn, I. H., Kostine, M., Kuijjer, M. L., & Bovée, J. V. (2019). Machine learning analysis of gene expression data reveals novel diagnostic and prognostic biomarkers and identifies therapeutic targets for soft tissue sarcomas. *PLoS Computational Biology*, *15*(2), e1006826.

Wang, Y., Mao, H., & Yi, Z. (2017). Protein secondary structure prediction by using deep learning method. *Knowledge-Based Systems*, *118*, 115–123.

Wu, C., Liu, Y., Yang, Y., Zhang, P., Zhong, W., Wang, Y., ... Li, H. (2020). Analysis of therapeutic targets for SARS-CoV-2 and discovery of potential drugs by computational methods. *Acta Pharmaceutica Sinica. B*, *10*(5), 766–788.

Xu, Z., Yang, L., Zhang, X., Zhang, Q., Yang, Z., Liu, Y., ... Liu, W. (2020). Discovery of potential flavonoid inhibitors against COVID-19 3CL proteinase based on virtual screening strategy. *Frontiers in Molecular Biosciences*, 7.

Zhang, H., Saravanan, K. M., Yang, Y., Hossain, M. T., Li, J., Ren, X., ... Wei, Y. (2020b). Deep learning based drug screening for novel coronavirus 2019-nCov. *Interdisciplinary Sciences, Computational Life Sciences*, *12*, 368–376.

Zhang, L., Lin, D., Sun, X., Curth, U., Drosten, C., Sauerhering, L., ... Hilgenfeld, R. (2020a). Crystal 449 structure of SARS-CoV-2 main protease provides a basis for design of improved α-ketoamide 450 inhibitors. *Nature*.

Zhang, L., Tan, J., Han, D., & Zhu, H. (2017). From machine learning to deep learning: Progress in machine intelligence for rational drug discovery. *Drug Discovery Today*, *22*(11), 1680–1685.

Zhang, R., Hristovski, D., Schutte, D., Kastrin, A., Fiszman, M., & Kilicoglu, H. (2021). Drug repurposing for COVID-19 via knowledge graph completion. *Journal of Biomedical Informatics*, *115*, 103696.

Zhavoronkov, A., Ivanenkov, Y. A., Aliper, A., Veselov, M. S., Aladinskiy, V. A., Aladinskaya, A. V., ... Aspuru-Guzik, A. (2019). Deep learning enables rapid identification of potent DDR1 kinase inhibitors. *Nature Biotechnology*, *37*(9), 1038–1040.

Zheng, C., Yu, W., Xie, F., Chen, W., Mercado, C., Sy, L. S., ... Jacobsen, S. J. (2019). The use of natural language processing to identify Tdap-related local reactions at five health care systems in the Vaccine Safety Datalink. *International Journal of Medical Informatics*, *127*, 27–34.

Zitnik, M., Agrawal, M., & Leskovec, J. (2018). Modeling polypharmacy side effects with graph convolutional networks. *Bioinformatics (Oxford, England)*, *34*(13), i457–i466.

ADDITIONAL READING

Cernile, G., Heritage, T., Sebire, N. J., Gordon, B., Schwering, T., Kazemlou, S., & Borecki, Y. (2021). Network graph representation of COVID-19 scientific publications to aid knowledge discovery. *BMJ Health & Care Informatics*, *28*(1), e100254. doi:10.1136/bmjhci-2020-100254 PMID:33419870

Dokland, T. (2006). *Techniques in microscopy for biomedical applications* (Vol. 2). World Scientific. doi:10.1142/5911

Kerfoot, E., Fovargue, L., Rivolo, S., Shi, W., Rueckert, D., Nordsletten, D., ... Razavi, R. (2016, August). Eidolon: visualization and computational framework for multi-modal biomedical data analysis. In *International Conference on Medical Imaging and Augmented Reality* (pp. 425-437). Springer. 10.1007/978-3-319-43775-0_39

Lytras, M. D., & Papadopoulou, P. (Eds.). (2017). *Applying big data analytics in bioinformatics and medicine*. IGI Global.

Palanisamy, V., & Thirunavukarasu, R. (2019). Implications of big data analytics in developing healthcare frameworks–A review. *Journal of King Saud University-Computer and Information Sciences*, *31*(4), 415–425. doi:10.1016/j.jksuci.2017.12.007

Peng, L., Peng, M., Liao, B., Huang, G., Li, W., & Xie, D. (2018). The advances and challenges of deep learning application in biological big data processing. *Current Bioinformatics*, *13*(4), 352–359. doi:10.2174/1574893612666170707095707

Raeven, R. H., van Riet, E., Meiring, H. D., Metz, B., & Kersten, G. F. (2019). Systems vaccinology and big data in the vaccine development chain. *Immunology*, *156*(1), 33–46. doi:10.1111/imm.13012 PMID:30317555

Szlezak, N., Evers, M., Wang, J., & Pérez, L. (2014). The role of big data and advanced analytics in drug discovery, development, and commercialization. *Clinical Pharmacology and Therapeutics*, *95*(5), 492–495. doi:10.1038/clpt.2014.29 PMID:24642713

Wise, C., Ioannidis, V. N., Calvo, M. R., Song, X., Price, G., Kulkarni, N., . . . Karypis, G. (2020). COVID-19 knowledge graph: accelerating information retrieval and discovery for scientific literature. *arXiv preprint arXiv:2007.12731*.

Zhang, G. L., Sun, J., Chitkushev, L., & Brusic, V. (2014). Big data analytics in immunology: A knowledge-based approach. *BioMed Research International*, *2014*. doi:10.1155/2014/437987 PMID:25045677

KEY TERMS AND DEFINITIONS

Bioinformatics: An interdisciplinary field of biology and computer science to perform data acquisition, storage, and analysis on biological data.

Deep Learning: A broad family of machine learning models based on neural networks. Typical deep learning models are deep neural networks, convolutional neural networks, recurrent neural networks, deep belief networks, and deep reinforcement learning.

Drug Discovery and Design: A process of identifying the active ingredient and design the drugs based on structure-activity-relationships.

Knowledge Graph: A graph-structured data model that can interpret semantics through analyzing the relationships between entities.

Machine Learning: A subject of artificial intelligence that aims at the task of computational algorithms, which allow machines to learning objects automatically through historical data.

NLP: A processing method of computational linguistics for human language based on algorithms.

Protein Structure Analysis: An analysis method to understand protein structure and the interactions with other proteins.

Vaccine Development: A process of finding a new antigen and then developing a vaccine accordingly.

Chapter 10
Developing a Web-Based COVID-19 Fake News Detector With Deep Learning

ABSTRACT

Fake news and misleading information have been determined as ongoing social challenges in the post-pandemic era. COVID-19-related misinformation has been posted online, which is a crucial impact on society. Despite technological abuses of spreading misinformation, artificial intelligence can help to terminate it. This chapter proposes a cloud-based architecture to detect misleading information on COVID-19-related news and articles. The system has been illustrated through misinformation extraction, fake news detection, and ground-truth testing. A web-based application has been presented with a dashboard-like user interface design using cloud computing. A bench of word embeddings and deep learning algorithms has been investigated for determining the optimal model. The anti-misinformation system can identify fake news in a second with a reliability study operated in a cloud computing environment. Potential limitations and suggestions are also discussed in terms of improving the system for industrial consideration.

INTRODUCTION

Deciphering what is fact and fiction sometimes is exhausting, particularly in the age of mass and social media. Fake news and misinformation have

DOI: 10.4018/978-1-7998-8793-5.ch010

been spreading even faster than the novel coronavirus since the beginning of 2020. Hundreds of posts containing misguiding information on COVID-19 are being left online, along with the huge impact on individual responses and disastrous consequences to the entire social system (Barua et al., 2020). Many social and psychological impacts from misinformation, such as xenophobia, depression, violence, human rights violation for specific clusters (i.e. females, minorities, and LGBT groups), and psychological health of healthcare workers, have been becoming one of the biggest challenges during the ongoing COVID-19 pandemic (Ali et al., 2021). The ability of spreading fake news and false information has increased exponentially in the digital era as irresponsible individuals and organizations generate and share misleading posts through social media. Despite the negative aspect of hyper-connected society, information technologies can also assist to terminate using the state-of-the-art techniques. The majority of news organizations have implemented the fact-check units, which can help editors to guarantee the trustworthiness of the news source by screening and checking material in the digital manner. Third-party fact-checkers have also been incorporated to strengthen the capability of the authenticity of news content in terms of automatic assessments of news organizations, in which journalists are required to obey the Trust Principles as the bottom line. On the other hand, public authorities and lawmakers have obligations to regulate the journalism systematically by issuing related policies and rules, facilitating news literacy, and unifying government supervision, whereas monitoring fake news and misinformation is the foundation of the evidence-based policy-making process and the efficient measure of the regulation. Moreover, the supervision from any other social participants also plays a critical role in fighting against fake news and rumors. Several advanced algorithms and auto-detection systems have been applied by technology companies to identify fake news and misinformation. Individuals can also avoid fake news and rumors through comparing news and posts from a variety of online resources.

However, checking facts manually is less efficient and annoying, thereby it cannot catch up with the rapid growth of information streaming updates, along with the popularity of social media in particular. Billons of contents in pieces have been posted and shared every day via internet, therefore it is impossible to review news and stories on the individual basis through a traditional way of the fact-checking process. Various cutting-edge technologies, such as machine learning, deep learning, and big data analytics, have been incorporated in fake news detection and tracking misinformation during the last decade as data science has becoming more popular and applicable

(Cardoso Durier da Silva et al., 2019). From the vision of interdisciplinary research, detecting fake news can be discussed from screening the false information it carries, analyzing its writing style, extracting its propagation patterns, and checking the credibility of its source (Zhou & Zafarani, 2020). As such, existing studies in this domain are concentrated on natural language processing (NLP), social network analysis (SNA) and knowledge engineering with artificial intelligence (AI) and big data analytics during the last few years (Oshikawa et al., 2018; Ahmed et al., 2019; Vishwakarma & Jain, 2020). To the industrial application perspective, Smart tools collaborating with data science methods have been proposed by internet technology companies for fighting against fake news and the spreading of misleading information. Facebook announced that the product team has improved the capability of fact-checking with machine learning, which can assist end-users to identify false stories and duplicated information (Lyons, 2018). Twitter also introduced its fake news detector, known as Fabula AI, which is one of the best deep learning-based fake news detectors in terms of its performance of stopping the spreading of misinformation by learning the differences between fake stories and real news through social media and internet (Pomputius, 2019). Besides, various prototypes of fake news detection and tracking applications have been developed by domain experts and data scientists. Janze & Risius (2017) developed an automatic detection tool for identifying fake news on social media platforms during the 2016 U.S. presidential election. Similarly, a machine learning-based fake news auto-detector has been implemented within a Facebook Messenger chatbot, achieving over 80% in accuracy with a real-world application (Della Vedova et al., 2018). Shu et al. (2019) introduced FakeNewsTracker, a smart tool for collecting, detecting, and visualizing fake news and misleading information, which can understand and track news pieces and social context with a range of effective techniques, such as data aggregation, data visualization, and predictive analysis.

With the rapid growth of digital life in the modern society, individuals are rending to acquire online news and information. Analyzing internet dependency relations (IDR) becomes one of the most popular topics in online activity study and cyber-based behavioral science (Ko et al., 2009). Experimental results from a survey with IDR-based analysis in early 2000s illustrated that users spent more hours on reading news online than on other activities per day (Patwardhan & Yang, 2003). On the other hand, people are more likely to spend time on social media as they may take advantages of reading and sharing posts for social interaction and communication, information searching and sharing, entertainment and relaxation, and pass time and surveillance about others (Whiting & Williams, 2013). Such the conclusions do not change overtime, which have been confirmed by reports from Pew Research Center. Compiled from the Pew's surveys and analyses in 2018, 37% of U.S. adults prefer to acquire news through online channels, in which 77% of them acknowledge the importance of online news (Geiger, 2019). Such numbers have been increased as of 2020, particularly, over half of Americans use a digital platform for news searching and acquiring, in which 11% of them prefer social media platforms (Shearer, 2021). Besides, individuals tend to scan articles' titles and headlines rather than text bodies due to the busy and fast rhythm of the modern lifestyle. The process of news reading has been affected by hyper-reading information from social media, thereby individuals are most likely to read only the headlines without clicking on the majority of articles (Waage, 2018). Moreover, news readers are more likely to trust the news headline when it aligns with their mindsets, whereas people are less likely to believe headlines that challenge their opinions (Moravec et al., 2018). As such, changes in the way of information acquisition, reading habit, and cognitive activity objectively reduce the cost of making and spreading fake news.

Fake news and misleading information regarding COVID-19 have been produced and disseminated on a large scale. In addition to the crisis of public health, false information about the pandemic has evolved into a new disaster, so-called "infodemic", which damages the structure of the public trust system. One of the most recent studies uncovers psychometric and communicational factors that affect belief in misinformation about COVID-19, including stigma, health status, heuristic information processing, trust, and subjective social class (Kim & Kim, 2020). Typical fake news and rumors are positively correlated to a broad range of topics, such as the conjecture of the man-made coronavirus in military labs, the panic caused by the overload capacity of hospitals, the discrimination against Asian communities and Chinese products, the surmise of the commercial scheme towards vaccine sells, the conspiracy theory of the 5G technology, and the attitude of political hoax about COVID-19 (Furini et al., 2020; Bruns et al., 2020; Casero-Ripollés, 2020). The tone of fake news, in general, is fatalistic, racist, and anti-intellectualism orientated (Ljungholm & Olah, 2020; Stanley et al., 2020). Therefore, ad-hoc text mining and classification tasks are demanded in order to implement the detector for COVID-19 related fake news and rumors. Moreover, short-message-like fake news may pollute the internet environment more significantly as it is more convincing and disseminating. During the COVID-19 outbreak, some individuals and organizations deliberately created misleading information by misrepresenting headlines of the fact, whereas regular people may receive information that is contrary to the original articles of facts. Therefore, instead of focusing on full text bodies, screening headlines of articles from online news and social media may provide the most efficient method for identifying fake news and misleading information, along with big data analytics, NLP, and text mining. Motivated by the current challenges of fake news detection in the context of COVID-19, this chapter is designed to illustrate how AI and big data can help in constructing an efficient detection system for monitoring COVID-19 related fake news and misleading information. The objectives of this chapter are:

- illustrating how to collect and clean fake news data from open-source databases and websites using web scraper APIs;
- developing a web-based fake news detector by utilizing a cloud-based system architecture, along with introducing the full cycle of the web app development;
- investigating various NLP approaches that can be applied to classifying fake news and misinformation in regards to the COVID-19 pandemic by examining several word embeddings and deep learning models;
- examining the alpha test results of the web application for detecting COVID-19 fake news and misinformation based on news from mainstream media and short-message-like data.

BACKGROUND

In the past few years, AI-based text classification approaches have been heavily applied in monitoring and screening fake news and false information. A machine learning-based credibility ranking approach has been proposed for detecting spam tweets by extracting content- and source-based features using regression analysis (Gupta & Kumaraguru, 2012). Conroy et al. (2015) provided a typology of multiple assessment approaches incorporating linguistic cue methods, machine learning, and network analysis to design a fake news detector. A Support Vector Machine (SVM)-based machine learning framework has been applied in detecting potential misleading information by analyzing features of satirical cues (Rubin et al., 2016). Such a framework can detect satirical news with a precision of 90% and a recall of 84%, which has been applied in minimizing the risk of causing misinformation. Ahmed et al. (2017) proposed an online fake news detector using the n-gram model and machine learning algorithms, where the optimal performance can be achieved by incorporating Term Frequency-Inverted Document Frequency (TF-IDF) as the feature extractor and SVM as the classifier, with 92% in accuracy. Ghosh & Shah (2018) performed a generalized fake news classification by combining approaches from information retrieval, text analysis, and neural networks. Such a hybrid model can achieve up to 82.4% in accuracy with the automatically detective and preventive attempts towards fake news problems. Various methods for false information detection have been investigated from a broad range of state-of-the-art technologies, such as NLP, SNA, and data mining (Bondielli & Marcelloni, 2019). Shu et al. (2020) proposed a data repository that contains news content, social context, and spatiotemporal features for analyzing and tracing fake news on social media. Such a data repository, named as FakeNewsNet, has been applied for classifying fake news using a variety of machine learning/deep learning models, such as SVM, logistic regression (LR), Naive Bayes (NB), Convolutional Neural Network (CNN), and Social Article Fusion (SAF), whereas CNN and SAF in general yield relatively higher performance in accuracy and F1 score.

Current studies in fake news detection are concentrated on using deep learning. A deep learning-based detector has been proposed through a three level hierarchical attention network, which is effective for detecting fake news automatically with an accuracy of 96.8% (Singhania & Rao, 2017). Yang et al. (2018) developed a CNN-based fake news detector that can be applied for identifying hidden patterns of fake news in textual contents and images. A multiple-level CNN architecture has been developed for fake news detection in cultural communication (Li et al., 2019). Besides, alternative deep learning algorithms, such as Recurrent Neural Network (RNN) and Long Short-term Memory (LSTM), have also been applied for detecting misinformation through internet and social media. Girgis et al. (2018) employed RNN and LSTM as the classifiers for identifying false information in online text. A RNN-based fake news detection system has been developed by using a bi-directional LSTM with the synonyms-based augmentation for the headline and the body part of the news (Suyanto, 2020). Moreover, some researchers aimed on the combination of deep learning models and word embeddings. Agarwal et al. (2020) proposed a blend of neural networks and word embeddings for predicting the nature of an article, along with a precision of 97.2%. Similarly, the FNDNet model for fake news detection has been developed for learning the discriminatory features of misleading information identification via multiple hidden layers followed by a CNN-based feature extractor, which can obtain 98.4% in accuracy (Kaliyar et al., 2020). Goldani et al. (2021)

proposed a fake news detector using CNN with margin loss and multiple embedding approaches. The optimal model can outperform the cutting-edge architectures by 7.9% and 2.1% on the test set of two well-known datasets in text classification, ISOT and LIAR, respectively. Instead of using a single deep learning model, some studies focuses on hybrid network architectures for implementing the text classification tasks in fake news detection, e.g. a bi-directional LSTM-RNN model (Bahad et al., 2019), the CNN-LSTM architecture (Umer et al., 2020), and the hybrid CNN-RNN approach (Nasir et al., 2021).

Nevertheless, training the ad-hoc models in regards to identifying COVID-19 fake news plays a critical role in the context of the infodemic, whereas the first stage towards such issues is to collect facts and fake news data. Existing studies aimed on collecting such information from news websites and social media platforms. Some studies have obtained the news sources and references from authority agencies and/or fact-checking websites. Elhadad et al. (2020) released an annotated COVID-19 misinformation and facts dataset, called Covid-19-FAKES, using machine learning with feature extraction, which can detect fake news through tweets. The annotated records in Covid-19-FAKES have been confirmed with official websites and Twitter accounts of authority organizations, such as WHO, UNICEF, and UN. A cross-domain dataset designed for COVID-19, namely FakeCovid, has been created by collecting fact-checked news articles from over 90 different websites with 40 languages from more than 100 countries after obtaining references from fact-checking websites, including Poynter and Snopes (Shahi & Nandini, 2020). Most studies relies on manually labeled articles of real and fake news. For example, a manually annotated dataset containing over 10,000 social media posts and articles of real and fake news on COVID-19 has been generated with multiple machine learning models, such as SVM, LR, decision tree, and gradient boost (Patwa et al., 2020). Qazi et al. (2020) provided a large Twitter dataset, named as GeoCoV19, containing over 520 million tweets with their geolocation, which has been applied for detecting and characterizing misleading information on COVID-19 (Memon & Carley, 2020).

To the text classification perspective, current studies in fake news classification are concentrated on using deep learning with word embeddings. Wani et al. (2021) proposed multiple text classifiers (i.e. CNN and LSTM with word embedding methods) for COVID-19 fake news detection by evaluating the model performance matrix per model, of which the best model can achieve an accuracy of 98.4%. An optimized architecture for automated identifying COVID-19 false information on Twitter has been develop based on a variety of machine learning/deep learning algorithms, such as LSTM, Gated Recurrent Units (GRU), NB, SVM, etc., along with two feature extractors (TF-IDF and N-gram), whereas the optimal model is the modified LSTM with the highest cross-validation accuracy at 94.2% (Abdelminaam et al., 2021). Some studies incorporated the efficiency of the classification tasks by training the deep learning models based on the headlines of articles. For instance, Wang et al. (2020) proposed a COVID-19 misinformation detection system for evaluating the credibility of article headlines from the mainstream media channels using CNN, LSTM, and Deep Belief Networks (DBNs). Such an approach can significantly reduce the time for model training and testing without decreasing the model performance. Alternatively, high-performance data flow pipelines have been employed in COVID-19 fake news detection in terms of implementing text classifiers into a parallel computing system, e.g. Madani et al. (2021) deployed a COVID-19 misinformation detector using Spark.

MAIN FOCUS OF THE CHAPTER

Issues, Controversies, Problems

Although general fake news and false information detection algorithms with machine learning and deep learning have been proposed and applied for a long while, the novel datasets and new information pool are essential for training the ad-hoc models as the text classifiers in regards to a specific problem or topic, e.g. detecting COVID-19 related fake news and rumors. However, one of the controversies lays on collecting enough data and labeling the records, along with the data cleaning process. Therefore, this chapter will emphasize on the data acquisition process throughout collecting, labeling, cleaning and aggregating information from open accessible databases and online resources. Fake news and facts related to headlines of COVID-19 articles will be extracted from third-party fact-checking networks and news intelligence platforms independently using content extractor APIs. Raw data will be cleaned and reshaped into a NLP-friendly format, followed by the preprocessing of the classification task. Several machine learning/deep learning algorithms will be applied to identifying COVID-19 fake news. The optimal model will be selected based on the model performance evaluation. However, using such models directly for classification tasks may cause the overfitting problem, therefore the optimal model will be then improved by fine-tuning the model structure, i.e. word embeddings as the feature extractor and machine learning/deep learning models as the classifier. Classical word embedding methods will be also introduced and investigated for the basic understanding of how such vectorizers works as the feature extractors for the text classification problem.

Features of the Chapter

One of the most important features of this chapter emphasizes on introducing the system architecture of the web-based application with the experimental design of the fake news detector. The proposed system architecture of fake news and misinformation detection consists of four major components, such as data preparation, feature extraction, fake news classification, and model deployment. The system starts from the data acquisition, extraction, cleaning, and preprocessing, followed by the initiative of the model training and testing process. To the experimental design perspective, the cleaned news dataset has been split into two subsets, i.e. 80% for training and 20% for testing. As per the common-used operation for text classification, several word embedding approaches have been chosen as the candidate feature extractors, including TF-IDF, Word2Vec, and Global Vector (GloVe), in which the most efficient methods will be selected for extracting text features based on the theoretical and experimental conclusions summarized from related studies. For text classification, classical machine learning models can be applied, however, the model performance is relatively lower (Chokshi & Mathew, 2021). Therefore, a variety of deep learning algorithms will be investigated in terms of the network structure for each model, such as DBN, RNN, LSTM, GRU, and CNN, whereas the candidate models will be selected as the fake news classifiers based upon the experimental results derived from recent studies. The optimal model will be determined by the model evaluation measurements, such as accuracy, AUC score, and F-1 measure. Finally, the optimal model will be deployed in a web-based user interface (UI) with Amazon Web Services (AWS) using Python Flask API containerized in Docker and powered by Elastic Container Service (ECS). Such a web-based application is expected to identify COVID-19 fake news and misinformation through news articles and/or message-like information in seconds with a high degree of reliability.

SOLUTIONS AND RECOMMENDATIONS

Data Extraction, Cleaning, and Preprocessing

Data acquisition plays the fundamental role in fake news detection. Real and fake news information should be annotated and categorized carefully without disturbing by any subjective aspirations. Two data sources are employed in this chapter. The Alliance Database can be acquired from Poynter, which is a non-profit journalism research institution owned by Tampa Bay Times and the International Fact-Checking Network, providing a fake news pool about COVID-19. Since the beginning of 2020, Poynter deployed the CoronaVirusFacts module in Alliance, which covers more than 100 fact-checkers in posting, sharing, and translating information in regards to the novel coronavirus globally, along with detecting and collecting related fake stories. For each record of misleading information in the database, Alliance provides various attributes, such as the affiliation of the fact-checker, the post date, geolocations, the content of the false information, and the comparison between real and fake news. The Alliance development team implements the database into a web-based UI with updating the fake news pool frequently. On the other hand, real news and true stories related to COVID-19 can be retrieved from an open-source database, namely Coronavirus News Dataset, provided by AYLIEN, which is a news intelligence platform. AYLIEN aims on assisting lawmakers, healthcare authorities, and governments in tracking and predicting the spread of COVID-19 in the form of publishing and sharing fact-checked information during the pandemic. By leveraging NLP and machine learning, newspapers have been transformed into a semi-structured data type, which can be obtained through the AYLIEN News API. The AYLIEN dataset collects newspaper articles in English from over 400 data sources, containing nearly 2 million real news records about COVID-19 from the late of 2019 to the middle of 2020. Each record provides the news headline, the publication date, the body part of the article, the source URL, etc. As such, aggregating Alliance and AYLIEN is necessary for the model training process as per the system architecture of the COVID-19 fake news and misinformation detection. However, neither Alliance nor AYLIEN is in the ready-to-use format, thereby requiring data extraction, cleaning, and preprocessing.

Despite adequate information available from Alliance, each attribute has been designed in a visual-only fashion on the website in the HTML format.

Users can check the fact through choosing a set of categories, including interested countries, news types, and fact-checking organizations, whereas the result has been well-organized block by block, thereby extracting such information is not straightforward. Manually extracting useful information from the website seems impossible due to the large scale of the data pool, therefore, web scraper will be the ideal solution towards the data extraction problem on Alliance. This chapter applies the web scraping approach using API in Python to capture data from Alliance, incorporating a range of attributes per fake news record, such has country, post date, news type, article headlines and contents, and fact-checking organizations between January 14 and August 11 in 2020. Overall, more than 8800 records have been extracted, in which news types originally annotated as false, misleading, partly false, mostly false, and no-evidence are labeled as the fake news for the next stage of model training and testing. On the other hand, the AYLIEN dataset can be downloaded directly from its official website at alien.com in the JSON object file format. Although extracting desired information from a JSON file is not difficult, its large volume and data structure challenge the data extraction and cleaning process for AYLIEN. Implementing such a task in a local computer seems less efficient due to the limited memory size and CPU capacity, thereby a high-performance computing environment is essential for reducing the time on data processing. In this chapter, a Spark-based data engine has been deployed on AWS with demanded Python modules installed. The JSON file is firstly unzipped into pieces, having 1,673,354 subfiles that contain a record of the news article. For each subfile, desire information, such as the publication date, the news headline, the source URL, has been extracted after removing unnecessary and broken data. Finally, the AYLIEN dataset has been reshaped into a CSV file containing records from November 1, 2019 to July 31, 2020. Such a cleaned dataset contains all fact-checked information that can be determined as the real news dataset.

Data preprocessing plays one of the most important role in the NLP domain, particularly to the efficient text classification perspective. News articles and headlines are typically unstructured data, thereby requiring to be refined by performing the preprocessing procedure before entering the model training process. The basic preprocessing operations for NLP are stop-words removal, tokenization, sentence segmentation, etc. Such techniques can be helpful in reducing the size of original data by removing the noise, thereby expecting better performances in model training and testing. As per the regular implementation of data preprocessing for NLP, both real news and false information have been involved by a series of preprocessing operations listed as follows:

Step 1. Lowercase. This is one of the basic preprocessing operations to transform a word into the lower case.
Step 2. Unnecessary data removal. The operation is to delete data such as the punctuation and the stop word, which do not provide much semantics to a sentence.
Step 3. Tokenization. The corpus is then proceeded by converting a word into a vector. Words in this sentence are split up into single items such as x_1,x_2,…,x_i . A list of all unique words is formatted as a dictionary that are tokenized as ["x_1", "x_2", …, "x_i"].

Based on the fact-checking capabilities of two data sources, each article in the Alliance dataset has been annotated as the fake news data, while articles in AYLIEN are labeled as real news. The headline of each annotated article has been extracted for the model training and testing process. Instead of training the fake news detection model using the full content of an article, such experimental design is less time-consuming without decreasing the model performance or eliminating the functionality of the system as per above-mentioned discussions. Moreover, only 8,800 real news records are randomly selected from the AYLIEN dataset in order to match the sample size of fake news. The sample size can be adjustable by increasing the number of real news records due to the large scale of AYLIEN, whereas the number of fake news samples is fixed with the limited records captured from Alliance.

Feature Extraction With Word Embeddings

In this stage, several popular and classical word embedding approaches are investigated in terms of their basic ideas. In general, a word embedding method converts words into vectors, representing each word in an n-dimensional dense vector, in which similar semantics are transformed into a similar vector. Conmen-used word embeddings include TF-IDF, Word2Vec, GloVe, etc. Such methods can be employed as the candidate feature extractors for the COVID-19 fake news detection.

TF-IDF

TF-IDF provides another vectorization based on the frequency method by account occurrence of a word in a document and the entire corpus. One of the most significant challenges in nature language processing is that common words such as 'is', 'are', 'a', 'the' etc. appear frequently in contrast with the important words in a sentence. Theoretically, the common words should be downplayed and other words should be emphasized by adding more weights. To do that, calculating the term frequency and the inverse document frequency is the essential stage of the initiative. Firstly, a term frequency can be calculated by

$$TF\left(t,d\right)=\frac{t}{d},$$

where t is the number of times a term appears in a document and d is the number of terms in the document. It measures the words relevant to the document for either common words or important tokens. The second stage is to denote more contribution of the word to the document by calculating the inverse document frequency as

$$IDF\left(N,n\right)=\log\left(\frac{N}{n}\right),$$

where N is the number of documents and n is the number of documents t has appeared in. The idea of the reasoning behind IDF is to distinguish common words and other tokens in terms of occurrences in a particular document and relevance to another. Importantly, if the common words appeared in all documents, i.e. $N=n$, IDF will return 0 because of the log term. Finally, TF-IDF is computed by putting TF and IDF together in a multiple term

$$TF-IDF\left(t,d,N,n\right)=TF\left(t,d\right)\times IDF\left(N,n\right)$$

Common terms can be penalized significantly but important terms will be assigned greater weights, which indicates a rare word is an important one for a single document from the context of the entire corpus. No significant difference between removing the stop words and using TF-IDF method exists, because IDF process exactly remove words without semantic values from the documents.

Word2Vec

Word2Vec is a combination of two prediction-based models, CBOW (Continuous bag of words) and Skip-gram model. Both of them map tokens in a neural network, which can learn weights that are word vector representations. Both CBOW and Skip-gram have their own advantages and disadvantages. CBOW is faster than Skip-gram, while Skip-gram works well with a small set of documents. CBOW has better representations for high frequent words, while Skip-gram is fit to represent rare words effectively. CBOW is a word embedding by predicting the probability of a word in a context. The current word *c* in the context *w* can be predicted as a probability *P*(*c*|*w*) by a neural network, which has three layers including an input layer, a hidden layer, and an output layer. The process can be interpreted by a neural network framework with input layer x_{ik} and hidden layer h_i given the input-hidden wrights $W_{V \times N}$ and a hidden-output wight $W'_{N \times V}$. The hidden layer neurons copy the weighted sum of inputs the next layer. The output layer is a *V* dimension vector y_i. The mean square error in CBOW is negative log likelihood of a word given a set of context, $-\log(p(w_o \mid w_i)$, where

$$p(w_o \mid w_i) = \frac{\exp\left(v'_{w_o} v_{w_i}\right)}{\sum_{w=1}^{W} \exp\left(v'_{w_o} v_{w_i}\right)},$$

where w_o is the output word and w_i are context words. Input-hidden weights and hidden-output weights are different. CBOW does not require a huge memory for computing, comparing to a co-occurrence matrix. Skip-gram model is similar to CBOW, however, it attempts to predict the context given a word. Input and hidden layers are similar to CBOW, while the output layer is different. The output is a combination of *V* dimension vectors $(y_1 y_2 \ldots y_c)$. For each context position, *C* distributions of *V* probabilities can be get by inout the target word into the neural network. Errors are computed by subtracting the first row of the target from the first row of the output, thus the size of error vectors will be determined by the number of target context words.

GloVe

GloVe, developed by Pennington et al. (need citation https://nlp.stanford.edu/pubs/glove.pdf), makes word embeddings more effective. It combines both the count-based matrix vectorization and the context-based skip-gram model. The idea of the GloVe is to capture word meanings by the ratios of co-occurrence probabilities. A co-occurrence probability of two words can be defined as

$$P_c(x_i \mid x_j) = \frac{C\left(x_i, x_j\right)}{C\left(x_i\right)},$$

where $C(x_i)$ is a simple count of a word x_i and $C(x_i,x_j)$ is the number of the co-occurrence between word x_i and word x_j. The relationship between two words, w_i and w_j, regarding to the third context word $\overline{w}_k$ is determined by a ratio of two co-occurrence probabilities, such as

$$F\left(w_i, w_j, \overline{w}_k\right) = \frac{P_c(\overline{w}_k \mid w_i)}{P_c(\overline{w}_k \mid w_j)},$$

where $F(.)$ is a exponential function of the linear difference between two words, therefore

$$F\left(w_i, w_j, \overline{w}_k\right) = \exp(\left(w_i - w_j\right)^T \overline{w}_k\right) = \frac{\exp\left(w_i^T \overline{w}_k\right)}{\exp\left(w_j^T \overline{w}_k\right)},$$

where $w_i^T \overline{w}_k$ can be a $log(.)$ term of the co-occurrence probability $log(P_c(\overline{w}_k \mid w_i))$ substituted by $C(w_i)$ and $C\left(w_i, \overline{w}_k\right)$. A bias term βi for wi can be added since $log(C(wi_i)$ is independent of the context word, the final form of $w_i^T \overline{w}_k$ can be represented by

$$w_i^T \overline{w}_k = log \frac{C\left(w_i, \overline{w}_k\right)}{C\left(w_i\right)} = log\left(C\left(w_i, \overline{w}_k\right)\right) - log\left(C\left(x_i\right)\right)$$

where $log\left(C\left(w_i, \overline{w}_k\right)\right) = w_i^T \overline{w}_k + \beta_i + \overline{\beta}_k$ and $\overline{\beta}_k$ is the bias of $\overline{w}_k$. We can obtain the loss function by minimizing the sum of the squared errors, such as

$$\sum_{w=1}^{W} f(C\left(w_i, w_j\right)(w_i^T \overline{w}_k \beta_i + \overline{\beta}_k - log\left(C\left(w_i, \overline{w}_k\right)\right))^2,$$

and the weight function is as follows.

$$f\left(x\right) = \{ \begin{matrix} (x / x_{max})^{\alpha} & \text{if } x < x_{max} \\ 1 & \text{otherwise} \end{matrix}$$

Deep Learning-based Text Classifiers

Several deep learning models have been applied for identifying news into real or fake as per reviewing the most recent studies in this field. In general, a deep learning-based text classification architecture consists of an embedding layer based on the word embedding matrix, hidden layers such as RNN, LSTM, GRU, etc., a flatten layer such as CNN, and the output layer. In this stage, common-used deep learning models are investigated in terms of the network architecture for each model, including DBN, RNN, LSTM, GRU, and CNN. Such models can be applied as the candidate text classifiers for the COVID-19 fake news detection.

DBN

Introduced by Hinton et al. (2006), DBN is a probabilistic generative graphical model and composed of multiple layers of latent variables with connections between the layers. DBN is based on a deep architecture of multiple stacks of Restricted Boltzmann machines (RBM), which provides the flexibility to DBNs through performing a non-linear transformation on input vectors and deliver its output vectors to the next RBM model as input in the sequence. In the network structure of DBN, it contains three RBM networks. The training process of DBN goes through layer-wise pre-training and fine-tuning, sequentially. Firstly, the RBM network of each layer is trained under unsupervised learning and feature vector are mapped to various feature spaces. Secondly, a back propagation (BP) network is used at the last layer of DBN, and then the feature vectors produced by RBM are further used to train classifier of entity relationship under supervision manner. In the training process, error information is delivered top-down to each layer of RBM through a back propagation network. The training process of RBM network helps DBN to avoid local optimum and long training time when initializing the weight parameters in the deep BP network.

RNN

The history of RNN can be tracked back to 1980s. As per the network structure of RNN, the nodes are connected as feedforward neural networks, which can cope with variable length sequences of inputs using internal memory. RNN permits the intermediate states to store information of previous inputs for each element in the sequence. The states is determined by the input and the previous state. In the architecture of an RNN, the hidden state at a certain time in the sequences is decided by a hyperbolic tangent function of input vector, parameter matrices and vector. However, the traditional fully connected RNN cannot avoid the gradient vanishing and exploding problems when using gradient descent method. The modified RNN architecture is used, such as independently recurrent neural network (IndRNN). Different from fully connected RNN, IndRNN can make each neuron only attain its own previous state and work independently.

LSTM

LSTM is a special form of RNN with feedback connections, which enable it to deal with entire sequences of data. The memory cell consists of a memory block and three gating units that can protect and control the cell state. These three gating units are forget gate layer, input gate layer, and output gate layer. Forget gate layer usually decides what information should be filtered out from the cell state. Input gate layer allows new information to be stored in the cell state. Output gate layer decided the output result. Peephole can help the communication between gate layer and cell state. Moreover, LSTM covers character sequences of different lengths, which means that linguistic features are not needed necessarily.

GRU

Based on RNN, GRU is a gating mechanism in recurrent neural network, proposed by Cho et al. (2014). Different from LSTM, GRU has two gates, including reset gate and update gate. The reset gate controls the short-term memory of the network and determines the influence on the candidate state. The reset gate will decide if the entire information from the pass hidden state is considered or ignored. The update gate controls the long-term memory of the network and the new information from the candidate state. The hidden state of the candidate is produced through calculating the entire information using a hyperbolic tangent function.

CNN

CNN has raised extensive attention as one of the most efficient classification approaches in recent years. It consists of multiple layers of artificial neurons with remarkable advantages in computer vision and natural language processing. CNN contains an input layer, an output layer, convolutional layers, pooling layers, fully connected layers, and normalization layers. Input layer usually converts the input data into a tensor with a certain shape. The convolutional layer transforms the input into an activation function with a proper shape. Connected to convolutional layers, pooling layers usually reduce the dimensions of the outputs of the neuron clusters into a single neuron. The fully connected layers help to connect each neuron in each layer. Weights define the calculation operation for each neuron. The last layer of the CNN is the classification layer, which conduct the predictions based on high-level features.

Model Evaluation and Deployment

This chapter applies two different word embedding methods, Word2Vec and GloVe, as the feature extractors. Word2Vec and GloVe have been proven to be the more efficient word embedding approaches, comparing to TF-IDF (Abdelminaam et al., 2021). On the other hand, two deep learning models, LSTM and CNN, have been chosen as the text classifiers due to the advantages of their model performance in contract with other deep neural networks for COVID-19 fake news detections (Wang et al., 2020). As such, four proposed algorithms are determined for the model training and testing process, followed by evaluating the model performance matrix, such as accuracy, AUC, and F-1 score, for each proposed algorithm. The evaluation results are shown in Table 1. Each model has been trained with the equal number of the sample size for both real and fake news. The model training and testing process has been implemented entirely in a cloud-computing environment.

Table 1. Model performance matrix for each proposed COVID-19 fake news detection algorithm

Feature Extractor	Classifier	Accuracy	AUC	F1
Word2Vec	CNN	86.71%	0.87	0.87
Word2Vec	LSTM	79.16%	0.79	0.79
GloVe	CNN	91.82%	0.92	0.92
GloVe	LSTM	92.87%	0.93	0.93

The optimal model can be determined by comparing the model performance matrix through Table 1. The ROC curves for CNN and LSTM are given in Figure 1. The visual results suggest that both CNN and LSTM can achieve a high degree in model performance. To the computational efficiency of learning processes perspectives, CNN has been examined as a faster model rather than LSTM in the COVID-19 fake news detection task (Wang et al., 2020). Thus, CNN will be chosen as the final classifier for implementing the web-based fake news detector for COVID-19 in the next stage.

Figure 1. ROC curves of CNN (right) and LSTM (left)

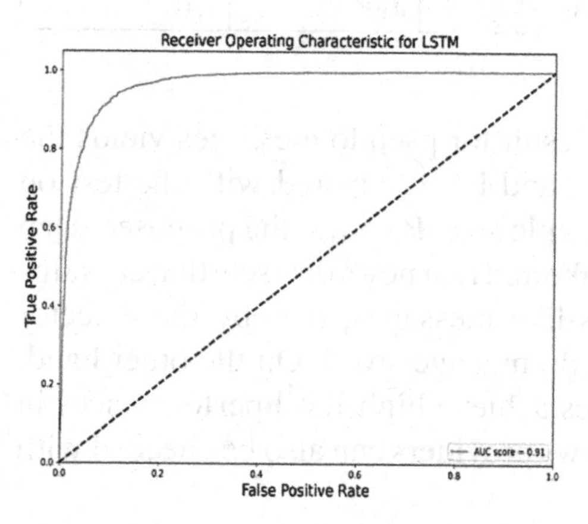

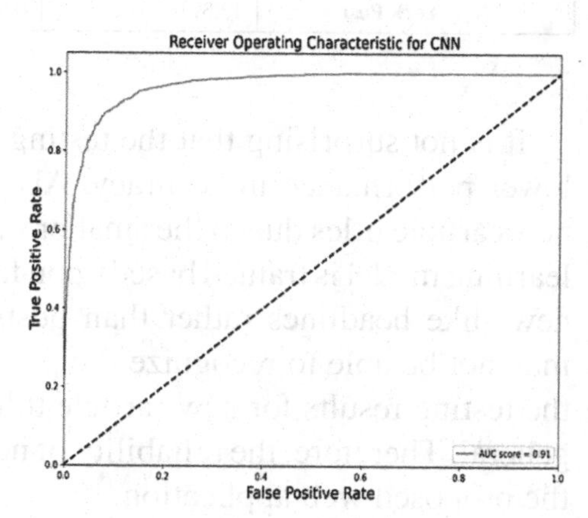

To store the data for the fake-news web application, a MongoDB database has been deployed with the Flask framework using a Flask extension, called MongoEngine. The configure information of MongoDB has been initialized in Flask in order to connect the database and the server. The interface of this app has been generated by HTML code in Flask. Additionally, this fake-news web app has been implemented in AWS elastic beanstalk with EC2 instance of t2.xlarge. To test the capability of the web-based COVID-19 fake news detector, an alpha testing procedure has been performed with 100 pseudo messages created by the developers and more than one million news headlines derived from COVID-19 related articles. Messages are deliberately generated by referring typical rumors from social media platforms, e.g. "don't go to Chinese restaurants as they produce Covid", "5G will get you infected", "don't get the vaccine because you will be dead", etc. Such messages are pre-defined as fake news. On the other side, news articles are randomly selected from news providers, such as The New York Times, Fox News, Reuters, and New York Post. Data in the news dataset is collected from the cloud using news APIs in Python. Such news titles are expected to be real. The test results are shown in Table 2.

Table 2. Alpha testing results of the web application for detecting COVID-19 fake news

Headline/Message Sources	Sample size	Accuracy	AUC	F1
Pseudo Messages	100	0.81	0.79	0.78
The New York Times	10,039	0.95	0.95	0.95
Fox News	19,911	0.95	0.93	0.94
Reuters	90,420	0.98	0.93	0.95
New York Post	13,391	0.93	0.93	0.93

It is not surprising that the testing result for pseudo messages yields the lower performance in accuracy, AUC, and F1, compared with the test on news article titles due to the smaller sample size. Besides, the proposed deep learning model is trained based upon fake and real news datasets that contains news-like headlines rather than posts-like messages, thereby the detector may not be able to recognize the pseudo menages well. On the other hand, the testing results for news article titles achieve high-level performances in general. Therefore, the reliability of news providers can also be checked with the proposed web application.

FUTURE RESEARCH DIRECTIONS

The future research directions can be stretched by improving the components of the proposed detector, throughout data collection to model deployment. Data can be collected from different platforms, such as social media. Besides, different data types, such as images, voice messages, and videos, should also been investigated and incorporated in terms of textual transformation of the data diversity. Moreover, alternative models can be examined for training the fake news detector, e.g. transfer learning for text classification (Slovikovskaya, 2019; Singhal et al., 2020), BERT-based word embedding methods (Al-Garadi et al., 2020; Kaliyar et al., 2021), and transformer-based models (Qazi et al., 2020; Kula et al., 2021). Lastly, the pre-trained model can be deployed through the big data tools, such as Hadoop, Spark, Cassandra, etc., for the higher performance in data processing (Karidi et al., 2019; Madani et al., 2021). On the other hand, existing studies suggest that future research should emphasizes on the decentralized architecture with deep learning by incorporating AI and blockchain. Huang et al. (2020) proposed an anti-rumor system for COVID-19 using deep learning and blockchain. Such a system starts from the journalists who provide the news sources, article compositions, and submissions, along with data stored in cloud and on-chain, followed by the article review by detecting fake news using deep learning. Hybrid crowed auditors will receive the article and AI-based review results on-chain and then perform the validation process through the QR code. Real news will be transacted to the next block. Such a system can fix mistakes from either human-based operations or machines from block to block. Similarly, Kolluri & Murthy (2021) developed a novel COVID-19 news verification system, namely CoVerifi, which allows users to vote on news content for determining whether or not an article is fake. Such a system can tackle the infodemic problem effectively under the framework of machine learning and blockchain.

CONCLUSION

The whole world has been impacted by misleading information significantly since the beginning of the COVID-19 pandemic, whereas more and more fake news and false information are generating every second as of today. Despite a variety of studies in terms of fake news detections using NLP, a systematic solution has not been provided due to the difficulties throughout data collection and validation, model training for text classification, and the test procedure for the application development. A web-based fake news detector is proposed in this chapter by investigating and examining all stages of the development. Such a detector is designed based on text classification algorithms by employing several word embeddings as the feature extractors and deep learning models as the classifiers. The whole operational procedures are performed in a cloud-based architecture, along with a testing process in AWS. Such a web-based fake news detector can identify misleading information in regards to COVID-19 effectively, thereby may help individuals to distinguish real information with fake news in a shorter time.

REFERENCES

Abdelminaam, D. S., Ismail, F. H., Taha, M., Taha, A., Houssein, E. H., & Nabil, A. (2021). Coaid-deep: An optimized intelligent framework for automated detecting covid-19 misleading information on twitter. *IEEE Access: Practical Innovations, Open Solutions*, *9*, 27840–27867. doi:10.1109/ACCESS.2021.3058066 PMID:34786308

Agarwal, A., Mittal, M., Pathak, A., & Goyal, L. M. (2020). Fake news detection using a blend of neural networks: An application of deep learning. *SN Computer Science*, *1*(3), 1–9. doi:10.100742979-020-00165-4

Ahmed, H., Traore, I., & Saad, S. (2017, October). Detection of online fake news using n-gram analysis and machine learning techniques. In *International conference on intelligent, secure, and dependable systems in distributed and cloud environments* (pp. 127-138). Springer. 10.1007/978-3-319-69155-8_9

Ahmed, S., Hinkelmann, K., & Corradini, F. (2019). Combining machine learning with knowledge engineering to detect fake news in social networks-a survey. In *Proceedings of the AAAI 2019 Spring Symposium* (Vol. 12, p. 8). AAAI.

Al-Garadi, M. A., Yang, Y. C., Lakamana, S., & Sarker, A. (2020). *A text classification approach for the automatic detection of twitter posts containing self-reported covid-19 symptoms*. Academic Press.

Ali, S., Khalid, A., & Zahid, E. (2021). Is CoViD-19 immune to misinformation? A brief overview. *Asian Bioethics Review*, *13*(2), 1–23. doi:10.100741649-020-00155-x PMID:33777228

Bahad, P., Saxena, P., & Kamal, R. (2019). Fake news detection using bi-directional LSTM-recurrent neural network. *Procedia Computer Science*, *165*, 74–82. doi:10.1016/j.procs.2020.01.072

Barua, Z., Barua, S., Aktar, S., Kabir, N., & Li, M. (2020). Effects of misinformation on COVID-19 individual responses and recommendations for resilience of disastrous consequences of misinformation. *Progress in Disaster Science*, *8*, 100119. doi:10.1016/j.pdisas.2020.100119 PMID:34173443

Bondielli, A., & Marcelloni, F. (2019). A survey on fake news and rumour detection techniques. *Information Sciences*, *497*, 38–55. doi:10.1016/j.ins.2019.05.035

Bruns, A., Harrington, S., & Hurcombe, E. (2020). 'Corona? 5G? or both?': The dynamics of COVID-19/5G conspiracy theories on Facebook. *Media International Australia*, *177*(1), 12–29. doi:10.1177/1329878X20946113

Cardoso Durier da Silva, F., Vieira, R., & Garcia, A. C. (2019, January). Can machines learn to detect fake news? a survey focused on social media. *Proceedings of the 52nd Hawaii International Conference on System Sciences.* 10.24251/HICSS.2019.332

Casero-Ripollés, A. (2020). Impact of Covid-19 on the media system. Communicative and democratic consequences of news consumption during the outbreak. *El Profesional de la Información*, *29*(2), e290223. doi:10.3145/epi.2020.mar.23

Cho, K., Van Merriënboer, B., Gulcehre, C., Bahdanau, D., Bougares, F., Schwenk, H., & Bengio, Y. (2014). *Learning phrase representations using RNN encoder-decoder for statistical machine translation.* doi:10.3115/v1/D14-1179

ChokshiA.MathewR. (2021). *Deep Learning and Natural Language Processing for fake news detection: A Survey*. doi:10.2139/ssrn.3769884

Conroy, N. K., Rubin, V. L., & Chen, Y. (2015). Automatic deception detection: Methods for finding fake news. *Proceedings of the Association for Information Science and Technology*, *52*(1), 1–4. doi:10.1002/pra2.2015.145052010082

Della Vedova, M. L., Tacchini, E., Moret, S., Ballarin, G., DiPierro, M., & de Alfaro, L. (2018, May). Automatic online fake news detection combining content and social signals. In *2018 22nd Conference of Open Innovations Association (FRUCT)* (pp. 272-279). IEEE. 10.23919/FRUCT.2018.8468301

Elhadad, M. K., Li, K. F., & Gebali, F. (2020, August). COVID-19-FAKES: A Twitter (Arabic/English) dataset for detecting misleading information on COVID-19. In *International Conference on Intelligent Networking and Collaborative Systems* (pp. 256-268). Springer.

Furini, M., Mirri, S., Montangero, M., & Prandi, C. (2020, July). Untangling between fake-news and truth in social media to understand the Covid-19 Coronavirus. In *2020 IEEE Symposium on Computers and Communications (ISCC)* (pp. 1-6). IEEE. 10.1109/ISCC50000.2020.9219663

Geiger, A. W. (2019). *Key findings about the online news landscape in America.* PEW Research Center.

Ghosh, S., & Shah, C. (2018). Towards automatic fake news classification. *Proceedings of the Association for Information Science and Technology*, *55*(1), 805–807. doi:10.1002/pra2.2018.14505501125

Girgis, S., Amer, E., & Gadallah, M. (2018, December). Deep learning algorithms for detecting fake news in online text. In *2018 13th International Conference on Computer Engineering and Systems (ICCES)* (pp. 93-97). IEEE. 10.1109/ICCES.2018.8639198

Goldani, M. H., Safabakhsh, R., & Momtazi, S. (2021). Convolutional neural network with margin loss for fake news detection. *Information Processing & Management*, *58*(1), 102418. doi:10.1016/j.ipm.2020.102418

Gupta, A., & Kumaraguru, P. (2012, April). Credibility ranking of tweets during high impact events. In *Proceedings of the 1st workshop on privacy and security in online social media* (pp. 2-8). 10.1145/2185354.2185356

Hinton, G. E., Osindero, S., & Teh, Y. W. (2006). A fast learning algorithm for deep belief nets. *Neural Computation*, *18*(7), 1527–1554. doi:10.1162/neco.2006.18.7.1527 PMID:16764513

Huang, T., Li, B., & Wang, X. (2020). Fake news monitoring and anti-rumor system using deep learning and blockchain. *GitHub*. https://github.com/pzhaoir/fake-news-monitoring-anti-rumor-systm/blob/main/En-Ch-Report-Fake%20News%20Monitoring%20and%20Anti.pdf

Janze, C., & Risius, M. (2017, January). *Automatic Detection of Fake News on Social Media Platforms*. Association for Information Systems.

Kaliyar, R. K., Goswami, A., & Narang, P. (2021). FakeBERT: Fake news detection in social media with a BERT-based deep learning approach. *Multimedia Tools and Applications*, *80*(8), 11765–11788. doi:10.100711042-020-10183-2 PMID:33432264

Kaliyar, R. K., Goswami, A., Narang, P., & Sinha, S. (2020). FNDNet–a deep convolutional neural network for fake news detection. *Cognitive Systems Research*, *61*, 32–44. doi:10.1016/j.cogsys.2019.12.005

Karidi, D. P., Nakos, H., & Stavrakas, Y. (2019, November). Automatic ground truth dataset creation for fake news detection in social media. In *International Conference on Intelligent Data Engineering and Automated Learning* (pp. 424-436). Springer. 10.1007/978-3-030-33607-3_46

Kim, S., & Kim, S. (2020). The Crisis of public health and infodemic: Analyzing belief structure of fake news about COVID-19 pandemic. *Sustainability*, *12*(23), 9904. doi:10.3390u12239904

Ko, C. H., Yen, J. Y., Liu, S. C., Huang, C. F., & Yen, C. F. (2009). The associations between aggressive behaviors and Internet addiction and online activities in adolescents. *The Journal of Adolescent Health*, *44*(6), 598–605. doi:10.1016/j.jadohealth.2008.11.011 PMID:19465325

Kolluri, N. L., & Murthy, D. (2021). CoVerifi: A COVID-19 news verification system. *Online Social Networks and Media*, *22*, 100123. doi:10.1016/j.osnem.2021.100123 PMID:33521412

Kula, S., Kozik, R., Choraś, M., & Woźniak, M. (2021, June). Transformer Based Models in Fake News Detection. In *International Conference on Computational Science* (pp. 28-38). Springer. 10.1007/978-3-030-77970-2_3

Li, Q., Hu, Q., Lu, Y., Yang, Y., & Cheng, J. (2019). Multi-level word features based on CNN for fake news detection in cultural communication. *Personal and Ubiquitous Computing*, 1–14.

Ljungholm, D. P., & Olah, M. L. (2020). Regulating fake news content during COVID-19 pandemic: Evidence-based reality, trustworthy sources, and responsible media reporting. *Review of Contemporary Philosophy*, *19*(0), 43–49. doi:10.22381/RCP1920203

Lyons, T. (2018). *Increasing our efforts to fight false news*. Facebook Newsroom.

Madani, Y., Erritali, M., & Bouikhalene, B. (2021). Fake News Detection Approach Using Parallel Predictive Models and Spark to Avoid Misinformation Related to Covid-19 Epidemic. In *Intelligent Systems in Big Data, Semantic Web and Machine Learning* (pp. 179–195). Springer. doi:10.1007/978-3-030-72588-4_13

Madani, Y., Erritali, M., & Bouikhalene, B. (2021). Using artificial intelligence techniques for detecting Covid-19 epidemic fake news in Moroccan tweets. *Results in Physics*, *25*, 104266. doi:10.1016/j.rinp.2021.104266

Memon, S. A., & Carley, K. M. (2020). *Characterizing covid-19 misinformation communities using a novel twitter dataset.* arXiv preprint arXiv:2008.00791.

Moravec, P., Minas, R., & Dennis, A. R. (2018). *Fake news on social media: People believe what they want to believe when it makes no sense at all.* Kelley School of Business Research Paper, (18-87).

Nasir, J. A., Khan, O. S., & Varlamis, I. (2021). Fake news detection: A hybrid CNN-RNN based deep learning approach. *International Journal of Information Management Data Insights*, *1*(1), 100007. doi:10.1016/j.jjimei.2020.100007

Oshikawa, R., Qian, J., & Wang, W. Y. (2018). *A survey on natural language processing for fake news detection.* arXiv preprint arXiv:1811.00770.

Patwa, P., Sharma, S., Pykl, S., Guptha, V., Kumari, G., Akhtar, M. S., . . . Chakraborty, T. (2020). *Fighting an infodemic: Covid-19 fake news dataset.* arXiv preprint arXiv:2011.03327.

Patwardhan, P., & Yang, J. (2003). Internet dependency relations and online consumer behavior: A media system dependency theory perspective on why people shop, chat, and read news online. *Journal of Interactive Advertising*, *3*(2), 57–69. doi:10.1080/15252019.2003.10722074

Pomputius, A. (2019). Putting Misinformation Under a Microscope: Exploring Technologies to Address Predatory False Information Online. *Medical Reference Services Quarterly*, *38*(4), 369–375. doi:10.1080/02763869.2019.1657739 PMID:31687908

Qazi, M., Khan, M. U., & Ali, M. (2020, January). Detection of fake news using transformer model. In *2020 3rd International Conference on Computing, Mathematics and Engineering Technologies (iCoMET)* (pp. 1-6). IEEE. 10.1109/iCoMET48670.2020.9074071

Qazi, U., Imran, M., & Ofli, F. (2020). GeoCoV19: A dataset of hundreds of millions of multilingual COVID-19 tweets with location information. *SIGSPATIAL Special*, *12*(1), 6–15. doi:10.1145/3404111.3404114

Rubin, V. L., Conroy, N., Chen, Y., & Cornwell, S. (2016, June). Fake news or truth? using satirical cues to detect potentially misleading news. In *Proceedings of the second workshop on computational approaches to deception detection* (pp. 7-17). 10.18653/v1/W16-0802

Shahi, G. K., & Nandini, D. (2020). *FakeCovid—A multilingual cross-domain fact check news dataset for COVID-19.* arXiv preprint arXiv:2006.11343.

Shearer, E. (2021). More than eight-in-ten Americans get news from digital devices. Pew Research Center.

Shu, K., Mahudeswaran, D., & Liu, H. (2019). FakeNewsTracker: A tool for fake news collection, detection, and visualization. *Computational & Mathematical Organization Theory*, *25*(1), 60–71. doi:10.100710588-018-09280-3

Shu, K., Mahudeswaran, D., Wang, S., Lee, D., & Liu, H. (2020). Fakenewsnet: A data repository with news content, social context, and spatiotemporal information for studying fake news on social media. *Big Data*, *8*(3), 171–188. doi:10.1089/big.2020.0062 PMID:32491943

Singhal, S., Kabra, A., Sharma, M., Shah, R. R., Chakraborty, T., & Kumaraguru, P. (2020, April). Spotfake+: A multimodal framework for fake news detection via transfer learning (student abstract). *Proceedings of the AAAI Conference on Artificial Intelligence*, *34*(10), 13915–13916. doi:10.1609/aaai.v34i10.7230

Singhania, S., Fernandez, N., & Rao, S. (2017, November). 3han: A deep neural network for fake news detection. In *International conference on neural information processing* (pp. 572-581). Springer. 10.1007/978-3-319-70096-0_59

Slovikovskaya, V. (2019). *Transfer learning from transformers to fake news challenge stance detection (FNC-1) task.* arXiv preprint arXiv:1910.14353.

Stanley, M. L., Barr, N., Peters, K., & Seli, P. (2020). Analytic-thinking predicts hoax beliefs and helping behaviors in response to the COVID-19 pandemic. *Thinking & Reasoning*, 1–14.

Suyanto, S. (2020, June). Synonyms-Based Augmentation to Improve Fake News Detection using Bidirectional LSTM. In *2020 8th International Conference on Information and Communication Technology (ICoICT)* (pp. 1-5). IEEE.

Umer, M., Imtiaz, Z., Ullah, S., Mehmood, A., Choi, G. S., & On, B. W. (2020). Fake news stance detection using deep learning architecture (CNN-LSTM). *IEEE Access: Practical Innovations, Open Solutions*, *8*, 156695–156706. doi:10.1109/ACCESS.2020.3019735

Vishwakarma, D. K., & Jain, C. (2020, June). Recent State-of-the-art of Fake News Detection: A Review. In *2020 International Conference for Emerging Technology (INCET)* (pp. 1-6). IEEE. 10.1109/INCET49848.2020.9153985

Waage, H. S. (2018). *Hyper-reading headlines: How social media as a news-platform can affect the process of news reading* (Master's thesis). University of Stavanger, Norway.

Wang, X., Zhao, P., & Chen, X. (2020). Fake news and misinformation detection on headlines of COVID-19 using deep learning algorithms. *International Journal of Data Science*, *5*(4), 316–332. doi:10.1504/IJDS.2020.115873

Wani, A., Joshi, I., Khandve, S., Wagh, V., & Joshi, R. (2021, January). Evaluating deep learning approaches for Covid19 fake news detection. In *Combating Online Hostile Posts in Regional Languages during Emergency Situation: First International Workshop, CONSTRAINT 2021, Collocated with AAAI 2021, Virtual Event, February 8, 2021, Revised Selected Papers* (p. 153). Springer Nature. 10.1007/978-3-030-73696-5_15

Whiting, A., & Williams, D. (2013). Why people use social media: A uses and gratifications approach. *Qualitative Market Research*, *16*(4), 362–369. doi:10.1108/QMR-06-2013-0041

Yang, Y., Zheng, L., Zhang, J., Cui, Q., Li, Z., & Yu, P. S. (2018). *TI-CNN: Convolutional neural networks for fake news detection.* arXiv preprint arXiv:1806.00749.

Zhou, X., & Zafarani, R. (2020). A survey of fake news: Fundamental theories, detection methods, and opportunities. *ACM Computing Surveys*, *53*(5), 1–40. doi:10.1145/3395046

ADDITIONAL READING

Aggarwal, C. C., & Zhai, C. (2012). A survey of text classification algorithms. In *Mining text data* (pp. 163–222). Springer. doi:10.1007/978-1-4614-3223-4_6

Deng, L., & Liu, Y. (Eds.). (2018). *Deep learning in natural language processing*. Springer. doi:10.1007/978-981-10-5209-5

Gulli, A., Kapoor, A., & Pal, S. (2019). *Deep learning with TensorFlow 2 and Keras: regression, ConvNets, GANs, RNNs, NLP, and more with TensorFlow 2 and the Keras API*. Packt Publishing Ltd.

Kowsari, K., Jafari Meimandi, K., Heidarysafa, M., Mendu, S., Barnes, L., & Brown, D. (2019). Text classification algorithms: A survey. *Information (Basel), 10*(4), 150. doi:10.3390/info10040150

Liu, J., Chang, W. C., Wu, Y., & Yang, Y. (2017, August). Deep learning for extreme multi-label text classification. In *Proceedings of the 40th international ACM SIGIR conference on research and development in information retrieval* (pp. 115-124). 10.1145/3077136.3080834

Minaee, S., Kalchbrenner, N., Cambria, E., Nikzad, N., Chenaghlu, M., & Gao, J. (2021). Deep Learning—based Text Classification: A Comprehensive Review. *ACM Computing Surveys, 54*(3), 1–40. doi:10.1145/3439726

Oshikawa, R., Qian, J., & Wang, W. Y. (2018). A survey on natural language processing for fake news detection. *arXiv preprint arXiv:1811.00770.*

Reis, J. C., Correia, A., Murai, F., Veloso, A., Benevenuto, F., & Cambria, E. (2019). Supervised learning for fake news detection. *IEEE Intelligent Systems, 34*(2), 76–81. doi:10.1109/MIS.2019.2899143

Varia, J. (2010). Migrating your existing applications to the aws cloud. *A Phase-driven Approach to Cloud Migration*, 1-23.

Wake, L. (2010). *Nlp: Principles in practice*. Ecademy Press.

KEY TERMS AND DEFINITIONS

Cloud Computing: An on-demand access that can compute resources of various services through the internet.

Deep Learning: A broad family of machine learning models based on neural networks. Typical deep learning models are deep neural networks, convolutional neural networks, recurrent neural networks, deep belief networks, and deep reinforcement learning.

Misinformation Detection: A task of monitoring the information that is unreliable and questioned to its accuracy and credibility.

NLP: A processing method of computational linguistics for human language based on algorithms.

Text Classification: A typical problem in information science that assigns a textual data to one or more classes or categories.

Web App Development: A process of creating a web application programs with customized functions that can benefit the users.

Web Scrapping: A process of extracting content information, such as texts, tables, images, from a certain website.

Word Embeddings: A representation vectors that is encoded based the similarity calculation in a vector space.

Chapter 11
Fighting Anti-Asian Hate in and After the COVID-19 Crisis With Big Data Analytics

ABSTRACT

Racism and physical attacks on Asian communities have spread in the U.S. and around the world. Xenophobia is a virus that may lead to an ongoing social problem in the post-pandemic era. Although existing studies have been done to classify anti-Asian haters, little is known on monitoring, tracking, and characterizing anti-Asian haters on social media platforms. In this chapter, a systematic examination of anti-Asian haters tracking and profiling methods is designed by using big data analytics with deep learning algorithms. Target haters are investigated and tracked by analyzing public opinions towards key topics in 2020, including the U.S. elections, stimulus checks, and economy opening strategies throughout data collection and preprocessing, text classification, sentiment analysis, data visualization, and association rule. Such a comprehensive study provides a variety of research opportunities in dealing with anti-Asian racism and xenophobia in and after the COVID-19 pandemic.

INTRODUCTION

A tsunami of hate, xenophobia, and scapegoats has been evolved into one of the biggest challenges in and after the COVID-19 pandemic, not just

DOI: 10.4018/978-1-7998-8793-5.ch011

throughout the United States, but around the world. As the outbreak of the novel coronavirus, millions of families have been harassed by a series of problems, such as health risks, financial distresses, security issues, and life uncertainties, whereas such recession exacerbates a tendency that the minority has been targeted from being possible hate. Due to an unproven speculation that as the COVID-19 starts from China, racism and crimes on Chinese and Asian communities have been dramatically increasing since the beginning of the pandemic. The tragedy in Georgia in regards to Atlanta spa shootings is one of the extreme manifestations in the form of fears of anti-Asian bias, while more racism and anti-Asian haters are invisible (Kao, 2021). Despite the announcement of the importance of protecting civil rights by lawmakers and FBI agencies, strong anxiety and negative emotions have been represented in terms of actions of discrimination, verbal abuse, and online harassment. Cyber-based anti-Asian voices, statements, and comments are everywhere on social media channels, such as Twitter, YouTube, Instagram, FaceBook, etc., along with the spreading of fake news and rumors.

Over the last few months, hatred incidents against Asian communities have quickly attracted scholars' attention, from sociological, psychological, and socio-economic perspectives. Tessler et al. (2020) examined the historical antecedents that link infectious diseases and Asian Americans, which illustrated the anxiety of such a group of people suffered by hate crimes and negative biases during the COVID-19 pandemic. The psychological impact of anti-Asian stigma due to COVID-19 has been investigated and discussed by examining how it affects mental health and recovery, through implementing evidence-based stigma reduction initiatives, and via coordinating federal response to anti-Asian racism including investment in mental health services and community-based efforts (Misra et al., 2020). Meanwhile, Gover et al. (2020) believed that pandemic-related health crises have been associated with the stigmatization, which can be explained by exploring the reproduction of inequality. Xenophobia has spread much faster than the virus itself in the form of the socio-economic impact of COVID-19 by covering the important global rise in hate towards Asian ethnic origins (Cheng, 2020). Potential impacts of pandemic-related racial discrimination related to public health implications have also been investigated by citing relevant cases from the history of previous disasters in the U.S., along with exploring theoretical and empirical evidence (Chen et al., 2020). Existing studies in this field indicate that the anti-Asian hate violence associated with COVID-19 is not accidental, but has historical origins and is systematical and institutional.

However, the situation seems inevitably getting worse and worse as the COVID-19 related issues have been highly politicized. A large scale of unconfirmed and false information has been posted on social media, where people with different visions are being divided due to the pandemic associated with political reasons. Moreover, the situation became even more confusing and dangerous when the phenomenon of the pandemic-related anti-Asian encountered the 2020 U.S. election. Particularly, for some political purpose, the former President, Donald J. Trump, deliberately declared that COVID-19 is "China Virus", whereas such an irresponsible remark has caused immeasurable harm to Asian communities (Masters-Waage et al., 2020). Nevertheless, provocative statements and voices have accelerated the division of the American people in the form of opposite opinions on anti-Asian racism, Black Lives Matter (BLM), stimulus checks, disease control measures, economy reopening strategies, vaccinations, and other topics related to the 2020 U.S. presidential election (Ho, 2021; Kim & Lee, 2021; Baccini et al., 2021; Yousefinaghani et al., 2021). The COVID-19 pandemic has been evolved into a political pandemic, which leads the world standing at a dangerous crossroads (Taskinsoy, 2020). This election season was unusual and sensitive when COVID-19 became one of the most important issues, combined with controversies in many other political sentiments beyond anti-Asian racism. These anti-Asian haters are homogenous in many characteristics due to the similar reactions and opinions on a specific topic of the 2020 presidential election.

On the other hand, social media platforms are one of the most convenient channels to release hate sentiments and racial animus. Hundreds of thousands pandemic-related tweets are generated every day, while a large number of anti-Asian hate posts have been detected in only three months since the beginning of the COVID-19 outbreak (Ziems et al., 2020). Similar to the fake news and rumors, the anti-Asian hate sentiment can be spreading exponentially in seconds through the internet, of which such a new infodemic becomes a significant social risk. The division among invidious who hold opposite opinions has been reflected in the commentaries of the mainstream media as news reporters and editorials have been divided according to their political tendencies. The Cable News Network and the Fox News Channel will stand on completely different positions when reporting the same news, particularly for the COVID-19 related topics (Mello et al., 2020), whereas such a phenomenon is also deepening the opposition of the anti-Asian hate in general. Therefore, understanding the complexity of the pandemic-related hate is not an easy task, but a comprehensive subject that involves a broad range of interdisciplinary studies, from sociology, infodemiology, to social system engineering, while big data and artificial intelligence can provide the most cutting-edge technologies in this research domain. Motivated by this blueprint, this chapter is design to explore how big data and AI can help in monitoring, tracking, and profiling anti-Asian racism and xenophobia. The objectives of this chapter are listed as follows:

- illustrating how to collect pandemic-related information from social media channels using data extraction tools and web scrapers, along with data cleaning and preprocessing for data preparation;
- investigating a variety of text classification in identifying anti-Asian haters using multiple word embedding models and deep learning algorithms;
- examining a data-driven approach in regards to constructing a xenophobia index for monitoring anti-Asian hate changes over time;
- analyzing anti-Asian haters' sentiments and visualizing their distributions on maps by performing sentiment analysis and visual mining;
- providing a novel big data method in tracking and profiling anti-Asian haters with a rule-based algorithm.

BACKGROUND

Most recent studies of pandemic-related public reactions have been proposed by using multiple big data approaches and AI models, which can be applied in data preprocessing, data collection, sentiment analysis, and visual mining. Most existing studies are concentrated on using data extracted from Twitter as it contains enough samples from tweets. Ziems et al. (2020) created a novel dataset, namely COVID-HATE, using the classical machine learning-based text classifier, which can identify anti-Asian haters and counter-haters from over 30 million tweets. The trained classifier can achieve an average AUC of 0.85 based on a hand-labeled dataset of more than 2,000 tweets. Similarly, a large scale classification of hate and counter speech from social media has been proposed using an ensemble learning algorithm with logistic regression (LR) functions. Such a classifier can detect both hate and counter-hate speech efficiently in terms of a higher model performance (Garland et al., 2020). Kabir & Madria (2020) developed a real-time analyzer, called Coronavis, which can be applied in visualizing topic modeling, analyzing subjectivity, and identifying human emotions for exploring the psychological and behavioral changes based on tweets shared during the COVID-19 pandemic. Such an analyzer can assist in managing the social crisis using text mining and Geographical Information System (GIS). Chakraborty et al. (2020) proposed a set of deep learning classifiers that can be used to perform sentiment analysis for COVID-19 tweets. Similarly, a cross-cultural polarity and emotion detection study has been proposed using sentiment analysis and deep learning on COVID-19 related tweets (Imran et al., 2020).

Classifying anti-Asian hater has attracted much attention from the research community in the form of detection and identification approaches using machine learning, deep learning, and Natural Language Processing (NLP). Vidgen et al. (2020) applied multiple text classification models, such as LSTM, AIBERT, XLNet, etc., to classify East Asian prejudice on social media based on pandemic-related tweets collected through Twitter Streaming API. Such a classifier can detect and categorize tweets into four classes, including hostility against East Asia, criticism of East Asian, meta-discussions of East Asian prejudice, and a neutral class, along with an F1 score of 0.83 for each category. Similarly, a transformer-based hate speech analyzer has been proposed by developing a BERT attention mechanism, which can detect novel keywords targeting the Asian community and older people (Vishwamitra et al., 2020). Nguyen et al. (2020) explored shifts in anti-Asian sentiment in the U.S. during the COVID-19 pandemic by examining racial sentiment changes using Support Vector Machine (SVM). Such a study investigates the general pattern and trend of anti-Asian tweets associated with the pandemic. Chaudhary et al. (2021) developed a conceptual framework for countering online pandemic-related hate speech using NLP. Such a survey study provides a thorough understanding of current works on utilizing NLP for prevention and intervention of cyber hate speech. A study that aims to investigate the impact of health, psychosocial, and social problems emanating from the COVID-19 pandemic has been proposed to understand public opinions, experiences, and issues with respect to COVID-19 by hybridizing NLP and thematic analysis, along with data collected from Twitter, Facebook, YouTube, and other online forums (Oyebode et al., 2021). Pei & Mehta (2021) developed a four dimensional category for detecting racism and xenophobia using deep leaning, which enables insightful meanings into the COVID-19 emergent topics related to anti-Asian expression on Twitter.

Other related studies have been proposed by analyzing multiple features and factors associated with anti-Asian racism and xenophobia, particularly when the 2020 U.S. presidential election meets the COVID-19 pandemic.

Most studies are concentrated on exploring a variety of political opinions of the election topics among people who hold conflict views of politicized terms. Various big data approaches have been involved in generating election-related data pools (Chen et al., 2021), performing social network analysis for social media users and communications (Ahmed et al., 2020), examining anti-Asian attitudes based on the multivariate regression analysis using data from a national survey (Dhanani & Franz, 2020), tracking anti-Asian haters from controversial terms using demographical and geographical features (Lyu et al., 2020), and exploring the relationship between public attitudes of Asian Americans and behavioral diversities of policy outcomes (Reny & Barreto, 2020). Nevertheless, little is known on how AI and big data can functionally work for classifying anti-Asian haters based on emotional and behavioral factors. Zhao et al. (2021) provides such an attempt for identifying racism and xenophobia using sentiment analysis and GIS. Several pandemic- and election-related features for both anti-Asian haters and non-haters have been measured by performing a series of state-of-the-art techniques, such as sentiment analysis, visual mining, and GIS, along with a binary classification task using a Deep Neural Network (DNN) model. Such a study can be determined as the first attempt in tracking and characterizing anti-Asian haters in the context of the 2020 U.S. election and the COVID-19 pandemic.

MAIN FOCUS OF THE CHAPTER

Issues, Controversies, Problems

The pandemic-related xenophobia is an ongoing social risk that leads a national-wide uncertainty as a bigger challenge in social, political, and psychological consequences. Existing studies in this domain have been represented in analyzing and investigating pandemic-related racism and xenophobia from multiple angels of public reactions, emotions, and behaviors using a variety of computational approaches, such as sentiment analysis, text classification, GIS, and social network analysis. Most studies apply sentiment analysis in identifying anti-Asian hate speech to extract racism and xenophobia factors, and for preprocessing of the social network analysis, NLP and deep learning approaches are more efficient as one of the most robust classifiers in detecting anti-Asian hate speech. Therefore, some studies have been drawn by using machine learning and deep learning models for text classification with transformer-based NLP approaches. Such studies are usually concentrated on identifying anti-Asian hate speech from social media by training text classifiers, and then apply the model into the whole dataset for the automatic labeling process. Nevertheless, little is known on the comprehensive understanding of the pandemic-related racism and xenophobia from monitoring, tracking, and characterizing anti-Asian haters, thereby such works are needed to be investigated and examined from both conceptual and empirical perspectives.

Features of the Chapter

This chapter aims on investigating how big data analytics and AI models help in monitoring, tracking, and characterizing pandemic-related anti-Asian racism and xenophobia throughout data acquisition and preprocessing, text classification, model training and selection, real-time xenophobia monitoring, and anti-Asian hater tracking and profiling. Data used in this chapter is collected from multiple social media channels, such as YouTube and Twitter, for different tasks, along with the information gathering and data cleaning process. To monitor racism and xenophobia, an optimal anti-Asian classifier has been trained before constructing the xenophobia index, which can be applied in detecting anti-Asian hate speech in real-time. Common-used word embedding methods, such as Embeddings from Language Models (ELMo) and Bidirectional Encoder Representations from Transformers (BERT), have been investigated in terms of the basic functionality, whereas BERT and GoogleNews-vectors have been applied in feature extraction, along with a set of machine learning and deep learning classifiers. To examine how big data helps in tracking and profiling anti-Asian haters, a novel dataset, namely COVID-HATE-TRACK, has been create based on tracking anti-Asian haters from an automatically labeled dataset based on tweets by using machine learning algorithms, followed by characterizing haters' profiles using sentiment analysis and Apriori algorithms.

SOLUTIONS AND RECOMMENDATIONS

Monitoring Racism and Xenophobia

Data Acquisition and Preprocessing

The pandemic-related anti-Asian haters are difficult to monitor and track, whereas their speech and voices can be found from social media channels, such as Twitter, YouTube, Facebook, etc., as many racial posts and comments are generated every day since the beginning of the COVID-19 outbreak. Monitoring racism and xenophobia requires a large scale of data pool for tracking, identifying, and analyzing anti-Asian haters. Such information can be collected from tweets and comments through APIs. This chapter provides an example in constructing the xenophobia index using data from YouTube comments. To collect the comments of pandemic-related videos, a web scraper has been applied to automatically scrape video titles and URLs from 16 YouTube channels of mainstream media channels, including Reuters, MSNBC, NBC News, ABC News, CBS News, Fox News, CNN, USA Today, Los Angeles Times, Breitbart, New York Post, POLITICO, Yahoo News, Huffpost, The New York Times, and NPR. Base on the keywords of "COVID-19" and "Coronavirus", the pandemic-related videos are filtered out in the scraping process. To perform comment query using YouTube Data API, the video IDs are extracted by the means of regular expression. Facepager application embedded with YouTube Data API is utilized for comment collection. For each comment, the attributes are categorized to comment information, users' information, date, and accumulation data. The comment information includes the comment ID and the text content. User name, user profile URL, user channel URL, user channel ID belong to the group of user information. The comment published date and updated data are also collected. The accumulation data of like count and reply count can also be collected in the scraping result for each comment. Through filtering, four data attributes are remained, including videoID, comment, date and channel.

In order to train a classifier for anti-Asian hate classification, a manually labeled dataset is created based on the labeling rules: a comment is with hateful hashtags, such as # Kungflu and #fuckChina; a comment has Asian-related information, including asian names of people, country, and etc; a comment expresses obvious insulting words. Otherwise, those comments that not meet with the labeling rules will be identified as non-hate comments.

In the labeling process, 3,163 comments are randomly extracted. With the consideration of the balance between anti-Asian hate and non-hate comments, 795 comments with anti-Asian-related terms, extracted through information retrieval, are added into the whole samples. Based on the labeling rules, a dataset with 3,756 comments is generated, which contains 778 anti-Asian hate and 2,981 non-hate comments, respectively. After a classifier is trained, the overall data can be detected as anti-Asian hate and non-hate. For those comments already with anti-Asian hate hashtags, they can be labeled as anti-Asian hate directly without going through classifier. For those comments only containing non-anti-Asian hate hashtags, HTML links, and emojis, they can be labeled as non-hate comments. Furthermore, two rounds of labeling, done by anti-Asian hate hashtags and classifier separately, can guarantee a reasonable classification for overall data labeling.

Extracting Features of Hate Speech

Similar to the fake news classification, textual features can be extracted by using a variety of word embedding methods. Most recent studies suggest two popular models, i.e. ELMo and BERT, which have been widely used in identifying anti-Asian hate speech. ELMo, introduced by Peters et al. (2018), is a novel way of deep contextualized word representation in a pre-trained bidirectional language model architecture. The empirical work in the paper demonstrates that it is extremely effective in practice. Comparing to traditional word embeddings such as Word2Vec and GloVe, ELMo assigns word representations a function of the entire input sentence. The main idea of ElMo word vectors is to compute a two-layer biLM (bidirectional language model), which has two layers stacked together. Both forward and backward language models are combined in a biLM, which is a joint maximization of log likelihood, such as

$$\sum_{i=1}^{N}(\log P(w_i \mid w_1,\cdots,w_{i-1};\Theta_T,\vec{\Theta}_M,\Theta_S) + \log P(w_i \mid w_{i+1},\cdots,w_N;\Theta_T,\overleftarrow{\Theta}_M,\Theta_S)),$$

where w_i is a word in a sequence of N token vector. Token representation Θ_T and Softmax layer Θ_S are both in the forward $\vec{V}$ direction and backward $\overleftarrow{V}$ direction, given a language model Θ_M, for example, LSTMs. All layers in a set of biLM representations $\mathbb{R}$ are decomposed into a single vector and specifically weighed for all biLM layers.

BERT is another state-of-the-art word embeddings. In 2018, Google AI Language launched a research paper talking about a new algorithm for many NLP tasks, including question answering and language inference (Devlin et al., 2018). This language representation model is based on the transformer architecture with deep bidirectionally pre-trained representations on a large corpus of unlabelled text including the entire Wikipedia (2,500 million words) and Book Corpus (800 million words). In contrast to previous models, BERT looks at a text sequence combined left-to-right and right-to-left training. The experimental results indicate that a bidirectionally trained model can make a deeper sense than any single-direction language models. One of the biggest challenges in a big neural network, i.e. BERT, is to setup millions parameters, thus, training the model on a small dataset will result in an overfilling problem. A model fine-tuning technique can solve the problem effectively by using a pre-trained BERT model on a large dataset and then a smaller one.

Model Training and Evaluation

The anti-Asian hate classification can be implemented as a binary problem between two classes of hate and non-hate. For classification problem, both machine learning and deep learning methods can achieve reasonable results. Based on model performance evaluation, a model selection process is designed among five major machine learning and deep learning methods, including LR, SVM, RF, CNN, and LSTM. To extract features from unstructured comment data, BERT embedding is used to pre-trained a 20-dimensional word representation, in order to prepare the training process for LR, SVM and Random Forest. Besides, GoogleNews-vectors, a vector pre-trained on billion of words of Google News, is utilized in the feature extraction process for model training of CNN and LSTM.

Table 1. Model performance comparison for determine the optimal algorithm of the xenophobia identification

Feature Extraction	Text Classifier	Accuracy	AUC	F1
BERT	LR	89%	0.90	0.65
	SVM	90%	0.88	0.70
	RF	81%	0.80	0.01
GoogleNews-vectors	LSTM	93%	0.98	0.84
	CNN	95%	0.99	0.88

Table 1 presents the model performance metrics, such as accuracy, AUC, and F1 score, for anti-Asian hate classification task. The training process is performed in cloud by deploying multiple clusters based on the computational needs. CNN reaches the highest accuracy of 95% and the third and third high accuracies were got by LSTM and SVM algorithms, as 93% and 90% separately. For AUC score, CNN still achieves the highest score of 0.99 and the score of LSTM is much closed to CNN. The top three highest F1 scores are also achieved by CNN, LSTM and SVM, with the scores of 0.88, 0.84, and 0.70, respectively. Therefore, the CNN classifier combined with GoogleNews-vectors as the feature extractor can be determined as the optimal algorithm of the xenophobia identification. In the next step, the overall 1,433,246 YouTube comments are classified and divided into 56,210 Hate comments and 1,377,036 non-hate comments.

Constructing the Xenophobia Index

During COVID-19 pandemic, most of major mainstream media update related news daily. Therefore, the news videos and corresponding commentary programs filled the YouTube channels, in which daily comments from different individuals were produced. It makes possible to create a dynamic time-series-based index to portray the tendency of anti-Asian hate over time. From this index, the hate signal in each day can be measured base on the attributes of time series. A calculation method, proposed by Shapiro et al. (2020), for signal computing for categorical data is illustrated as follows,

$$s(t) = \frac{n_p(t) - n_n(t)}{n_p(t) + n_m(t) + n_n(t)}$$

where, $s(t)$ represents the signal at time t. $n_p(t)$, $n_m(t)$, and $n_n(t)$ represent the daily number of the positive, the neutral, and the negative, respectively. In this case, the anti-Asian hate classification problem can be dealt with as a binary problem, which means that only two categories, hate and non-hate, will be classified. In a Twitter research conducted by Ziems et al (2020), the researchers concluded that counterhate activity is not much active than hate activity through their sentiment analysis experiment. Therefore, based on this conclusion, the number of counterhate comments could be ignored. In the Formula 1, it means that the number of neutral can be regarded as 0, which leads to a new signal calculation for anti-Asian hate classification task as shown in the following equation,

$$s(t) = \frac{m_h(t)}{m_h(t) + m_n(t)}$$

where $m_h(t)$ and $m_n(t)$ are the number of hate and non-hate comments, separately. The daily hate signal $s(t)$ is equal to the ratio of the number of the hate comments to the total number of comments from two categories. If no hate comments appear in a certain day, the daily hate signal will be equal to 0. In another case, when the number of hate comments is closed to the number of non-hate comments, the daily hate signal will be approximately equal to 0.5.

The generated anti-Asian hate index is shown as Figure 1A. The indexes for 16 different mainstream media are in 16 different colors. It is obvious that diverse mainstream media has various hate signal pattern. The highest hate ratios were reached by four mainstream media, including Breitbart, MSNBC, NPR, and New York Post. The mainstream media channels having the most spikes are Los Angeles Times, USA Today, and The New York Times. The spikes are mainly occurred from January to March and June to November. However, it can be seen that the anti-Asian hate index is not continuous for some mainstream media channels, such as NPR and POLITICO, since COVID-19-related news is not updated daily and hate comments are not frequently posted in those channels. This issue can be solved if the 16 channels are analyzed as a whole. In Figure 1B, it describes the overall anti-Asian hate index from January to December in 2020, which indicates the variation of hate signal to the historic events. Based on the COVID-19 spread in the United States analyzed by Google statistics for COVID-19, the overall time series can be divided in to four phrases according to the daily change of COVID-19 infections in the United States. Phase I indicates the COVID-19 outbreak has been spreading rapidly in East Asian and Europe. Although the first COVID-19 confirmed case was found in the State of Washington on January 21, 2020, numbers of confirmed cases did not spike nationwide. The xenophobia ratio is at a relatively high level in Phrase I due to the misinformation and fake news, which confuse most individuals to believe that the novel virus is originally from China. In Phrase II, a remarkable spike of the xenophobia index occurred on March 18, the same day after Trump posted a tweet on coronavirus using term “Chinese Virus”. In June 20, Trump held a rally in Tulsa, followed by another spike of the index. Moreover, two significant spikes showed up after Trump's rallies on August 17 and September 13, respectively. The trend of the xenophobia seems decreased, while spikes of the index are significantly affected by political factors.

Figure 1. Visualizing the xenophobia index. (A) anti-Asian hate indicators of mainstream media channels; (B) the xenophobia index changes with political factors.

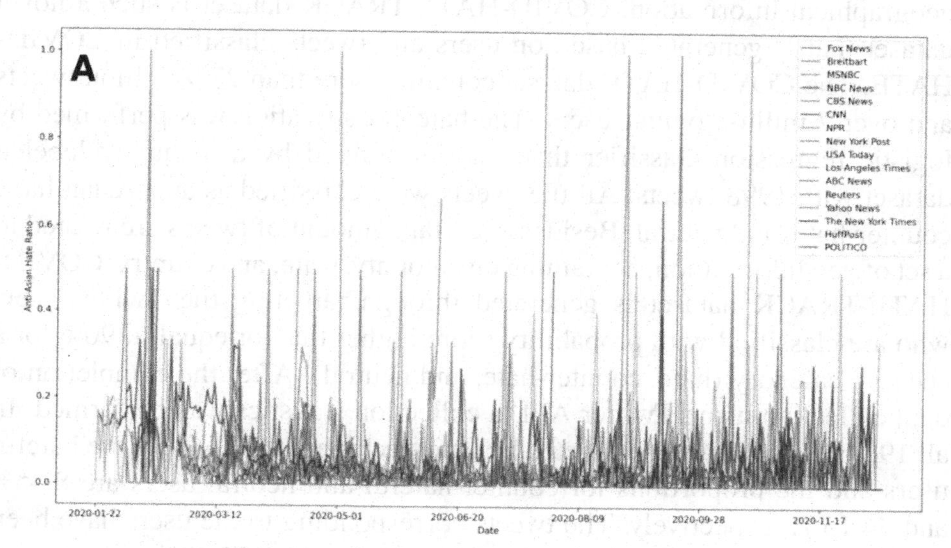

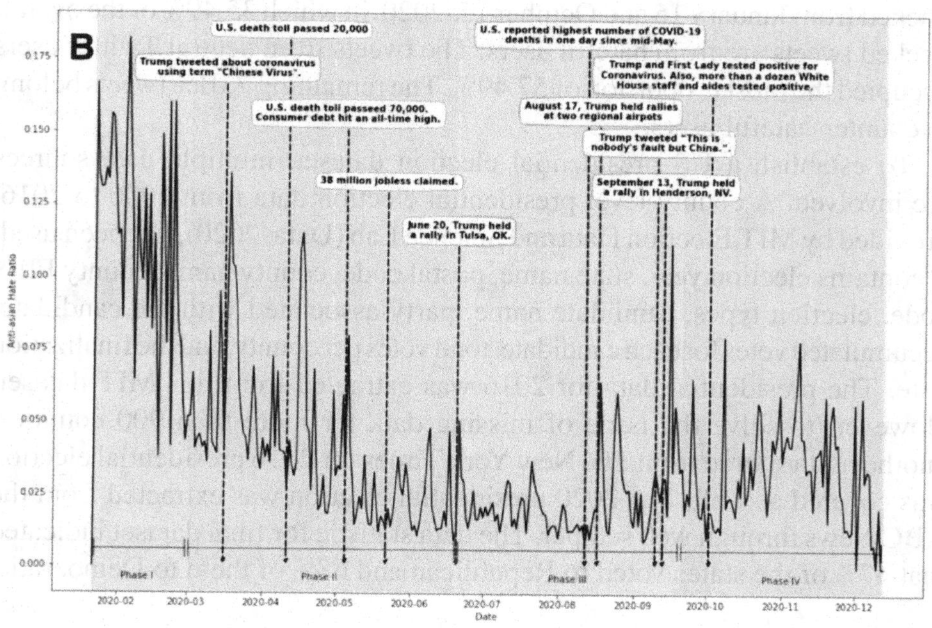

Tracking and Profiling Anti-Asian Haters

Tracking Haters Back From Their Posts

Similar to constructing the xenophobia index, analyzing anti-Asian racism also needs a solid dataset for tracking and profiling such a group of people in a multi-dimensional vision of characteristics, such as sentiments and geographical information. COVID-HATE-TRACK dataset is such a novel dataset that is generated based on users and tweets classified in COVID-HATE. The COVID-HATE dataset contains more than 27.9 million tweets and over 7 million twitter users. The hate classification was performed by logistic regression classifier that was pre-trained by a manually labeled dataset with 1998 tweets. All the tweets were classified as anti-Asian hate, counter-hate, and neutral. Besides, a certain amount of tweets are related to a set of geo information, containing city, county, state, and country. COVID-HATE-TRACK dataset is generated through targeting the Twitter users who are classified with probability score higher than or equal to 90% for a specific category (hate, counter-hate, and neutral). After the completion of data collection using Twitter API, a collection statistics was performed. In all 192,611 Twitter users targeted by hate tracking, 50.17% users are hateful users and the proportions for counter-hateful and neutral users are 8.45% and 44.13%, respectively. The tweets corresponding to the users have been tracked from January 15 and October 15, 2020, in which 35.49% of the overall tracked tweets are from hateful users. The tweets from neutral Twitter users occupied the most proportion of 57.49%. The remaining 7.01% tweets belong to counter-hateful users.

To establish a US presidential election dataset, multiple data sources are involved. A county-level presidential election data from 2000 to 2016, provided by MIT Election Data and Science Lab (Data, 2020), has been used. It contains election year, state name, postal code, county name, county FIPS code, election types, candidate name, party associated with the candidate, accumulated votes for each candidate, total votes per county, and the finalization date. The presidential data for 2016 was extracted from this MIT dataset. However, to solve the issue of missing data for more than 900 counties, another data source from The New York Times for 2016 presidential election was covered as well. The 2020 presidential election was extracted from the NBC News through web scraper. The data statistic for final dataset indicated that 47% of the states voted to Republican and 68% of them to Democratic, while in 2020 25% states supported Republican and the Democratic was voted by 51% states. Due to the missing data for 2016 presidential election result, only 4,585 counties were included, while the amount of county-level results is 4,632 for 2020. The proportions of county votes for Republican and Democratic are 68% and 29%, respectively, in 2016. In 2020, the proportions were changed to 68% for Republican and 29% for Democratic, separately.

Analyzing Haters' Sentiments and Visualizing Their Distributions

The Twitter users can be politically classified for candidate parties based on sentiment analysis toward Trump and Biden in 2020 and Trump and Hillary in 2016. A sentiment analysis tool, called valence aware dictionary and sentiment reasoned (VADER), was utilized to create sentiment-based features toward these two parties. A compound score was calculated through VADER module function to measure whether the sentence has positive, negative, or neutral view. Based on this tool, four sentiment variables were generated with tweets from the dataset for tracking anti-Asian haters. The compound score was calculated base on the Trump, Biden, Republican, and Democratic, which are the key words for the US presidential election in 2020. For each of the four features, it has four categorical values, containing like, dislike, swing, and does-not-care. If a user only posted tweets positive attitude toward a specific election keyword, this user will be identified as "like" on certain keyword, vice versa. If a user changes his/her mind to give opposite attitude toward the same keyword, this user will be labeled as "swing". For those users who never post tweets having sentiment preference toward any specific keyword, they will be regarded as "does-not-care" on the corresponding feature. Based on the mentioned labeling rules, 207,460 tweets from the dataset for tracking anti-Asian haters are labeled and 34,885 Twitter users are classified as like, dislike, swing, and does-not-care. The statistics indicate that for Trump 27.37% Twitter users are identified as like, 33.77% as swing, and 18.92% as does-not-care. Moreover, for Biden 20.07% user are identified as like, 15.31% as swing, and 51.40% as does-not-care. In the statistics on candidate's parties of Republicans and Democrats, 5.72%, 3.72%, 1.10%, and 88.90% Twitter users are labeled as like, dislike, swing, and does-not-care, respectively for Republicans. For Democrats, the proportions for like, dislike, swing, and does-not-care are 7.30%, 4.82%, 1.62%, and 86.08%, separately.

The COVID-HATE dataset includes geographical information for certain amount of tweets, with which each tweet can be located by longitude and latitude values. Through data visualization analysis performed on the COVID-HATE dataset, it can be known that large numbers of anti-Asian hate spike happened after "China Virus" tweet was posted by Trump. The tendency of anti-Asian increasing is significantly remarkable starting from the coasts, such as the areas of California, Texas, New York and Florida. To retrieve more geographical information, a python-written tool, called Reverse Geocoder, is used with tweets for geospatial analysis. All the retrieved information is collected in the dataset of anti-Asian hater tracking. Data analysis on the collected anti-Asian hater tracking shows that one user may post tweets from different locations since the user may travel to another place and use mobile device to post the tweets. In this case, the user will be labeled with multiple geolocation information for all the places where the tweets posted by the user are located. Through applying Reverse Geocoder, geographical information for over 11,600 Twitter users are completed collected, which includes country, state or province, and county or township. Among all the Twitter users, based on the geolocation information of their posted tweets, 11,188 Twitter users are located in United States, while 4,228 users lie on the overseas. From data analysis on the dataset of tracking anti-Asian haters, the top numbers of pandemic-related anti-Asian haters originated from 6 different states, such as Kansas, California, Texas, New York, Florida and Pennsylvania. Visualizing the hotspot of anti-Asian haters can be drawn by creating county-level choropleth maps and heatmaps, as shown in Figure 2. Figure 2A provides an overview of the distribution of the anti-Asian hater on Twitter. Hotspot areas scattered in four states are illustrated in Figure 2B-E. Obviously, the anti-Asian haters are more likely to locate in urban areas. Besides, the higher population density an area has, the higher possibility of anti-Asian hate occurrences. From the presidential election perspectives, 3,630 Twitter users who posted hate tweets are from the states supporting for the Republican, while 5,474 Twitter hate users are located in the states that vote for the Democrats. The similar tendency also happens to the voting counties. 3,997 haters are from a Republican county, however, 5,128 anti-Asian hate users are in a county supporting Democrats.

Figure 2. Visualizing anti-Asian hate Twitter users in the U.S. (A) the nationwide county-level choropleth map; (B) the heatmap of hotspot areas in California; (C) the heatmap of hotspot areas in Texas; (D) the heatmap of hotspot areas in New York; (E) the heatmap of hotspot areas in Florida.

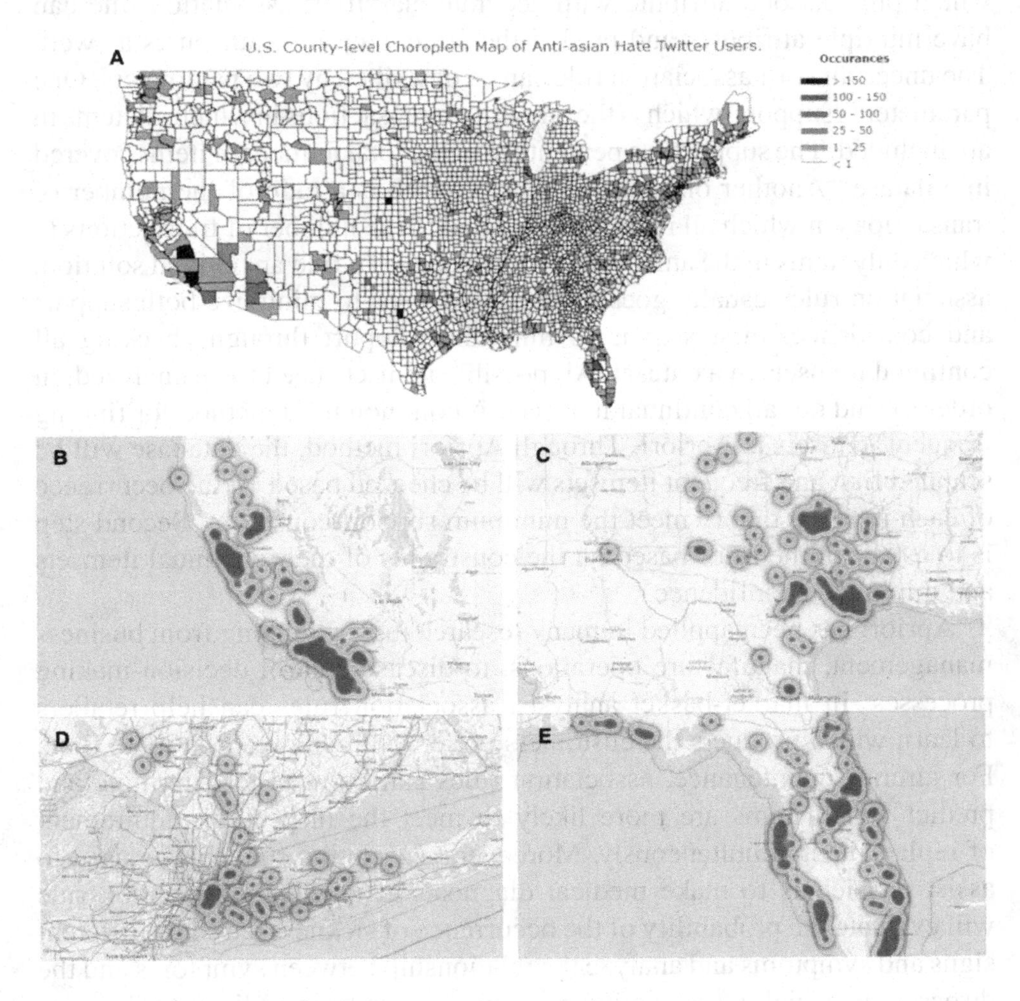

Profiling Anti-Asian Haters

Profiling haters can be a first step towards preventing the spreading of hate speech. Some characteristics, such as age, ethnicity, gender, religion, sentiments, and socioeconomic status, can represent the community of the haters and help to find out the existing conflicts in a society. To implement this issue in practice, association rule, which is a data mining approach used for exploring the relation and interpretation in a large dataset, is involved to profile anti-Asian hater. Association rules can analyze transactional behaviors according to information of training transactions. Different from classification which predicts one attribute with accurate classifier, association rule can have multiple attributes and predict the combination of attributes as well. The uncertainty of association rule can be described by two parameters. One parameter is support, which is the amount of transactions in which all items in are included. The support is a percentage of the total number of items covered in a dataset. Another one is confidence, which is a ratio of the number of transactions in which all items are included to the number of transactions in which only items in the antecedent are included. To find an optimal solution, association rules usually goes through two steps to minimize both support and confidence. First step is to minimize support through checking all continual itemsets in a dataset. All possible itemsets need to be analyzed, in order to find out all continual itemsets. A common used method for finding frequent itemsets is Apriori. Through Apriori method, the database will be scanned first and frequent itemsets will be checked based on the occurrence of each item, so that to meet the minimum support constraint. Second step is to establish the rules based on the constraints of their continual itemsets and minimized confidence.

Apriori has been applied in many research topics, ranging from business management, manufacture operations, to disease control decision-making processes. In market-basket analysis, association rules can help retailers to learn which products the customers prefer to purchase at the same time. For aircraft maintenance, association rules can be used to determine and predict which items are more likely to meet the inspection requirement of replacement simultaneously. Moreover, association rule can be used to assist physicians to make medical diagnosis efficiently. Association rule will evaluate the probability of the occurrence of sickness based on different signs and symptoms and analyze the relationship between symptoms and the diseases. Especially, association rule mining has been utilized to discover frequent symptom patterns of COVID-19 patients (Tandan et al., 2021). Such ideologies can be applied in profiling anti-Asian haters. This chapter provides an example for characterizing pandemic-related racisms by producing additional two indicators, i.e. sentiments on stimulus checks and economy reopening strategies, other than political sentiments and geographical factors. To simply the computational process, only 500 anti-Asian hate users have been randomly selected from the COVID-HATE-TRACK dataset. The Apriori algorithm is then performed with 4 input variables, such as sentiment indicators for Trump, Biden, stimulus checks, and economy reopening.

To better compute major measures from the Apriori algorithm, the dataset has been converted into a list of features for each record. For instance, a list of {anti-Asian hater, Trump supporter, economy reopening supporter} indicates that such a person is an anti-Asian hater who supports Trump and economy reopening without any disease control restrictions. Some records have been removed from the experimental dataset due to the difficulty of detections on sentiment factors. Finally, 476 lists of features are captured for profiling the characteristics of anti-Asian haters. The most important part in generating rules based on Apriori algorithm is to figure out which side a parameter should be set up. To find the answer to the question that a person who is an anti-Asian hater also supports for which political leaders or opinions, the anti-Asian hater indicator should be kept on the left-hand-side (lhs), while others will be on the right-hand-side (rhs). Numbers of rules have been generated via Apriori, whereas only one rule can achieve a 100% in confidence, that is "{anti-Asian hater} => {economy reopening supporter}", which means 100% of the individuals who produce anti-Asian tweets also support for the economy reopening without any restrictions of the disease control measures. Such a rule-based analysis uncovers the relationship between anti-Asian racism and economic impacts, rather than any other political issues. Obviously, the sample size is limited, thereby the conclusion may change with capturing more records and characteristics of anti-Asian haters.

FUTURE RESEARCH DIRECTIONS

Despite the effectiveness in monitoring, tracking and profiling anti-Asian racism and xenophobia, several challenges and difficulties are still remaining, laying the foundations for many research directions in the future. Various features, such as life styles, culture orientations, socioeconomic status, etc., can be captured from tweets and YouTube comments, which are beneficial for extracting novel sentiment-based factors that will be applied in many tasks, e.g. classifying haters through multi-dimensionally personal attributes. Moreover, the NLP-based models and the sentiment-based approaches are limited in English language, whereas analyzing data from multi-languages across the globe can extend the feature extraction effectively as one of the future research directions. Besides, this chapter only employs data from two social media platform with textual data inputs, while detecting anti-Asian racism and xenophobia can be proposed through extracting features from images, voices, and videos from multiple social media channels, online forums, and other multi-purpose instant messaging apps. Such suggestions may be helpful for better understanding the pandemic-related racism and xenophobia in the future.

CONCLUSION

Anti-Asian hate is an ongoing social problem that has changed the world significantly. Although a variety of research studies have been proposed in terms of emotional and behavioral analysis, sentiment analysis, and internet-based hate speech classifications, it is essential to monitor, track, and profile pandemic-related racism and xenophobia using big data analytics and deep learning models, as illustrated in this chapter. The xenophobia index can be applied in monitoring the anti-Asian hate comments in real-time, along with the usage of cutting-edge techniques in NLP and text classification. The deep learning model can be applied to automatically label pandemic-related posts efficiently with a 95% accuracy. Tracking and profiling anti-Asian haters have also been investigated and examined by incorporating sentiment-based features and geographical factors in regards to the 2020 U.S. presidential election, which is one of the most significant political issues during the pandemic. Again, the effectiveness of the experimental study in this chapter suggests that big data is one of the most powerful analytical tools in deciphering the complexities of the social change in the context of the COVID-19 pandemic.

REFERENCES

Ahmed, W., Seguí, F. L., Vidal-Alaball, J., & Katz, M. S. (2020). Covid-19 and the "film your hospital" conspiracy theory: Social network analysis of twitter data. *Journal of Medical Internet Research*, *22*(10), e22374. doi:10.2196/22374 PMID:32936771

Baccini, L., Brodeur, A., & Weymouth, S. (2021). The COVID-19 pandemic and the 2020 US presidential election. *Journal of Population Economics*, *34*(2), 739–767. doi:10.100700148-020-00820-3 PMID:33469244

Chakraborty, K., Bhatia, S., Bhattacharyya, S., Platos, J., Bag, R., & Hassanien, A. E. (2020). Sentiment Analysis of COVID-19 tweets by Deep Learning Classifiers—A study to show how popularity is affecting accuracy in social media. *Applied Soft Computing, 97*, 106754. doi:10.1016/j.asoc.2020.106754 PMID:33013254

Chaudhary, M., Saxena, C., & Meng, H. (2021). *Countering Online Hate Speech: An NLP Perspective.* arXiv preprint arXiv:2109.02941.

Chen, E., Deb, A., & Ferrara, E. (2021). # Election2020: The first public Twitter dataset on the 2020 US Presidential election. *Journal of Computational Social Science*, 1–18. doi:10.100742001-021-00117-9 PMID:33824934

Chen, J. A., Zhang, E., & Liu, C. H. (2020). Potential impact of COVID-19–related racial discrimination on the health of Asian Americans. *American Journal of Public Health, 110*(11), 1624–1627. doi:10.2105/AJPH.2020.305858 PMID:32941063

Cheng, S. O. (2020). Xenophobia due to the coronavirus outbreak–A letter to the editor in response to "the socio-economic implications of the coronavirus pandemic (COVID-19): A review". *International journal of surgery (London, England), 79*, 13–14. doi:10.1016/j.ijsu.2020.05.017 PMID:32407798

Data, M. E. (2020). *Science Lab, 2018," County Presidential Election Returns 2000-2016*. Harvard Dataverse.

Devlin, J., Chang, M. W., Lee, K., & Toutanova, K. (2018). *Bert: Pre-training of deep bidirectional transformers for language understanding*. arXiv preprint arXiv:1810.04805.

Dhanani, L. Y., & Franz, B. (2020). Unexpected public health consequences of the COVID-19 pandemic: A national survey examining anti-Asian attitudes in the USA. *International Journal of Public Health, 65*(6), 747–754. doi:10.100700038-020-01440-0 PMID:32728852

Garland, J., Ghazi-Zahedi, K., Young, J. G., Hébert-Dufresne, L., & Galesic, M. (2020). *Countering hate on social media: Large scale classification of hate and counter speech.* doi:10.18653/v1/2020.alw-1.13

Gover, A. R., Harper, S. B., & Langton, L. (2020). Anti-Asian hate crime during the COVID-19 pandemic: Exploring the reproduction of inequality. *American Journal of Criminal Justice, 45*(4), 647–667. doi:10.100712103-020-09545-1 PMID:32837171

Ho, J. (2021, January). Anti-Asian racism, Black Lives Matter, and COVID-19. *Japan Forum, 33*(1), 148–159. doi:10.1080/09555803.2020.1821749

Imran, A. S., Daudpota, S. M., Kastrati, Z., & Batra, R. (2020). Cross-cultural polarity and emotion detection using sentiment analysis and deep learning on COVID-19 related tweets. *IEEE Access: Practical Innovations, Open Solutions, 8*, 181074–181090. doi:10.1109/ACCESS.2020.3027350 PMID:34812358

Kabir, M., & Madria, S. (2020). *Coronavis: A real-time covid-19 tweets analyzer.* arXiv preprint arXiv:2004.13932.

Kao, A. C. (2021). Invisibility of Anti-Asian Racism. *AMA Journal of Ethics, 23*(7), 507–511. doi:10.1001/amajethics.2021.507

Kim, M. J., & Lee, S. (2021). Can stimulus checks boost an economy under covid-19? evidence from south korea. *International Economic Journal, 35*(1), 1–12. doi:10.1080/10168737.2020.1864435

Lyu, H., Chen, L., Wang, Y., & Luo, J. (2020). Sense and sensibility: Characterizing social media users regarding the use of controversial terms for covid-19. *IEEE Transactions on Big Data.*

Masters-Waage, T. C., Jha, N., & Reb, J. (2020). COVID-19, Coronavirus, Wuhan Virus, or China Virus? Understanding How to "Do No Harm" When Naming an Infectious Disease. *Frontiers in Psychology, 11*, 3369. doi:10.3389/fpsyg.2020.561270 PMID:33362626

Mello, H. L. (2020). Innovation diffusion, social capital, and mask mobilization: Culture change during the COVID-19 pandemic. In *COVID-19* (pp. 134–151). Routledge. doi:10.4324/9781003142065-14

Misra, S., Le, P. D., Goldmann, E., & Yang, L. H. (2020). Psychological impact of anti-Asian stigma due to the COVID-19 pandemic: A call for research, practice, and policy responses. *Psychological Trauma: Theory, Research, Practice, and Policy, 12*(5), 461–464. doi:10.1037/tra0000821 PMID:32525390

Nguyen, T. T., Criss, S., Dwivedi, P., Huang, D., Keralis, J., Hsu, E., Phan, L., Nguyen, L. H., Yardi, I., Glymour, M. M., Allen, A. M., Chae, D. H., Gee, G. C., & Nguyen, Q. C. (2020). Exploring US shifts in anti-Asian sentiment with the emergence of COVID-19. *International Journal of Environmental Research and Public Health, 17*(19), 7032. doi:10.3390/ijerph17197032

Oyebode, O., Ndulue, C., Adib, A., Mulchandani, D., Suruliraj, B., Orji, F. A., Chambers, C. T., Meier, S., & Orji, R. (2021). Health, Psychosocial, and Social Issues Emanating From the COVID-19 Pandemic Based on Social Media Comments: Text Mining and Thematic Analysis Approach. *JMIR Medical Informatics*, *9*(4), e22734. doi:10.2196/22734 PMID:33684052

Pei, X., & Mehta, D. (2021). *Beyond a binary of (non) racist tweets: A four-dimensional categorical detection and analysis of racist and xenophobic opinions on Twitter in early Covid-19*. doi:10.1109/BigData52589.2021.9671945

Peters, M. E., Neumann, M., Iyyer, M., Gardner, M., Clark, C., Lee, K., & Zettlemoyer, L. (2018). *Deep contextualized word representations*. doi:10.18653/v1/N18-1202

Reny, T. T., & Barreto, M. A. (2020). Xenophobia in the time of pandemic: Othering, anti-Asian attitudes, and COVID-19. *Politics, Groups & Identities*, 1–24. doi:10.1080/21565503.2020.1769693

Shapiro, A. H., Sudhof, M., & Wilson, D. J. (2020). Measuring news sentiment. *Journal of Econometrics*. Advance online publication. doi:10.1016/j.jeconom.2020.07.053

Tandan, M., Acharya, Y., Pokharel, S., & Timilsina, M. (2021). Discovering symptom patterns of COVID-19 patients using association rule mining. *Computers in Biology and Medicine*, *131*, 104249. doi:10.1016/j.compbiomed.2021.104249 PMID:33561673

TaskinsoyJ. (2020). The world is at a dangerous crossroads on" China virus" and US" political virus". doi:10.2139/ssrn.3713745

Tessler, H., Choi, M., & Kao, G. (2020). The anxiety of being Asian American: Hate crimes and negative biases during the COVID-19 pandemic. *American Journal of Criminal Justice*, *45*(4), 636–646. doi:10.100712103-020-09541-5 PMID:32837158

Vidgen, B., Botelho, A., Broniatowski, D., Guest, E., Hall, M., Margetts, H., . . . Hale, S. (2020). *Detecting East Asian prejudice on social media*. doi:10.18653/v1/2020.alw-1.19

Vishwamitra, N., Hu, R. R., Luo, F., Cheng, L., Costello, M., & Yang, Y. (2020, December). On Analyzing COVID-19-related Hate Speech Using BERT Attention. In *2020 19th IEEE International Conference on Machine Learning and Applications (ICMLA)* (pp. 669-676). IEEE. 10.1109/ICMLA51294.2020.00111

Yousefinaghani, S., Dara, R., Mubareka, S., Papadopoulos, A., & Sharif, S. (2021). An Analysis of COVID-19 Vaccine Sentiments and Opinions on Twitter. *International Journal of Infectious Diseases.*

Zhao, P., Chen, X., & Wang, X. (2021). Classifying COVID-19-related hate Twitter users using deep neural networks with sentiment-based features and geopolitical factors. *International Journal of Society Systems Science, 13*(2), 125–139. doi:10.1504/IJSSS.2021.116373

Ziems, C., He, B., Soni, S., & Kumar, S. (2020). *Racism is a virus: Anti-asian hate and counterhate in social media during the covid-19 crisis.* arXiv preprint arXiv:2005.12423.

ADDITIONAL READING

Aggarwal, C. C., & Zhai, C. (2012). A survey of text classification algorithms. In *Mining text data* (pp. 163–222). Springer. doi:10.1007/978-1-4614-3223-4_6

Deng, L., & Liu, Y. (Eds.). (2018). *Deep learning in natural language processing*. Springer. doi:10.1007/978-981-10-5209-5

Garg, S., & Ramakrishnan, G. (2020). Bae: Bert-based adversarial examples for text classification. *arXiv preprint arXiv:2004.01970*. doi:10.18653/v1/2020.emnlp-main.498

Gulli, A., Kapoor, A., & Pal, S. (2019). *Deep learning with TensorFlow 2 and Keras: regression, ConvNets, GANs, RNNs, NLP, and more with TensorFlow 2 and the Keras API.* Packt Publishing Ltd.

Kowsari, K., Jafari Meimandi, K., Heidarysafa, M., Mendu, S., Barnes, L., & Brown, D. (2019). Text classification algorithms: A survey. *Information (Basel), 10*(4), 150. doi:10.3390/info10040150

Levy, O., & Goldberg, Y. (2014, June). Dependency-based word embeddings. In *Proceedings of the 52nd Annual Meeting of the Association for Computational Linguistics* (Volume 2: Short Papers) (pp. 302-308).

Liu, J., Chang, W. C., Wu, Y., & Yang, Y. (2017, August). Deep learning for extreme multi-label text classification. In *Proceedings of the 40th international ACM SIGIR conference on research and development in information retrieval* (pp. 115-124). 10.1145/3077136.3080834

Minaee, S., Kalchbrenner, N., Cambria, E., Nikzad, N., Chenaghlu, M., & Gao, J. (2021). Deep Learning—based Text Classification: A Comprehensive Review. *ACM Computing Surveys*, *54*(3), 1–40. doi:10.1145/3439726

Sun, C., Qiu, X., Xu, Y., & Huang, X. (2019, October). How to fine-tune bert for text classification? In *China National Conference on Chinese Computational Linguistics* (pp. 194-206). Springer. 10.1007/978-3-030-32381-3_16

Wake, L. (2010). *Nlp: Principles in practice*. Ecademy Press.

KEY TERMS AND DEFINITIONS

Association Rule: A machine learning method that can find out correlative relationships from large-scale data.

Deep Learning: A broad family of machine learning models based on neural networks. Typical deep learning models are deep neural networks, convolutional neural networks, recurrent neural networks, deep belief networks, and deep reinforcement learning.

GIS: A system that can convert all types of data into a map to present all descriptive information and understand geographic context.

NLP: A processing method of computational linguistics for human language based on algorithms.

Sentiment Analysis: A process of analyzing and extracting the subjective opinions of texts.

Social Media Analysis: A process of analyzing and tracking social media through related data collected from social network.

Text Classification: A typical problem in information science that assigns a textual data to one or more classes or categories.

Web Scrapping: A process of extracting content information, such as texts, tables, images, from a certain website.

Chapter 12
Big Data Thinking of Economy, Investment, and Business in COVID-19

ABSTRACT

The COVID-19 pandemic has disrupted the global economy and has changed lifestyles. Stock markets crashed in the early stage of the pandemic in the United States. Small businesses have been affected by COVID-19, and the impact will remain for a while. It is highly uncertain in making formulation of right decisions with the evolution of the disease and its variants. Pandemic impacts are hidden behind the complex social system, which can be uncovered by using big data analytics. In this chapter, a systematic survey is given based on current studies in regards to pandemic-related social and economic changes using big data analytics by incorporating economic development, investment management, and business operation issues. This chapter provides a way of big data thinking for research communities, economic policymakers, and individuals to understand economics and social science in the context of COVID-19 and the post-pandemic era.

INTRODUCTION

From December 2019 to January 2020, when the US-China trade war and Brexit were turning white-hot, anxiety has become the dominated sentiment due to the significant impact on the global economy. Despite a moderated

DOI: 10.4018/978-1-7998-8793-5.ch012

economic growth predicted by IMF, the overall market sentiment seemed cautiously optimistic. However, only a few days later, the COVID-19 outbreak changed the story line unexpectedly. As fear, panic, uncertainty, and rational assessment are combined together, revenue and profits of hundred thousands of companies are expected to be lower than ever since the beginning of the pandemic. Such the inference is initially reflected in the capital market, when the US stock market was crashed suddenly on February 20, 2020. $5 trillion in values of the S&P 500 index have been lost in the week from 24th to 28th of February, while global markets shrank nearly $6 trillion in wealth in the same week (Mazur et al., 2021). The tourism industry was affected by COVID-19 significantly, according to the International Air Transportation Association (IATA) predictions that the air travel business would lose hundred billion dollars as the Chinese travelers were restricted by the stay-at-home policies and travel bans (Suau-Sanchez et al., 2020). The spillover of COVID-19 has also impacted the global economy throughout a variety of industrial sectors and market participants, such as hospitality, sports, oil-dependent countries, import-dependent countries, banks and FinTech, etc. (Ozili & Arun, 2020). In a word, the COVID-19 pandemic has caused a huge amount of economic losses and is evolving into the next great recession.

The sharp reduction in economic activity associated with public disease control measures will also result in millions of job losses. The labor market is undergoing tremendous pressures due to the efforts of the stay-at-home policies, immigration restrictions, nonessential business closures, and reductions of demand in products and services. As such, economists have estimated over 20% decline of GDP in the U.S. on an annual basis, while more than 95% of the domestic economy is impacted by the mandatory closures in almost all states (Walmsley et al., 2021). The negative impact on economy rapidly transformed in investment markets, along with a sharp sell-off in Treasury markets (Schrimpf et al., 2020). Aggressive actions have been taken as the interest rate is moving back to zero, issued by the U.S. Federal Reserve, which responded with a set of economic stimulus plans in terms of the actions targeting on relieving cash-flow stress for small and medium-size businesses. Such a monetary policy is successful in stimulating economic growth without significant effects on inflation, as markets no longer believe that even quantitative easing can bring inflation to target, which leaves few options for monetary policy apart from low interest rates (Feldkircher et al., 2021; Lilley & Rogoff, 2020). However, the COVID-19 outbreak seriously affects global economy through various economic issues, such as economic growth and uncertainty, bubble and burst, as well as risks in investment markets, thereby the macroeconomic impact from the COVID-19 pandemic has been becoming significant and prominent.

The COVID-19 pandemic has also revolutionized the way of life, consequently impacting the decision-making process in business management and industrial operation. The fragility of the global supply chain has been highlighted in the pandemic, whereas such a problem is affected by severely restricted international traveling and in-person commercial activities. With the spread of the global pandemic, various industrial sectors are meeting huge operational crisis, such as shortages of raw materials, increased costs in human resource, financing difficulties, etc., while cutting-edge technologies can help in combating the health and economic crisis in terms of innovations for creative solutions in business management and operational processes, along with the spirit of resilience, strategic agility, and entrepreneurship (Liu et al., 2020). On the other hand, the consumer behaviors have been strongly disrupted by the lockdown and social distancing policies, thereby new buying, working, and learning habits, such as online shopping, offline products/ services delivering, and remote learning, will also emerge via technological advances, demographical diversities, and innovative business models, which have been impacted by the COVID-19 pandemic in blurring the boundaries between working, leisure, and education (Sheth, 2020). Moreover, prominent changes in lifestyle behaviors have been investigated during the COVID-19 pandemic, along with substantial changes on outdoor activities, physical events, stress management, social support, and restorative sleep (Balanzá-Martínez et al., 2021). Such insights will also benefit in understanding the COVID-19 pandemic from the social and economical perspectives.

The social system is a precision machine that incorporates many aspects of economy, from both macroeconomics and microeconomics perspectives, which can be impacted by internal and external shocks. The COVID-19 pandemic is such a shock, along with the impact from black swans and gray rhinos, which has fundamentally changed the world. It is urgent to respond to the pandemic crisis in terms of economic development, business operation, and crisis management, from the government, the entrepreneur, to the individual. This chapter stands to reasons that decision-makers can make the optimal choices by leveraging digitalized methods and data-driven approaches in managing economic crisis associated with the COVID-19 pandemic. Business operators enable to identify and analyze the shape of the novel reality, while big data analytics and computational methods can be applied to solve the problem effectively. The objectives of this chapter are:

- illustrating how big data works in monitoring the macroeconomic health and policy-making measures in the context of COVID-19;
- investigating the role of big data analytics in providing insights for better decision-making in asset and investment management under the market crash caused by the pandemic;
- examining how COVID-19 changes the way of thinking and doing in business operations and management, and how to apply big data and AI to deal with the current challenges.

BACKGROUND

Over the past decade, big data and artificial intelligence (AI) technologies have been widely applied as matching, searching, and forecasting approaches in analyzing economic theories, firms and consumers, productivity growth, price discrimination, market competition, and inequality (Mihet & Philippon, 2019). The economic system is a complex system that is composed of various heterogeneous, dispersed and interconnected factors, which involve a wide range of complex behaviors that can be investigated by using big data analytics and AI models, e.g. economic security simulation, economic early warning modeling, economic calibration and visualization, etc. (Li et al., 2020). The relationship between economic growth and other factors is ambiguous, whereas both positive and negative effects of impacts on economic growth can be examined by analyzing how the individuals' attitude toward the future, along with utilizing big data (Ghosh et al., 2020). State-of-the-art techniques are increasingly applied in exploring the relevant patterns of economic activities, of which big data analytics and AI models have been frequently used in discussing a variety of core topics in macroeconomics, such as GDP growth modeling, unemployment estimation, inflation monitoring, and monetary and fiscal policy-making (Choi, 2019; Simionescu & Zimmermann, 2017; Medeiros et al., 2021). The current debate over computational approaches in economics and social science has been quickly transformed into a detailed discussion of interactions between big data and computational behavioral macroeconomics, which lead to new thinking styles and research paradigms (D'Orazio, 2017). Moreover, novel analytical approaches, such as social network and semantic analysis, have been proposed to improve the monitoring and forecasting models of macroeconomic indicators (Elshendy & Fronzetti Colladon, 2017). Such approaches have enlarged the research boundary beyond the classical economics, while cutting-edge techniques, such as statistical inference, time-series forecasting, machine learning (ML), and deep learning (DL) can be employed as some of the most powerful tools in analyzing and monitoring macroeconomic indicators. Macroeconomic policymakers and observers can use big data to monitor the economic performance, especially when uncertainty evolves into a crisis.

As the blood of industrial enterprises, finance is the artery of the economy. The health of the financial markets is one of the most critical issues in economic strength, social stability, and individual wealth. The financial market is a complex system that consists of many active traders and the high-frequency volatility of price returns, while valuable insights can be gained from usages of big data analytics for better decision making, e.g. forecasting financial market volatility using ML and big data analytics (Yang et al., 2020). Anomaly detection in financial markets has attracted attention as many institutional and individual investors store, process, and analyze historical data of assets to make rational strategies and informed decisions, where a variety of data-driven approaches can be employed, e.g. supervised learning, clustering, and statistical inference (Ahmed et al., 2017). Moreover, extracting meaningful representations and recognizing patterns from financial big data can be operated by using DL models, which allow producing more efficient predictions than standard methods in finance (Hasan et al., 2017). In today's digitalized age, the boundary between the investment market, social media, and FinTech blurs, in which online financial platforms are significantly correlated to other industries, such as material manufacturing and telecommunication, whereas social media is related to service providers and technology-intense companies (Liu et al., 2018). Such insights can be obtained by extracting and analyzing daily information from online search engines, cyber-based finance platforms, mobile devices, and social media channels, along with web crawling techniques, big data processing, ML/DL algorithms, and natural language processing (NLP), which have been formed into the concept of big data finance (Fischer & Krauss, 2018; Bukovina, 2016; Xing et al., 2018). Such a concept has been applied in various research topics, such as financial risk management, asset management, credit risk assessment, and financial fraud detection (Cerchiello & Giudici, 2016; Campos et al., 2017; Pérez-Martín et al., 2018; Tang & Karim, 2019). Herein, the innovation of blockchain technology has prompted a new type of investment market, i.e. Bitcoin, which can interact with big data finance in the form of the new frontier of research domains (Wang et al., 2020).

Applying big data analytics and AI approaches has been becoming popular in industry in the form of new analytical tools that deliver evidence-based decision-making processes. Various industrial sectors are under pressure of sustainable integrations throughout the overall business operations and management, while data-driven approaches provide unique insights to improve the sustainable business models and performance (Raut et al., 2019). Hence, big data analytics plays a critical role in modern business operations management. From technical, strategical, and architectural perspectives, different types of methods can be applied in dealing with multiple operational management topics, such as inventory management, revenue forecasting, marketing analysis, transportation planning, supply chain management, and risk analysis (Choi et al., 2018). The services-oriented architecture has been widely referenced to investigate how big data analytics enhances business intelligence through big-data-as-a-service, where such an ideology is feasible for developing business intelligence and enterprise information systems (Sun et al., 2015). The COVID-19 crisis has impacted the modern business operations significantly and fundamentally due to the challenges and risks associated with the pandemic. Business operators and practitioners are seeking to make sense of how big data analytics can help to deal with the uncertainty during the COVID-19 outbreak and in the age of post-pandemic. One of the most critical and painful lessons learned from the COVID-19 pandemic is the weakness of the global supply chain, whereas a variety of innovative approaches using big data analytics has been proposed to strengthen supply chains in different industrial sections, such as manufacturing, healthcare, automobile, airline, and agriculture, along with applications of big data analytics (Bag et al., 2021; Belhadi et al., 2021; Sharma et al., 2020). Besides, understanding consumer behaviors is critical in business management and marketing analytics, where big data analytics and AI methods have contributed more powerful tools in analyzing buying patterns and shopping ways (Hofacker et al., 2016). New business models have been inspired by the current reality caused by COVID-19, thereby understanding such new patterns of consumer behaviors will benefit business operations and management.

MAIN FOCUS OF THE CHAPTER

Issues, Controversies, Problems

Much of the research communities' attention is currently focused on the impact of the COVID-19 pandemic on the economy, particularly in terms of economic conditions, financial market dynamics, and business operations. The crisis caused by COVID-19 has meant substantial alternations to macroeconomic policy-making, investment, international trade, supply chain, consumption, lifestyle patterns and changes. Due to the ongoing nature of the pandemic, researches have been increasing to emphasize on economy development, financial market volatility, business operations and management, consumer behaviors, and social changes, while such topics have been incorporated with big data analytics, which can be applied in providing insights for evidence-based decision-making processes. Therefore, understanding how big data and AI works for monitoring the economic condition, guiding asset and wealth management, and assisting business operations seems extremely important in the context of COVID-19.

Features of the Chapter

This chapter provides a systematic and comprehensive survey on current applications of big data analytics that deal with a broad range of social and economic changes incorporating economic development, investment management, and business operation issues associated with COVID-19. Existing studies in this field are divided into three layers, ranging from the macroeconomic monitoring and measures, asset and investment management, to business operations and social changes, along with the discussion of the role of big data in between. Various ongoing and urgent topics, such as pandemic recession, economic uncertainty, sustainable development, labor demand and supply, energy consumption, financial market volatility, housing and rental, consumer behaviors, risk control, business operation and management, and social changes, have been incorporated with examining how big data works for providing insightful evidence in the decision-making of economic policies, investment strategies, and business operations.

SOLUTIONS AND RECOMMENDATIONS

Economic Monitoring and Measures

Recession, Bubble, and Economic Uncertainty

The COVID-19 pandemic has impacted the economic development in the form of recession, bubbles, and uncertainty, while such issues have been discussed and measured with big data and AI. Coulombe et al. (2021) applied ML methods to analyze UK economy in terms of catching the COVID-19 recession. A large dataset containing 112 monthly macroeconomic and financial indicators has been used, along with time-series models, such as random walk and factor augmented autoregressive, and ML methods, such as random forests, ridge regression and neural networks. Diebold (2020) discussed economic activity using the Aruoba-Diebold-Scotti (ADS) Index of Business conditions (ADS) under the circumstances of Great Recession and Pandemic Recession. ADS applied linear models for locally-linear estimations. He et al. (2020) generated the real-time index to characterize the pandemic-related impact on service industry. The study discussed the influence of COVID-19 on the Chinese economy, based on value creation, cost, leverage, and inventory, through big data analysis. Narayan (2020) applied hourly exchange rate data for different currencies to analyze bubble activity and bubble type behavior. ADF regression model and t-statistics were used to calculate and evaluate the bubbles. Altig et al. (2020) investigated several forward-looking measures in regards to the economic uncertainty, including stock market volatility, economic Twitter chatter, subjective uncertainty about business growth, newspaper-based uncertainty measure, etc., in the context of COVID-19. The study concluded that large economic uncertainty happened corresponding to the pandemic, whereas the significant uncertainty is not the sign of economic recovery. Shoss et al. (2020) conducted a study aimed on the COVID-19 health crisis and pandemic-related economic threats. The correlation results suggest a negative relation between search interest in recession and search interest in helping and a lightly positive relation between COVID-19 search interest and helping. Moreover, the study pointed out that health disaster factors at macro can micro-level can portray the helping behavior during the COVID-19 pandemic.

Economic Growth vs. Sustainable Development

The balance between economic growth and sustainable development plays a crucial role in making strategies for both national and local governance. Current studies are mainly focusing on discussing the relationships between GDP growth, energy consumption, environmental shocks, and sustainable cities using a broad range of cutting-edge technologies, such as big data, AI, and Internet of Things (IoT), along with the impact from the COVID-19 pandemic. Magazzino et al. (2021a) discussed the possible correlation between the use of renewable energy and GDP increasing in Brazil using artificial neural network (ANN). The empirical results suggest that the renewable energies can represent the tendency of economic growth in Brazil and economic growth positively influences GDP growth. Mele & Magazzino (2021) uncovered relationships among environmental pollution, economic growth, and COVID-19 death in India using annual pollution data. A causality analysis was conducted with a predictive linear regression model. For each time-series variable, a set of stationarity tests, including ADF, ERS, PP, and KPSS, has been incorporated. Toda-Yamamoto test was used to evaluate the causal relationship between each certain pollution variable and the economic growth. Similarly, Magazzino et al. (2021b) applied DL methods based on daily data of New York state to explore the relationship between COVID-19 deaths and air pollution. The empirical results based on AUPRC analysis suggest that pollution factors, such as $PM_{2.5}$ and NO_2 to deaths, have impacted on accelerating COVID-19 deaths and economic growth, along with the environmental improvement discussions in New York. Lovric et al. (2021) designed a ML-based approach to analyze the air quality change during the COVID-19 pandemic in Graz, Austria. The feature data included concentrations of nitrogen dioxide, particulate matter, ozone and total oxidant. A random forest regression model was trained to predict the values of air concentrations during the pandemic. Scott et al. (2020) performed another ML-based analysis based on data-driven IoT systems. A structural equation was involved to model the collected data from multiple data sources, such as CompTIA, Deloitte, DNV GL, SCC and etc. Ezugwu et al. (2021) proposed a framework based on ML methods and the IoT technique to fulfill an interaction between city population and the environmentally negative influence caused by COVID-19. Such an AI-based framework can help national health system to release the medical burden during the pandemic. Lyons & Lăzăroiu (2020) investigated smart sustainable cities by estimating and analyzing data from

5,200 respondents collected from IoT sensors, along with structural equation modeling and ML.

Labor Demand and Supply

The structure of labor demand and supply has been changed with the spread of the COVID-19 outbreak, whereas big data analytics and AI models have been widely applied to deliver insights. Yu et al. (2020) examined the impact of COVID-19 on the labor force participation rate among different countries by coupling with big data on the occurrence of historical pandemic. The empirical results indicate that the country with a higher uncertainty avoidance index will go through an obvious decline in labor force participation rate. Gulyas & Pytka (2020) used the data of unemployment insurance records before May 2020 in Austria. A machine learning approach was applied to estimate earning losses based on individual worker and job characteristics, which indicate that the pandemic-related job suspension may potentially cause the lower loss in wages and incomes. Campello et al. (2020) proposed a big data method that can be applied to analyze the different dimensions of the COVID-19 impact on the US job market, with data from LinkUp, a job market data provider. Time series analysis was applied to analyze the job posting dynamics. Geo-based analysis was involved to examine the geographical distribution of job postings during 2017 to 2019. Bauer et al. (2021) investigated the search behavior in labor market during the COVID-19 pandemic, with leveraging big data. Data visualization was performed to compare the changes in difference sectors, such as construction, consumer goods, finance, etc., before and after the COVID-19 crisis. Besides, time series analysis were employed for search intensity analysis of job seekers. Brinca et al. (2021) assessed labor demand and supply shocks during the COVID-19 pandemic through utilizing a Bayesian structural vector autoregression, along with analyzing monthly records of working hours and real wages. Forsythe et al. (2020) used job vacancy data obtained through Burning Glass Technologies and the Bureau of Labor Statistics (BLS) employment data to examine the impact of COVID-19 on labor market. Campos-Vazquez et al. (2021) examined interactive relationship between labor demand and job searching by extracting and analyzing information from website. The results derived by big data indicate that the decrease in the number of job advertisements may not cause the structural change in labor demand during the pandemic in Mexico. Similarly, Shuai et al. (2021) discussed a set of urgent topics in labor market dynamics among COVID-19 tendency, labor market, and government policies, with using job advertisement data and time series analysis, which is applied to analyze the job posting and regression method that is implemented to examine the advertised wages.

Asset and Investment Management Under the Crash

Oil Market

The oil market has been shocked by the pandemic due to the reduced energy consumption and the limited supply of oil production and logistics. Such a shock has been quickly reflected in the international crude oil market, causing a huge amount of losses for investors in the oil market. Leveraging big data and AI allows investors to analyze and predict market trends, thereby reducing the risk. Lyke investigated the reaction of the US oil and gas firms to the COVID-19 pandemic by analyzing stock return and volatility, along with the EGARCH model that is used to model sentiment-stock return. Devpura & Narayan (2020) established an oil price volatility model to analyze macroeconomic stability and perform risk management, where the pandemic has the impact on oil price volatility. Qin et al. (2020) proved the interrelationship between COVID-19 and oil price using quarterly oil price data from 1996 to 2020. A capital asset pricing model (ICAPM) was used to portray the transmission between COVID-19 and oil price. Residual-based bootstrap (RB)-based modified-likelihood ratio (LR) statistics were utilized to prove the hypothesis in this study. Wu et al. (2021) collected oil-related news and perform feature extraction through convolutional neural network for model training. The news was collected based on four categories, including oil price, oil production, oil consumption, and oil inventory. The experimental results demonstrated that news information can be used to forecast the oil price. Hawash et al. (2020) discussed the impact of COVID-19 on oil and gas sector digital transformation. A proposed conceptual model was used to analyze the potential relationship between digital transformation and value change for oil and gas sector. Oyewola et al. (2021) performed image transformation for time-series-base crude oil price with the combination between Directed Acyclic Graph and Convolutional Neural Netowrk. Such a method can convert crude oil price into 2-D images, with an accuracy of 99.16% for model prediction of the impact of the pandemic on crude oil price image. Huang & Zheng (2020) examined the relationship between investor sentiment and crude oil futures price during the COVID-19 pandemic. A regression method was used to model the structural changes among the relationships by implicating time series analysis to portray the tendency of WTI crude oil futures price and OVX over time.

Stock Market

Similar to the oil market, the stock market was crashed in the beginning of the pandemic. Current studies mainly focus on uncovering the relationship between stock market and COVID-19, predicting price returns, and examining internal and external shocks by using various big data and AI approaches, such as time series forecasting, ML, DL, and NLP. Onali (2020) examined the relationship between COVID-19 cases and deaths and stock markets using VAR and GARCH models for analyzing impose responses of shocks from COVID-19 deaths to stock market returns from April 2019 to April 2020. Baek et al. (2020) discussed the impact of COVID-19 on the US stock market volatility based on Markov Switching AR model. The study implemented feature selection using ML to target the potential economic indicators. Khattak et al. (2021) used Least Absolute Shrinkage and Selection Operator (LASSO) to examine the internal and external shocks to the European market during the COVID-19 pandemic. Costola et al. (2020) proposed a NLP-based study on the effect of news on stock prices during the COVID-19 pandemic. The results suggest a positive correlation between news sentiment and market returns based on analyzing the news data containing 203,886 pandemic-related news articles from MarketWatch, Reuters, The New York Times. Lee (2020) discussed the relationship between COVID-19 sentiment and US stock market using big data. This study utilized the Daily News Sentiment Index (DNSI) and Google Trends data during the pandemic. Time series regression model was applied to predict US stock returns using DNSI. Xue et al. (2021) applied Word2Vec to analyze the response of stock markets to COVID-19. The results show that the rising of daily confirmed cases has positive correlation with the decline of the stock market. Ghosh et al. (2021) performed stock price prediction as a binary classification problem in the context of COVID-19. Macroeconomic indicators were involved as explanatory variables for model training, while Kernel Principal Component (KPCA) analysis, Stacking and Deep Neural Network (DNN) models were incorporated for feature extraction.

Real Estate Market

The real estate market is another investment object that is severely affected by the COVID-19 pandemic, which has changed the structure of the housing and rental markets profoundly. Big data can be used to track and analyze the features of the real estate market, thereby instructing investors to observe the market change intuitively and effectively in the context of COVID-19. Grybauskas et al. (2021) collected 18,992 property listings from Vilnius during the COVID-19 pandemic through performing a web scraper. To predict apartment revisions, 15 different ML algorithms were applied. Based on the calculation of SHAP values, the study concludes that the time-on-the-market variable is the most reliable variable for price revision prediction. Kuk et al. (2021) investigated the COVID-19 impact on rental housing market in metropolitan areas of the United States by collecting and analyzing rental data from Craigslist from March to June, 2020. A regression model was established to estimate the impact of COVID-19 cases. Kristof (2021) applied data from STR and Inside Airbnb from January 2018 to September 2020. A regression analysis was used to analyze the impact of COVID-19 on the housing rental industry. Text mining methods were also applied for analyzing new use-cases. Nhamo et al. (2020) used a multi-source-critical document analysis to study the impact of COVID-19 on hotels and Airbnbs, along with applying data visualization and time series analysis in discussing the confirmed booking and to portray the cancelled gross booking over time. Zhu (2021) established a real estate virtual e-commerce model and performed big data methods, which can be applied for prediction in real estate industry and provide reliable methods to target customers. Garzoli et al. (2021) combined recent information from 71 financial and macro predictors, containing 38 financial variables and 33 macroeconomic indicators. A mean squared error criteria was used to evaluate the accuracy of model prediction in the context of COVID-19.

Consumer Behaviors, Business Operations, and Social Changes

Consumer Behavior Changes by COVID-19

Consumer behaviors have been changed by COVID-19 significantly, while such changes affect business operators to make decisions via analyzing the new patterns in the market. Current studies suggest that changes in consumer behaviors have been captured through applying big data analytics, which can be used in marketing analytics for better business management and operations. Yedjou et al. (2021) employed ML to analyze the impact of various food types on COVID-19 cases and deaths. The data used in this article was a COVID-19-related diet dataset from Kaggle. A correlation-based feature selection method was used to extract important features that contribute significantly to the impact on COVID-19 cases and deaths. Garcia-Ordas et al. (2020) studied the impact of COVID-19 on people mortality. Principal component analysis (PCA) was performed to extract important features. A K-Means clustering method was used to classify the different countries based on features, including distribution of fat, energy, the ingested amount, and protein in 23 types of food. Safara (2020) established a ML-based model to predict consumer behavior during the COVID-19 pandemic, which can release the burden for government and retail industry. The data involved in this study was collected from DigiKala, an online shopping websit. Correlation analysis was used to measure the impact of each feature on prediction of COVID-19 consumer purchase volume. Akour et al. (2021) proposed a behavior model to analyze performances and action of college students for mobile learning platforms during the COVID-19 pandemic. The data involved in this study was collected from a designed survey. Six ML models, such as BayesNet, Logistic LWL[d], AdaBoostM1, OneR, and J48, were performed to train the classifiers for prediction. Sheth (2020) analyzed the consumer behavior influence by lockdown and social distancing during the COVID-19 pandemic. The study discussed the changes of habits caused by pandemic and managerial implications from the effect of COVID-19 on consumer behavior. Pantano et al. (2021) collected 15,000 tweets to analyze the impact of COVID-19 on consumer behavior, with the use of content analysis. The results indicated the cultural differences and preferences from consumers of various countries. Chauhan and Shah (2020) analyzed customer behavior and sentiment during the pandemic in India. The data was collected from a survey, which can be analyzed and visualized to explore shopping behavior, purchase decisions, media consumption habits across different genders and age groups. Brandtner et al. (2021) collected consumer satisfaction results and evaluation comments to examine the influence of COVID-19 on consumer satisfaction during the pandemic. Text mining analysis was performed to identify the influenced product groups. The consumer sentiment data was scraped from related review platforms. This study built a service quality assessment model based on five categorical data. Shang et al. (2021) investigated the impact of COVID-19 on the user behaviors and environmental benefits of bike sharing during the pandemic by developing a novel calculation method for trip distances and trajectories. Topological indices was used to examine the user behavior change of bike sharing caused by the pandemic.

Crisis Management and Risk Analysis in COVID-19

The COVID-19 pandemic produces big risks to small and mid-size businesses, whereas the effective crisis management can reduce risks and losses caused by the pandemic using data-driven approaches with big data, along with technological innovations. Loh and Teoh (2021) examined the big data analytics adoption on the problem of Malaysian manufacturing Small and Medium Enterprises (SMEs) by gathering the data from 185 manufacturing SMEs. Partial Least Square (PLS) methods were involved to analyze the relationship among major factors. Park (2021) implemented big data analytics to handle COVID-19 crisis management by collecting data from news media, such as Saudi Press Agency, and social media, including Twitter and YouTube from February to August 2020. Semantic network analysis was used to identify major social issues from the data. Chang & McAleer (2020) discussed the preparations of different countries for COVID-19 pandemic and risk analysis of COVID-19 through analyzing the Global Health Security Index. Three Pythagorean means were calculated and compared to evaluate the correlations of the numerical score rankings. Wang et al. (2020) discussed the COVID-19-related risk management conducted by universities in China. This study pointed out quick response actions from Chinese universities to COVID-19 pandemic. Furthermore, this study suggested the current challenges of risk management towards to COVID-19. McAleer (2020) discussed the risk management of different countries based on scores from Global Health Security Index 2019. This study also explained the benefit of vaccine development concerning risk management. Herold et al. (2021) analyzed the usages of digitalization in L&SC based on the framework of institutional logics. Data analysis was performed to find out the main factors have direct impact on digitalization. Looy (2021) discussed the relationship between business process management and digital innovation. The study proposed TOE framework based on technology innovation framework. Akpan (2020) suggested the benefits of technologies for small business and innovation in the time of COVID-19 global pandemic. This study emphasized big data analysis, model prediction, and visualization analysis as powerful tools to make proper decisions for complex business issues.

Business Operations Under The Pandemic Shock

Business operations have been impacted by a series of disease control measures, such as lockdown orders, travel restrictions, and quarantine policies. Current studies are concentrated on discussing how big data helps business operators to overcome the difficulties in global supply chain, tourism and traveling, and hotel management. Bage et al. (2021) discussed the power of big data to help restore the supply chains influenced by COVID-19 pandemic. Partial least squares structural equation modeling method was performed to test hypothetical model. The data involved in this study was gathered from manufacturing industries. Gossling et al. (2020) discussed the impact differences between COVID-19 and previous pandemics and further analyzed the changes on society, economy and tourism, caused by the pandemic. Time series analysis and data visualization were used to understand the impact on global tourism and global travel. Gallego & Font (2020) utilized Skyscanner data on air passenger searches and picks from November 2018 to December 2020 to examine the demand changes of air passengers caused by the COVID-19 crisis through big data technique. Wu et al. (2020) collected daily hotel price data from multiple sources using a web scraper. Rate fluctuation tendencies of hotel room were analyzed through ANOVA test, correlation analysis, and descriptive analysis. Besides, fluctuations in hotel room rates were also discussed based on star rating and district. Jiang & Wen (2020) examined the impact of COVID-19 on hotel marketing and management practices from three different aspects, including AI, robotics, and health care. The study suggests that the hotel should use AI solution to deal with pandemic circumstances.

Human Resource Management With Social Changes

Social changes caused by COVID-19 have affected business operators and industrial managers in the form of new patterns in human resource management, while such changes can be incorporated with applying big data analytics and AI. Hanyi et al. (2021) collected 1.56 million tweets from March 2020 to July 2020 and DL methods to analyze individuals' emotional responding to work-from-home during the pandemic. Six classifiers were trained through BERT models based on transformer learning. Besides, discontinuity growth model was also involved in this study for hypotheses testing. Zhang et al. (2021) scraped more than 1 million tweets based on keywords, including "work from home", "remote work", and "telework", to analyze the attitudes and feelings of the people. Sentiment analysis, content analysis, and topic modeling were performed through natural language processing techniques. Chong et al. (2020) established a two-stage moderated-mediation model based on conservation of resources theory to predict the correlation between daily COVID-19 task setbacks and people's exhaustion. A Monte Carlo simulation was used to measure the conditional indirect effects for the modeling results. Bughin et al. (2021) applied ML techniques to portray vaccination intent across five European countries, including Germany, France, Spain, Italy, and Sweden, for COVID-19. Random forest algorithm was used to train prediction model for vaccination intent. Such a study may further contribute to optimal decision-making in human resource management in the post-pandemic era.

FUTURE RESEARCH DIRECTIONS

Despite many challenges and difficulties caused by COVID-19, the research on economic topics may produce some new theories. The implicit hypotheses from current studies should be examined in the context of the COVID-19 crisis, thereby presenting a unique timing to test existing theories and providing numerous research opportunities in the future. Moreover, some related topics have not been incorporated in this chapter due to limited literatures and unverified results, e.g. monetary policies in COVID-19, pandemic-related investment strategies of Bitcoin, educational model shifts and shocks from the pandemic, etc. Such topics may be continuously investigated as big data plays the critical role in the future research directions. Besides, it is urgent for research communities, public authorities, and business participants to work together in establishing a comprehensive economic monitoring system, with a hyperconnected framework that incorporates as many as economic information pools and business data lakes as possible. Such a system can offer timely updated guidelines in terms of indicators of economic healthy, operational conditions, and business management, thereby requiring the integration and control by using big data and AI, which will be one of the most exciting research domains in the future.

CONCLUSION

This chapter contributes to the ongoing discourse on the impacts of COVID-19 on the economic development, asset and investment management, and business operations. Particularly, this chapter investigates the role of big data analytics in the above studies during the COVID-19 crisis. From the big data perspective, the challenges derived from the pandemic require an innovative way of thinking, thereby more researches are needed to emphasize on how data-driven approaches can be applied in monitoring economic conditions, predicting market trends, analyzing new patterns of consumer behaviors, reducing operational risks, and improving business management. As such, this chapter provides a roadmap for making better decisions by research communities, economic policymakers, and individuals in the context of COVID-19 and the post-pandemic era.

REFERENCES

Ahmed, M., Choudhury, N., & Uddin, S. (2017, July). Anomaly detection on big data in financial markets. In *2017 IEEE/ACM International Conference on Advances in Social Networks Analysis and Mining (ASONAM)* (pp. 998-1001). IEEE. 10.1145/3110025.3119402

Akour, I., Alshurideh, M., Al Kurdi, B., Al Ali, A., & Salloum, S. (2021). Using machine learning algorithms to predict people's intention to use mobile learning platforms during the COVID-19 pandemic: Machine learning approach. *JMIR Medical Education*, *7*(1), e24032. doi:10.2196/24032 PMID:33444154

Akpan, I. J., Soopramanien, D., & Kwak, D. H. (2020). Cutting-edge technologies for small business and innovation in the era of COVID-19 global health pandemic. *Journal of Small Business and Entrepreneurship*, 1–11. do i:10.1080/08276331.2020.1820185

Altig, D., Baker, S., Barrero, J. M., Bloom, N., Bunn, P., Chen, S., Davis, S. J., Leather, J., Meyer, B., Mihaylov, E., Mizen, P., Parker, N., Renault, T., Smietanka, P., & Thwaites, G. (2020). Economic uncertainty before and during the COVID-19 pandemic. *Journal of Public Economics*, *191*, 104274. doi:10.1016/j.jpubeco.2020.104274 PMID:32921841

Baek, S., Mohanty, S. K., & Glambosky, M. (2020). COVID-19 and stock market volatility: An industry level analysis. *Finance Research Letters*, *37*, 101748. doi:10.1016/j.frl.2020.101748 PMID:32895607

Bag, S., Dhamija, P., Luthra, S., & Huisingh, D. (2021). How big data analytics can help manufacturing companies strengthen supply chain resilience in the context of the COVID-19 pandemic. *International Journal of Logistics Management*. Advance online publication. doi:10.1108/IJLM-02-2021-0095

Bag, S., Gupta, S., Choi, T. M., & Kumar, A. (2021). Roles of Innovation Leadership on Using Big Data Analytics to Establish Resilient Healthcare Supply Chains to Combat the COVID-19 Pandemic: A Multimethodological Study. *IEEE Transactions on Engineering Management*.

Balanzá-Martínez, V., Kapczinski, F., de Azevedo Cardoso, T., Atienza-Carbonell, B., Rosa, A. R., Mota, J. C., & De Boni, R. B. (2021). The assessment of lifestyle changes during the COVID-19 pandemic using a multidimensional scale. *Revista de Psiquiatría y Salud Mental*, *14*(1), 16–26. doi:10.1016/j.rpsm.2020.07.003 PMID:32962948

Bauer, A., Hartl, T., Hutter, C., & Weber, E. (2021, December). Search Processes on the Labor Market during the Covid-19 Pandemic. In CESifo Forum (Vol. 22, No. 4, pp. 15-19). Institut für Wirtschaftsforschung (Ifo).

Belhadi, A., Kamble, S., Jabbour, C. J. C., Gunasekaran, A., Ndubisi, N. O., & Venkatesh, M. (2021). Manufacturing and service supply chain resilience to the COVID-19 outbreak: Lessons learned from the automobile and airline industries. *Technological Forecasting and Social Change*, *163*, 120447. doi:10.1016/j.techfore.2020.120447 PMID:33518818

Brandtner, P., Darbanian, F., Falatouri, T., & Udokwu, C. (2021). Impact of COVID-19 on the customer end of retail supply chains: A big data analysis of consumer satisfaction. *Sustainability*, *13*(3), 1464. doi:10.3390u13031464

Brinca, P., Duarte, J. B., & Faria-e-Castro, M. (2021). Measuring labor supply and demand shocks during covid-19. *European Economic Review*, *139*, 103901. doi:10.1016/j.euroecorev.2021.103901 PMID:34538878

Bughin, J. R., Cincera, M., Reykowska, D., & Ohme, R. (2021). Big Data is Decision Science: the Case of Covid-19 Vaccination. In *Handbook of Research on Applied Data Science and Artificial Intelligence in Business and Industry* (pp. 126–150). IGI Global. doi:10.4018/978-1-7998-6985-6.ch006

Bukovina, J. (2016). Social media big data and capital markets—An overview. *Journal of Behavioral and Experimental Finance*, *11*, 18–26. doi:10.1016/j.jbef.2016.06.002

Campello, M., Kankanhalli, G., & Muthukrishnan, P. (2020). *Corporate hiring under COVID-19: Labor market concentration, downskilling, and income inequality (No. w27208)*. National Bureau of Economic Research.

Campos, J., Sharma, P., Gabiria, U. G., Jantunen, E., & Baglee, D. (2017). A big data analytical architecture for the Asset Management. *Procedia CIRP*, *64*, 369–374. doi:10.1016/j.procir.2017.03.019

Campos-Vazquez, R. M., Esquivel, G., & Badillo, R. Y. (2021). How has labor demand been affected by the COVID-19 pandemic? Evidence from job ads in Mexico. *Latin American Economic Review*, *30*, 1–42. doi:10.47872/laer-2021-30-1

Cerchiello, P., & Giudici, P. (2016). Big data analysis for financial risk management. *Journal of Big Data*, *3*(1), 1–12. doi:10.118640537-016-0053-4

Chang, C. L., & McAleer, M. (2020). Alternative global health security indexes for risk analysis of COVID-19. *International Journal of Environmental Research and Public Health*, *17*(9), 3161. doi:10.3390/ijerph17093161 PMID:32370069

Chauhan, V., & Shah, M. H. (2020). An empirical analysis into sentiments, media consumption habits, and consumer behaviour during the Coronavirus (COVID-19) outbreak. *Purakala*.

Choi, K. S. (2019). K-SuperCast: A big data based GDP forecasting model. *The Korean Data & Information Science Society*, *30*(4), 723–743. doi:10.7465/jkdi.2019.30.4.723

Choi, T. M., Wallace, S. W., & Wang, Y. (2018). Big data analytics in operations management. *Production and Operations Management*, *27*(10), 1868–1883. doi:10.1111/poms.12838

Chong, S., Huang, Y., & Chang, C. H. D. (2020). Supporting interdependent telework employees: A moderated-mediation model linking daily COVID-19 task setbacks to next-day work withdrawal. *The Journal of Applied Psychology*, *105*(12), 1408–1422. doi:10.1037/apl0000843 PMID:33271029

Costola, M., Nofer, M., Hinz, O., & Pelizzon, L. (2020). *Machine learning sentiment analysis, Covid-19 news and stock market reactions (No. 288)*. Leibniz Institute for Financial Research SAFE.

Coulombe, P. G., Marcellino, M., & Stevanović, D. (2021). Can machine learning catch the Covid-19 recession? *National Institute Economic Review*, *256*, 71–109. doi:10.1017/nie.2021.10

D'Orazio, P. (2017). Big data and complexity: Is macroeconomics heading toward a new paradigm? *Journal of Economic Methodology*, *24*(4), 410–429. doi:10.1080/1350178X.2017.1362151

Devpura, N., & Narayan, P. K. (2020). Hourly oil price volatility: The role of COVID-19. *Energy Research Letters*, *1*(2), 13683. doi:10.46557/001c.13683

Diebold, F. X. (2020). *Real-time real economic activity: Exiting the great recession and entering the pandemic recession (No. w27482)*. National Bureau of Economic Research. doi:10.3386/w27482

Elshendy, M., & Fronzetti Colladon, A. (2017). Big data analysis of economic news: Hints to forecast macroeconomic indicators. *International Journal of Engineering Business Management*, *9*, 1847979017720040. doi:10.1177/1847979017720040

Ezugwu, A. E., Hashem, I. A. T., Oyelade, O. N., Almutari, M., Al-Garadi, M. A., Abdullahi, I. N., Otegbeye, O., Shukla, A. K., & Chiroma, H. (2021). A Novel Smart City-Based Framework on Perspectives for Application of Machine Learning in Combating COVID-19. *BioMed Research International*, *2021*, 2021. doi:10.1155/2021/5546790 PMID:34518801

Feldkircher, M., Huber, F., & Pfarrhofer, M. (2021). Measuring the effectiveness of US monetary policy during the COVID-19 recession. *Scottish Journal of Political Economy*, *68*(3), 287–297. doi:10.1111jpe.12275 PMID:33821043

Fischer, T., & Krauss, C. (2018). Deep learning with long short-term memory networks for financial market predictions. *European Journal of Operational Research*, *270*(2), 654–669. doi:10.1016/j.ejor.2017.11.054

Forsythe, E., Kahn, L. B., Lange, F., & Wiczer, D. (2020). Labor demand in the time of COVID-19: Evidence from vacancy postings and UI claims. *Journal of Public Economics*, *189*, 104238. doi:10.1016/j.jpubeco.2020.104238 PMID:32834178

Gallego, I., & Font, X. (2021). Changes in air passenger demand as a result of the COVID-19 crisis: Using Big Data to inform tourism policy. *Journal of Sustainable Tourism*, *29*(9), 1470–1489. doi:10.1080/09669582.2020.1773476

García-Ordás, M. T., Arias, N., Benavides, C., García-Olalla, O., & Benítez-Andrades, J. A. (2020, December). Evaluation of country dietary habits using machine learning techniques in relation to deaths from COVID-19. *Health Care*, *8*(4), 371. PMID:33003439

Garzoli, M., Plazzi, A., & Valkanov, R. I. (2021). *Backcasting, Nowcasting, and Forecasting Residential Repeat-Sales Returns: Big Data meets Mixed Frequency.* Swiss Finance Institute Research Paper, (21-21).

Ghosh, I., & Chaudhuri, T. D. (2021). FEB-stacking and FEB-DNN models for stock trend prediction: A performance analysis for pre and post covid-19 periods. *Decision Making: Applications in Management and Engineering*, *4*(1), 51–84. doi:10.31181/dmame2104051g

Ghosh, T., Parab, P. M., & Sahu, S. (2020). Analyzing the Importance of Forward Orientation in Financial Development-Economic Growth Nexus: Evidence from Big Data. *Journal of Behavioral Finance*, 1–9.

Gössling, S., Scott, D., & Hall, C. M. (2020). Pandemics, tourism and global change: A rapid assessment of COVID-19. *Journal of Sustainable Tourism*, *29*(1), 1–20. doi:10.1080/09669582.2020.1758708

Grybauskas, A., Pilinkienė, V., & Stundžienė, A. (2021). Predictive analytics using Big Data for the real estate market during the COVID-19 pandemic. *Journal of Big Data*, *8*(1), 1–20. doi:10.118640537-021-00476-0 PMID:34367876

Gulyas, A., & Pytka, K. (2020). The Consequences of the Covid-19 Job Losses: Who Will Suffer Most and By How Much? *Covid Economics*, *1*(47), 70–107.

Gyódi, K. (2021). Airbnb and hotels during COVID-19: different strategies to survive. *International Journal of Culture, Tourism and Hospitality Research*.

Hasan, A., Kalıpsız, O., & Akyokuş, S. (2017, October). Predicting financial market in big data: deep learning. In *2017 International Conference on Computer Science and Engineering (UBMK)* (pp. 510-515). IEEE. 10.1109/UBMK.2017.8093449

Hawash, B., Abuzawayda, Y. I., Mokhtar, U. A., Yusef, Z. M., & Mukred, M. (2020). Digital Transformation In The Oil And Gas Sector During Covid-19 Pandemic. *International Journal of Management*, *11*(12), 725–735. doi:10.34218/IJM.11.12.2020.067

He, P., Niu, H., Sun, Z., & Li, T. (2020). Accounting index of COVID-19 impact on Chinese industries: A case study using big data portrait analysis. *Emerging Markets Finance & Trade*, *56*(10), 2332–2349. doi:10.1080/1540496X.2020.1785866

Herold, D. M., Ćwiklicki, M., Pilch, K., & Mikl, J. (2021). The emergence and adoption of digitalization in the logistics and supply chain industry: An institutional perspective. *Journal of Enterprise Information Management*, *34*(6), 1917–1938. doi:10.1108/JEIM-09-2020-0382

Hofacker, C. F., Malthouse, E. C., & Sultan, F. (2016). Big data and consumer behavior: Imminent opportunities. *Journal of Consumer Marketing*, *33*(2), 89–97. doi:10.1108/JCM-04-2015-1399

Huang, W., & Zheng, Y. (2020). COVID-19: Structural changes in the relationship between investor sentiment and crude oil futures price. *Energy Research Letters*, *1*(2), 13685. doi:10.46557/001c.13685

Iyke, B. N. (2020). COVID-19: The reaction of US oil and gas producers to the pandemic. *Energy Research Letters*, *1*(2), 13912. doi:10.46557/001c.13912

Jiang, Y., & Wen, J. (2020). Effects of COVID-19 on hotel marketing and management: A perspective article. *International Journal of Contemporary Hospitality Management*, *32*(8), 2563–2573. doi:10.1108/IJCHM-03-2020-0237

Khattak, M. A., Ali, M., & Rizvi, S. A. R. (2021). Predicting the European stock market during COVID-19: A machine learning approach. *MethodsX*, *8*, 101198. doi:10.1016/j.mex.2020.101198 PMID:33425689

Kuk, J., Schachter, A., Faber, J. W., & Besbris, M. (2021). The COVID-19 Pandemic and the Rental Market: Evidence From Craigslist. *The American Behavioral Scientist*, *65*(12). doi:10.1177/00027642211003149

Lee, H. S. (2020). Exploring the initial impact of COVID-19 sentiment on US stock market using big data. *Sustainability*, *12*(16), 6648. doi:10.3390u12166648

Li, M., Li, T., Quan, D., & Li, W. (2020). Economic system simulation with big data analytics approach. *IEEE Access: Practical Innovations, Open Solutions*, *8*, 35572–35582. doi:10.1109/ACCESS.2020.2969053

Lilley, A., & Rogoff, K. (2020). Negative interest rate policy in the post Covid-19 world. *VOX, CEPR Policy Portal*, *17*, 28.

Liu, P., Xia, X., & Li, A. (2018). Tweeting the financial market: Media effect in the era of Big Data. *Pacific-Basin Finance Journal*, *51*, 267–290. doi:10.1016/j.pacfin.2018.07.007

Liu, Y., Lee, J. M., & Lee, C. (2020). The challenges and opportunities of a global health crisis: The management and business implications of COVID-19 from an Asian perspective. *Asian Business & Management*, *19*(3), 1. doi:10.105741291-020-00119-x

Loh, C. H., & Teoh, A. P. (2021, May). The Adoption of Big Data Analytics Among Manufacturing Small and Medium Enterprises During Covid-19 Crisis in Malaysia. In *Ninth International Conference on Entrepreneurship and Business Management (ICEBM 2020)* (pp. 95-100). Atlantis Press. 10.2991/aebmr.k.210507.015

Lovrić, M., Pavlović, K., Vuković, M., Grange, S. K., Haberl, M., & Kern, R. (2021). Understanding the true effects of the COVID-19 lockdown on air pollution by means of machine learning. *Environmental Pollution*, *274*, 115900. doi:10.1016/j.envpol.2020.115900 PMID:33246767

Lyons, N., & Lăzăroiu, G. (2020). Addressing the COVID-19 crisis by harnessing Internet of Things sensors and machine learning algorithms in data-driven smart sustainable cities. *Geopolitics, History, and International Relations*, *12*(2), 65–71. doi:10.22381/GHIR12220209

Magazzino, C., Mele, M., & Morelli, G. (2021a). The relationship between renewable energy and economic growth in a time of Covid-19: A Machine Learning experiment on the Brazilian economy. *Sustainability*, *13*(3), 1285. doi:10.3390u13031285

Magazzino, C., Mele, M., & Sarkodie, S. A. (2021b). The nexus between COVID-19 deaths, air pollution and economic growth in New York state: Evidence from Deep Machine Learning. *Journal of Environmental Management*, *286*, 112241. doi:10.1016/j.jenvman.2021.112241 PMID:33667818

Mazur, M., Dang, M., & Vega, M. (2021). COVID-19 and the march 2020 stock market crash. Evidence from S&P1500. *Finance Research Letters*, *38*, 101690. doi:10.1016/j.frl.2020.101690 PMID:32837377

McAleer, M. (2020). Prevention Is Better Than the Cure: Risk Management of COVID-19. *Journal of Risk and Financial Management*, *13*(3), 1–5. doi:10.3390/jrfm13030046

Medeiros, M. C., Vasconcelos, G. F., Veiga, Á., & Zilberman, E. (2021). Forecasting inflation in a data-rich environment: The benefits of machine learning methods. *Journal of Business & Economic Statistics*, *39*(1), 98–119. doi:10.1080/07350015.2019.1637745

Mele, M., & Magazzino, C. (2021). Pollution, economic growth, and COVID-19 deaths in India: A machine learning evidence. *Environmental Science and Pollution Research International*, *28*(3), 2669–2677. doi:10.100711356-020-10689-0 PMID:32886309

Mihet, R., & Philippon, T. (2019). The economics of Big Data and artificial intelligence. In *Disruptive Innovation in Business and Finance in the Digital World*. Emerald Publishing Limited. doi:10.1108/S1569-376720190000020006

Min, H., Peng, Y., Shoss, M., & Yang, B. (2021). Using machine learning to investigate the public's emotional responses to work from home during the COVID-19 pandemic. *The Journal of Applied Psychology*, *106*(2), 214–229. doi:10.1037/apl0000886 PMID:33818121

Narayan, P. K. (2020). Did bubble activity intensify during COVID-19? *Asian Economics Letters*, *1*(2), 17654.

Nhamo, G., Dube, K., & Chikodzi, D. (2020). Impacts and implications of COVID-19 on the global hotel industry and Airbnb. In *Counting the Cost of COVID-19 on the global tourism industry* (pp. 183–204). Springer. doi:10.1007/978-3-030-56231-1_8

OnaliE. (2020). COVID-19 and stock market volatility. Available at SSRN 3571453.

Oyewola, D. O., Augustine, A. F., Dada, E. G., & Ibrahim, A. (2021). Predicting Impact of COVID-19 on Crude Oil Price Image with Directed Acyclic Graph Deep Convolution Neural Network. *Journal of Robotics and Control*, *2*(2), 103–109. doi:10.18196/jrc.2261

OziliP. K.ArunT. (2020). *Spillover of COVID-19: impact on the Global* Economy. doi:10.2139/ssrn.3562570

Pantano, E., Priporas, C. V., Devereux, L., & Pizzi, G. (2021). Tweets to escape: Intercultural differences in consumer expectations and risk behavior during the COVID-19 lockdown in three European countries. *Journal of Business Research*, *130*, 59–69. doi:10.1016/j.jbusres.2021.03.015

Park, Y. E. (2021). Developing a COVID-19 Crisis Management Strategy Using News Media and Social Media in Big Data Analytics. *Social Science Computer Review*. doi:10.1177/08944393211007314

Pérez-Martín, A., Pérez-Torregrosa, A., & Vaca, M. (2018). Big Data techniques to measure credit banking risk in home equity loans. *Journal of Business Research, 89*, 448–454. doi:10.1016/j.jbusres.2018.02.008

Qin, M., Zhang, Y. C., & Su, C. W. (2020). The essential role of pandemics: A fresh insight into the oil market. *Energy Research Letters, 1*(1), 13166. doi:10.46557/001c.13166

Raut, R. D., Mangla, S. K., Narwane, V. S., Gardas, B. B., Priyadarshinee, P., & Narkhede, B. E. (2019). Linking big data analytics and operational sustainability practices for sustainable business management. *Journal of Cleaner Production, 224*, 10–24. doi:10.1016/j.jclepro.2019.03.181

Safara, F. (2020). A computational model to predict consumer behaviour during COVID-19 pandemic. *Computational Economics*, 1–14. doi:10.100710614-020-10069-3 PMID:33169049

SchrimpfA.ShinH. S.SushkoV. (2020). *Leverage and margin spirals in fixed income markets during the Covid-19 crisis*. doi:10.2139/ssrn.3761873

Scott, R., Poliak, M., Vrbka, J., & Nica, E. (2020). COVID-19 response and recovery in smart sustainable city governance and management: Data-driven Internet of Things systems and machine learning-based analytics. *Geopolitics. History and International Relations, 12*(2), 16–22.

Shang, W. L., Chen, J., Bi, H., Sui, Y., Chen, Y., & Yu, H. (2021). Impacts of COVID-19 pandemic on user behaviors and environmental benefits of bike sharing: A big-data analysis. *Applied Energy, 285*, 116429. doi:10.1016/j.apenergy.2020.116429 PMID:33519037

Sharma, R., Shishodia, A., Kamble, S., Gunasekaran, A., & Belhadi, A. (2020). Agriculture supply chain risks and COVID-19: mitigation strategies and implications for the practitioners. *International Journal of Logistics Research and Applications*, 1-27.

Sheth, J. (2020). Impact of Covid-19 on consumer behavior: Will the old habits return or die? *Journal of Business Research, 117*, 280–283. doi:10.1016/j.jbusres.2020.05.059 PMID:32536735

Shoss, M. K., Horan, K. A., DiStaso, M., LeNoble, C. A., & Naranjo, A. (2021). The conflicting impact of COVID-19's health and economic crises on helping. *Group & Organization Management*, *46*(1), 3–37. doi:10.1177/1059601120968704

Shuai, X., Chmura, C., & Stinchcomb, J. (2021). COVID-19, labor demand, and government responses: Evidence from job posting data. *Business Economics (Cleveland, Ohio)*, *56*(1), 29–42. doi:10.105711369-020-00192-2 PMID:33311717

Simionescu, M., & Zimmermann, K. F. (2017). *Big data and unemployment analysis* (No. 81). GLO Discussion Paper.

Suau-Sanchez, P., Voltes-Dorta, A., & Cugueró-Escofet, N. (2020). An early assessment of the impact of COVID-19 on air transport: Just another crisis or the end of aviation as we know it? *Journal of Transport Geography*, *86*, 102749. doi:10.1016/j.jtrangeo.2020.102749 PMID:32834670

Sun, Z., Zou, H., & Strang, K. (2015, October). Big data analytics as a service for business intelligence. In *Conference on e-Business, e-Services and e-Society* (pp. 200-211). Springer. 10.1007/978-3-319-25013-7_16

Tang, J., & Karim, K. E. (2019). Financial fraud detection and big data analytics–implications on auditors' use of fraud brainstorming session. *Managerial Auditing Journal*, *34*(3), 324–337. doi:10.1108/MAJ-01-2018-1767

Van Looy, A. (2021). A quantitative and qualitative study of the link between business process management and digital innovation. *Information & Management*, *58*(2), 103413. doi:10.1016/j.im.2020.103413

Walmsley, T. L., Rose, A., & Wei, D. (2021). Impacts on the US macroeconomy of mandatory business closures in response to the COVID-19 Pandemic. *Applied Economics Letters*, *28*(15), 1293–1300. doi:10.1080/13504851.2020.1809626

Wang, C., Cheng, Z., Yue, X. G., & McAleer, M. (2020). Risk Management of COVID-19 by Universities in China. *Journal of Risk and Financial Management*, *13*(2), 1–6. doi:10.3390/jrfm13020036

Wang, X., Chen, X., & Zhao, P. (2020). The relationship between Bitcoin and stock market. *International Journal of Operations Research and Information Systems*, *11*(2), 22–35. doi:10.4018/IJORIS.2020040102

Wu, B., Wang, L., Wang, S., & Zeng, Y. R. (2021). Forecasting the US oil markets based on social media information during the COVID-19 pandemic. *Energy*, *226*, 120403. doi:10.1016/j.energy.2021.120403 PMID:34629690

Wu, F., Zhang, Q., Law, R., & Zheng, T. (2020). Fluctuations in Hong Kong hotel industry room rates under the 2019 Novel Coronavirus (COVID-19) outbreak: Evidence from big data on OTA channels. *Sustainability*, *12*(18), 7709. doi:10.3390u12187709

Xing, F. Z., Cambria, E., & Welsch, R. E. (2018). Natural language based financial forecasting: A survey. *Artificial Intelligence Review*, *50*(1), 49–73. doi:10.100710462-017-9588-9

Xue, F., Li, X., Zhang, T., & Hu, N. (2021). Stock market reactions to the COVID-19 pandemic: The moderating role of corporate big data strategies based on Word2Vec. *Pacific-Basin Finance Journal*, *68*, 101608. doi:10.1016/j.pacfin.2021.101608

Yang, R., Yu, L., Zhao, Y., Yu, H., Xu, G., Wu, Y., & Liu, Z. (2020). Big data analytics for financial Market volatility forecast based on support vector machine. *International Journal of Information Management*, *50*, 452–462. doi:10.1016/j.ijinfomgt.2019.05.027

Yedjou, C. G., Alo, R. A., Liu, J., Enow, J., Ngnepiepa, P., Long, R., ... Tchounwou, P. B. (2021). Chemo-Preventive Effect of Vegetables and Fruits Consumption on the COVID-19 Pandemic. *Journal of Nutrition & Food Sciences*, *4*(2). PMID:33884222

Yu, Z., Xiao, Y., & Li, Y. (2020). The response of the labor force participation rate to an epidemic: Evidence from a cross-country analysis. *Emerging Markets Finance & Trade*, *56*(10), 2390–2407. doi:10.1080/1540496X.2020.1787149

Zhang, C., Yu, M. C., & Marin, S. (2021). Exploring public sentiment on enforced remote work during COVID-19. *The Journal of Applied Psychology*, *106*(6), 797–810. doi:10.1037/apl0000933 PMID:34138587

Zhu, D. (2021). Research and Analysis of a Real Estate Virtual E-Commerce Model Based on Big Data Under the Background of COVID-19. *Journal of Organizational and End User Computing*, *33*(6), 1–16. doi:10.4018/JOEUC.20211101.oa28

ADDITIONAL READING

Aldridge, I., & Avellaneda, M. (2021). *Big data science in finance*. John Wiley & Sons.

Dwivedi, Y. K., Hughes, D. L., Coombs, C., Constantiou, I., Duan, Y., Edwards, J. S., Gupta, B., Lal, B., Misra, S., Prashant, P., Raman, R., Rana, N. P., Sharma, S. K., & Upadhyay, N. (2020). Impact of COVID-19 pandemic on information management research and practice: Transforming education, work and life. *International Journal of Information Management*, *55*, 102211. doi:10.1016/j.ijinfomgt.2020.102211

Feyisa, H. L. (2020). The World Economy at COVID-19 quarantine: contemporary review. *International Journal of Economics, Finance and Management Sciences, 8*(2), 63-74.

Jia, Q., Guo, Y., Wang, G., & Barnes, S. J. (2020). Big data analytics in the fight against major public health incidents (including COVID-19): A conceptual framework. *International Journal of Environmental Research and Public Health*, *17*(17), 6161. doi:10.3390/ijerph17176161 PMID:32854265

Jing, J. (2020). Big data analysis and empirical research on the financing and investment decision of companies after COVID-19 epidemic situation based on deep learning. *Journal of Intelligent & Fuzzy Systems*, 1-10.

Keshky, E., El Sayed, M., Basyouni, S. S., & Al Sabban, A. M. (2020). Getting through covid-19: The pandemic's impact on the psychology of sustainability, quality of life, and the global economy–a systematic review. *Frontiers in Psychology*, *11*, 3188. doi:10.3389/fpsyg.2020.585897 PMID:33281683

Kumar, N., Kumar, G., & Singh, R. K. (2021). Analysis of barriers intensity for investment in big data analytics for sustainable manufacturing operations in post-COVID-19 pandemic era. *Journal of Enterprise Information Management*. Advance online publication. doi:10.1108/JEIM-03-2021-0154

Lim, G. C., & McNelis, P. D. (2008). *Computational macroeconomics for the open economy* (Vol. 2). MIT Press.

Miranda, M. J., & Fackler, P. L. (2004). *Applied computational economics and finance*. MIT press.

Riswantini, D., Nugraheni, E., Arisal, A., Khotimah, P. H., Munandar, D., & Suwarningsih, W. (2021). Big Data Research in Fighting COVID-19: Contributions and Techniques. *Big Data and Cognitive Computing*, *5*(3), 30. doi:10.3390/bdcc5030030

KEY TERMS AND DEFINITIONS

Business Operations and Management: A business function that can transform basic materials and labor into products and services.

Computational Macroeconomics: A research subject that involves computational tools to solve macroeconomic problems.

Consumer Behavior Analysis: A study of investigating consumer purchase performance and customer engagement through qualitative and quantitative ways.

Crisis Management: A process in which an organization handle unexpected event that likely leads to risks.

Economic Impact Analysis: An analysis method of examining the impact of an event on economy of a community.

Predictive Analysis: An analytical technique that makes simulations and forecasting in regards to uncertainties and unknown events using a variety of mathematical processes, such as statistical modeling, data mining, machine learning, etc.

Risk Analysis: A process of analyzing and detecting potential problems that could cause negative impact.

Social System: A network of relationships connecting the existing entities in a group.

About the Authors

Peng Zhao is a data science professional with experience in industry, teaching, and research. He has a broad range of practical data science experience in different industries, including finance, mobile device, consumer intelligence, big data technology, insurance, and biomedical industries. He is a leading machine learning expertise in a Big Data & AI company in New Jersey. He also manages a data scientist team providing a variety of data consulting services to individuals, businesses, and non-profit organizations.

Xi Chen is a lecturer in the School of Humanity and Law at Beijing University of Civil Engineering and Architecture. She is also a research assistant in Beijing Research Base for Architectural Culture. Her current research interests include English academic writing, settlement evolution and urbanization in China and the U.S., urban studies, inter-cultural analysis, globalization, and cognitive linguistics.

Index

G

H

I

K

L

M

N

O

P

R

S

Printed in the United States
by Baker & Taylor Publisher Services

Printed in the United States
by Baker & Taylor Publisher Services